Science and the Garden

# Science
## and the
# Garden

## The Scientific Basis
## of Horticultural Practice

Second Edition

Edited by

# David S. Ingram
# Daphne Vince-Prue
# Peter J. Gregory

Published for the
Royal Horticultural Society
by Blackwell Publishing

Royal
Horticultural
Society

Main cover image is reproduced courtesy of David De Lossy/Getty Images. The other cover
images are (from top to bottom) courtesy of: Patrick Echlin, Multi-imaging Centre, University
of Cambridge; Peter Beales Roses; Debbie White, Royal Botanic Garden, Edinburgh; Brian
Thomas, University of Warwick; and Debbie White, Royal Botanic Garden, Edinburgh.

*Library of Congress Cataloging-in-Publication Data*
Science and the garden : the scientific basis of horticultural practice / edited by
David S. Ingram, Daphne Vince-Prue, Peter J. Gregory. – 2nd ed.
p. cm.
Includes bibliographical references and index.
ISBN-13: 978-1-4051-6063-6 (pbk. : alk. paper)
ISBN-10: 1-4051-6063-2 (pbk. : alk. paper)    1. Horticulture. I. Ingram, David S.
II. Vince-Prue, Daphne.   III. Gregory, P. J.
SB318.S29 2008
635.01′5–dc22
2007043768

A catalogue record for this book is available from the British Library.

Set in 10/11.5 pt Times by Graphicraft Limited, Hong Kong
Printed and bound in Singapore by Markono Print Media Pte Ltd

1   2008

# Contents

# Foreword

Since its establishment in 1804, the Royal Horticultural Society (RHS) has set out to foster and encourage the advancement of horticultural science and to stimulate a wider understanding of both the principles and practices involved. This new and important book had its origins in the RHS Science and Horticultural Advice Committee, and the balance of authors involved demonstrates the vitality of the relations between the RHS scientific staff and the wider science community involved with the RHS.

The book progresses logically from consideration of the basic structures and functions of garden plants, through nomenclature and genetics to the environmental factors affecting growth, to methods of propagation and production, to pest and disease control, and finally to post-harvest management and storage. It is principally a book of 'why' with strong emphasis on the underlying science but it also, where appropriate, deals with 'how' and gives the rationale behind practical advice. Although written with the student in mind it will also appeal to gardeners, growers and scientists who will appreciate the width of expertise deployed by the authors in covering the subjects and in bringing an objective perspective to the impact of biotechnology on horticulture.

It is humbling to consider the wonder of plants and to appreciate their structure and function which combine engineering and chemical manufacture beyond the dreams of man: highly efficient, fully co-ordinated and multi-functional organisms with no moving parts and fuelled by natural resources. It is said that environmental factors drive evolutionary responses and this is clearly demonstrated by the infinite variations found in plants to exploit environmental niches. Gardeners need to appreciate that this adaptation to environment is not for decoration and their delectation but for the plant's function and survival. Understanding the science underpinning the differences between form, function and survival strategies of different plant groups will greatly aid their cultivation and the enjoyment of gardening. The relationship between plants and the environment in which they grow is dynamic. The plant can obviously respond immediately to short-term stress caused by factors such as variation in water supply and temperature, but it also must often be sensitive to regular seasonal changes to trigger major physiological processes such as the change from vegetative growth to flowering. Much is already known about these processes and is incorporated into horticultural practice, but the further understanding of these mechanisms, and of the trigger signals that initiate them, is an exciting area of science that will have great implications for the gardener and the commercial grower.

Gardening, fortunately, is not solely the application of science. Indeed gardening is a combination of practical, aesthetic and philosophic ideals. The chapters on Selecting and Breeding Plants (Chapter 5), Shape and Size (Chapter 11), and Colour, Scent and Sound in the Garden (Chapter 12) tackle these wider issues as well as considering the underlying science, whereas the chapters on controlling the undesirables (Chapters 15 and 16) start off with an interesting perspective on bio-diversity:

In the imagined Arcadian wilderness before gardening was invented there were no undesirables, only a rich biodiversity. Today's gardeners find this richness excessive and relabel some of it pests, diseases and weeds. A pest, disease or weed is simply biodiversity being over-assertive, thus limiting or preventing the growth, flowering or fruiting of cultivated plants. It may therefore be necessary to take some corrective action, to either prevent or reduce the problem. Before taking such action, however, it is important to be able to identify the organism or environmental factors involved.

These rich veins of common sense run throughout the book and temper the hard science with an awareness of the needs, desires and hopes of gardeners. The book is the culmination of much hard work in writing down and then editing the collective expert scientific knowledge of the authors, who were brought together by their interest in and involvement with the RHS. It is a substantial achievement and will give real benefit and pleasure to all interested in horticulture and gardening.

Building on its strong and still highly relevant foundations, the second edition of *Science and the Garden* has been substantially edited to sharpen its focus and expanded to address many of the key issues facing the natural world. Concerns about diversity, sustainability, conservation and climate are now widespread. However, the gardener can instinctively appreciate the complexity of these interacting and fluid forces, and will welcome the very readable, coherent and scientific approach to these issues which will shape our future world.

John MacLeod
RHS Professor of Horticulture

## Biographical details

John MacLeod was, for many years, involved in applied agronomic research on the Agricultural Development and Advisory Service Experimental Farms before becoming Director of the Experimental Centres and then, in 1990, Director of the National Institute of Agricultural Botany in Cambridge.

A member of the Royal Horticultural Society Science and Horticultural Advice Committee since 1993, he became Chairman in 2000. He is now the Royal Horticultural Society Professor of Horticulture.

# Preface to Second Edition

We hope that this second edition of *Science and the Garden* will be of interest and value to students of horticulture, professional horticulturists and home gardeners. In preparing it we have completely re-edited all the chapters from the first edition, and increased their number, removing unnecessary material, adding new information and re-ordering or re-writing where appropriate. If, in our thorough editing, we have introduced errors, the fault is ours, not that of the authors. We have also added four new chapters, dealing with matters that we felt were not adequately covered in the first edition, namely Diversity in the Plant World (Chapter 1), Conservation and Sustainable Gardening (Chapter 18), Gardens and the Natural World (Chapter 19) and Gardens for Science (Chapter 20).

We are aware of the difficulty of finding an appropriate style and level to cater for a wide readership. We have therefore tried to use straightforward language throughout, but have taken care not to over-simplify or 'dumb-down' scientific information which we believe to be of fundamental importance to the practice of horticulture.

We thank Blackwell Publishing and the Royal Horticultural Society for commissioning a second edition, Dr Malcolm Vincent, Chief Scientist of the Joint Nature Conservation Committee, for his valuable comments on drafts of Chapters 18 and 19, and Mrs Barbara Haynes for her editorial support and advice on content and presentation.

David S. Ingram
(Burton in Lonsdale & Edinburgh)
Daphne Vince-Prue
(Goring-on-Thames)
Peter J. Gregory
(Dundee)
May 2007

# Preface to First Edition

*Science and the Garden* has been written primarily for students of horticulture, but we expect that it will also be of interest to amateur gardeners and professional growers who would like to understand more about the science that underlies horticultural practices.

Most conventional gardening books concentrate on how and when to carry out horticultural tasks such as pruning, seed sowing and taking cuttings. In contrast, the aim of the present book is to explain in straightforward terms some of the science that underlies these practices. We address such diverse questions as: why are plants green? Why should one cut beneath a leaf node when taking cuttings? Why do plants need so much water? Why is light so important and what effect does it have on plant growth? How do plants detect drying soils and how is growth modified to improve their survival chances? Why are plants more resistant to freezing in the autumn than in spring? How do plants detect seasonal changes in their environment? Why do chrysanthemums flower in the autumn and onions produce their bulbs in the summer?

The first part of the book is concerned with some fundamental principles. Chapter 1 (Chapters 2 and 3 in the second edition) describes the structural features of the plant, and introduces biochemical and physiological processes such as photosynthesis and water and solute transport, which are expanded on in later chapters in relation to particular aspects of horticultural practice. Chapter 2 (Chapter 4 in the second edition) introduces the often difficult question of how plants are named. Plant names are a problem for many gardeners and this chapter explains the structure of plant nomenclature in simple terms; it outlines the rules for naming plants, discusses why names sometimes change and, most importantly for the gardener, what is being done to achieve stability in plant nomenclature.

It is often thought that genetic modification (GM) is the 'new' thing in horticulture, but the fact is that most plants grown in gardens (except weeds) have been genetically manipulated in the sense that their genes differ from those of their wild relatives. 'Designing Plants' (Chapter 3; Chapter 5 in the second edition) explains how new plants have been developed through cross-breeding and selection processes that have been going on for centuries. The chapter concludes with a look towards the future by showing how new plants can be 'designed' by introducing specific genes using GM technology.

The remainder of the book is more immediately concerned with the practices of horticulture. With the exception of a few aquatics, gardening depends on the soil and Chapter 4 (Chapters 6 and 7 in the second edition) describes the different types of soil, explains how to recognise them and introduces the science underlying soil management practices. Water conservation is an important consideration in many gardens and may well become more important with climate change. The selection of suitable plants is itself a form of water conservation by the gardener and Chapter 5 (Chapter 8 in the second edition), 'Choosing a Site', describes how certain plants are adapted to grow in dry conditions. All gardens have shady areas and Chapter 5 also explains how plants are able to detect shade from trees and neighbouring buildings and modify their growth accordingly. It ends with advice on how to choose plants for particular situations using scientific principles.

'Raising Plants from Seed' (Chapter 6; Chapter 9 in the second edition) and 'Vegetative Propagation' (Chapter 7; Chapter 10 in the second edition) are basic horticultural practices. These two chapters discuss the science underlying embryo development, seed maturation and ripening, dormancy and how it may be broken, and the storage of seeds. They also look at vegetative propagation, such as taking cuttings, layering and micropropagation, with special emphasis on the physiological processes underlying these practices, most notably the hormonal control of growth and development. The science of grafting is also considered.

Once plants have been propagated and the site has been selected with due consideration for soil and aspect, the choice of a particular plant for that situation is usually determined by factors such as colour, size and shape, topics that are covered in Chapter 8 (Chapters 11 and 12 in the second edition). The choice of suitable plants also depends on factors such as the time of flowering and, for edible crops, the yields of storage organs such as potato tubers and onion bulbs. The time at which plants enter dormancy and increase their resistance to freezing conditions often determines their ability to grow and even to survive in a particular locality. These processes are largely governed by seasonal factors, although they may be modified by local conditions. Chapter 9 (Chapter 13 in the second edition), 'Seasons and Weather', focuses on such seasonal factors as day length and temperature and explains how these are sensed by the plant and how the information is translated into the observed displays. Gardening in the greenhouse is a specialised form of gardening, requiring knowledge of how the physical conditions of the greenhouse interact with the physiology of the plant if optimum yields are to be achieved. Such interactions and their implications for successful greenhouse management are discussed in detail in Chapter 10 (Chapter 14 in the second edition).

As all gardeners will be aware, no matter how great their horticultural skills, pests, diseases and weeds are a constant problem and can often cause disaster. Chapter 11 (Chapters 15 and 16 in the second edition), 'Controlling the Undesirables', describes how such organisms can be recognised, how they affect plant growth and what strategies are available for combating them. Throughout this chapter the emphasis is on integrated management of pests, diseases and weeds and the use of methods that are environmentally friendly.

Harvesting the flowers, fruit and vegetables that are the product of many hours' labour can be the most satisfying of tasks for the successful gardener. Fittingly then the final chapter in the book (Chapter 17 in the second edition) considers the physiological basis of the maturation process, and discusses the best ways of harvesting and storing flowers, fruit and vegetables to ensure maximum quality, storage life and flavour.

The book has been edited and written by past and present members of the Royal Horticultural Society's Scientific and Horticultural Advice Committee, past and present members of the scientific staff of the Society, and other specialists. The contents reflect the particular interests of the authors, and their judgement as to the scientific information that is likely to be of greatest importance to gardeners and horticulturists.

We would like to thank all those who have contributed to the volume by writing particular chapters, by commenting on draft chapters and by giving their general support. We also wish to thank Mrs Joyce Stewart, Royal Horticultural Society Director of Horticulture, for allowing her staff to participate in the project and for arranging a grant from the Royal Horticultural Society towards the cost of printing the colour plates. DSI also thanks Napier University for its support during his tenure of a Visiting Professorship there, Mrs Janet Prescott for managing the project and contributing significantly to the editing process, and Mrs Jane Stevens, of St Catharine's College, Cambridge, for her assistance during the final stages of editing the volume.

David S. Ingram
(Cambridge)
Daphne Vince-Prue
(Goring-on-Thames)
Peter J. Gregory
(Reading)

# List of Contributors

**Guy Barter** began his horticultural career in commercial horticulture before joining the Royal Horticultural Society, as superintendent of the Trials Department at the RHS Garden, Wisley. After a period as a horticultural journalist for *Gardening Which?*, he returned to work for the RHS Gardening Advisory Service of which he is currently head.

**Anna Dourado** was an International Plant Genetic Resources Institute (IPGRI) Intern at the Asian Vegetable Research and Development Center (AVRDC), Taiwan, and a Visiting Lecturer at Massey University, New Zealand before becoming a Lecturer in Vegetable Production at the University of Bath. She has been Senior Lecturer in Crop Production at the Royal Agricultural College, Cirencester, and Head of Horticultural Science, Advice and Trials at the Royal Horticultural Society. She is now a horticultural consultant, part-time lecturer in horticulture and a freelance writer.

**Peter J. Gregory** was Lecturer, then Reader, in Soil Science at the University of Reading. He was then Principal Scientist at CSIRO, Australia, before becoming Professor of Soil Science and Pro-Vice-Chancellor (Research) at the University of Reading. He is now Chief Executive of the Scottish Crop Research Institute, near Dundee.

**Andrew Halstead** is Principal Entomologist at the RHS Garden, Wisley and has 35 years' experience of diagnosing and giving advice on garden pest problems. He is a Fellow of the Royal Entomological Society and a past President of the British Entomological and Natural History Society.

**David S. Ingram OBE, VMH, FIHort, FIBiol, FRSE** was Research Fellow in Plant Pathology at the Universities of Glasgow and Cambridge before becoming Lecturer then Reader in Plant Pathology at the University of Cambridge. He then became Regius Keeper of the Royal Botanic Garden Edinburgh and Royal Horticultural Society Professor of Horticulture, and was Chairman of the Science and Horticultural Advice Committee of the Royal Horticultural Society until 2000. He was Master of St Catharine's College, Cambridge, from 2000 to 2006, Chairman of the Darwin Initiative for the Survival of Species from 1999 to 2005 and is now Honorary Professor at Edinburgh and Glasgow Universities, Senior Visiting Fellow at the ESRC Genomics Forum, Edinburgh, an independent member of the Joint Nature Conservation Committee and Honorary Fellow, Royal Botanic Garden Edinburgh and Myerscough College, Preston. Professor Ingram was awarded the Victoria Medal of Honour of the Royal Horticultural Society in 2004.

**David S. Johnson** has spent his career working on all aspects of post-harvest research on fresh horticultural crops at the East Malling site in Kent (currently East Malling Research). In 2007 he received the Marsh Horticultural Research Award presented by the Marsh Christian Trust in conjunction with the Royal Horticultural Society.

**Stephen L. Jury** is the Herbarium Curator and Principal Research Fellow in the School of Plant Sciences at the University of Reading. He has a great interest in the taxonomy of cultivated plants and has served on the Advisory Panel on Nomenclature and Taxonomy of the Royal Horticultural Society, and is currently also a member of its Science and Horticultural Advice Committee.

**Ray Mathias** is a consultant in science communication and education working with research councils, schools and other organisations to support public, teacher and student engagement with science and engineering. Previously he was, for 10 years, Head of Science Communication and Education at the John Innes Centre with responsibility for external and media liaison and science in society activities. For over 20 years he was an active researcher working in plant biotechnology, first at the Plant Breeding Institute in Cambridge, and then at the John Innes Centre, Norwich. He has worked on the application of biotechnology in cereals and brassica crops and on the development of alternative non-food oilseed crops.

**Jon Pickering** is a horticultural graduate and received his doctorate from the University of Reading following research on the horticultural uses of green waste compost. He was formerly a horticultural scientist at the RHS Garden, Wisley where he provided advice on topics related to soils, plant nutrition, growing media and composting. He currently works for a consultancy firm involved with large-scale composting operations and sustainable waste-management technologies.

**Chris Prior** began his scientific career as a research fellow in the University of Papua New Guinea, working on diseases of cocoa. He then worked on biological control of a range of tropical pests with CAB International. He joined the Royal Horticultural Society in 1996 as Senior Plant Pathologist and was Head of Horticultural Science from 2001 until his departure in 2007.

**Michael Saynor** graduated from the University of Leeds in 1965. He then joined the Agricultural Development and Advisory Service (ADAS), formerly the National Agricultural Advisory Service, based in Cambridge. After postings to the west country and the south east, he left ADAS in 1991 to become an independent consultant.

**Simon Thornton-Wood** was a Tropical Botanist at the Natural History Museum, London, before becoming Horticultural Taxonomist for the National Trust. He later became Head of Botany at the RHS Garden, Wisley, progressing through a series of roles to RHS Director of Science and Learning in 2006.

**Daphne Vince-Prue** was Lecturer in Horticulture and Reader in Botany at the University of Reading before becoming a Scientific Advisor to the Agriculture and Food Research Council. She later became Head of the Physiology and Chemistry Division at the Glasshouse Crops Research Institute. Following her retirement she was for many years a member of the Science and Horticultural Advice Committee of the Royal Horticultural Society. Dr Vince-Prue was awarded a Gold Veitch Memorial Medal of the Royal Horticultural Society in 2002.

**Timothy Walker** has spent most of his professional life at the University of Oxford Botanic Garden where he is currently *Horti Praefectus*. He is particularly interested in the practice and theory of plant conservation. He is the Ernest Cook College Lecturer in Plant Conservation at Somerville College and a Research Lecturer in the Department of Plant Sciences at the University of Oxford.

# 1

# Diversity in the Plant World

## SUMMARY

In this chapter the stages in evolution of the diversity of plant life on Earth are outlined and the essential characteristics of the most successful land plants summarised, as an introduction to Chapters 2 and 3. The characteristics, origins and occurrence in the garden of primitive plants are then summarised. Finally, flowering land-plant forms, their occurrence in the wild and their value in the garden are presented in tabular form.

## INTRODUCTION

The most remarkable thing about plants is that they are green (Fig. 1.1), a property that makes it possible for them to generate the energy required to sustain almost the entire living world.

This is an extravagant claim, but to appreciate its significance it is necessary first to consider what happens to the average motor car if, like the one in Fig. 1.2, it is neglected for long enough: the bodywork rusts and the non-ferrous components disintegrate and decay. Indeed, it is the usual experience that all inanimate things, left to themselves, eventually reach a state of disorder: books turn to dust, buildings crumble and machines rust. This general tendency is expressed in the second law of thermodynamics, which states, in essence, that in an isolated system the degree of disorder and chaos – the *entropy* – can only increase.

## CREATING ORDER OUT OF DISORDER

When one thinks about living things, however, it is immediately apparent that they are able to create order out of disorder, assembling atoms and molecules to form tissues and bodies of great complexity and sophistication (Fig. 1.1). Now, living things must obey the laws of physics and chemistry, just as a motor car must, so how is this creation of order out of disorder possible, thermodynamically? The answer is that the cells of living things are not isolated systems in a thermodynamic sense, as a motor car is, for they are constantly deriving energy from another external source, the sun. It is necessary to go back in time to find out how this came about.

The Earth first condensed from dust and ashes about 4600 million years ago, and life must have appeared some time during the first thousand million years of the planet's existence. The molecules that made life possible may have arrived from another planet in, for example, a comet, but current theories suggest they were probably generated here on Earth. The earliest life forms would have derived their organic molecules (those containing carbon) from their

1

**Fig. 1.1** 'The most remarkable thing about plants is that they are green.' Giant redwoods growing in Younger Botanic Garden, Argyll, a Specialist Garden of the Royal Botanic Garden Edinburgh. *Photograph by David S. Ingram.*

slime on the surface of marshy soils and wet lawns even today (Fig. 1.3). These primitive organisms were so successful that they have remained virtually unchanged throughout almost the entire course of evolutionary time. The Cyanobacteria possess the ability to capture the electromagnetic radiation of the sun and incorporate it into a chemical energy source. This is made possible by the presence of light-absorbing pigments which give the blue-green algae their characteristic colour, the most significant being the green pigment *chlorophyll* **a** (Fig. 2.2). For the first time in evolutionary history there had appeared on earth *autotrophic* organisms, which were able to make their own food, in contrast to *heterotrophic* organisms, which must derive their food from external sources. The great diversity of plants alive today sprang from these humble beginnings.

The evolution of photosynthesis had another, significant, consequence. As the number of photosynthetic organisms increased they altered the Earth's atmosphere. This is because the most efficient form of photosynthesis, the one employed by most primitive plants, involves the splitting of water molecules ($H_2O$) to release oxygen ($O_2$) (see Fig. 2.1). This increased the oxygen level in the atmosphere, which had two important effects. First, some of the 'new' oxygen in the outer layer of the atmosphere was converted to *ozone* ($O_3$), which absorbs the *ultraviolet (UV) radiation* from sunlight, which is very damaging to living organisms. This meant that organisms could survive in the surface layers of water, and even on land. The current depletion of the *ozone layer* as a result of human activity is a serious reversal of a critical stage in the evolution of life on Earth (see Chapter 18). Second, the increase in the level of oxygen made possible the process of *aerobic respiration*, whereby carbon molecules formed by photosynthesis are broken down to release energy required for building bodies in far greater quantities than are released by *anaerobic respiration*, which occurs in the absence of oxygen.

Before the atmosphere became enriched by oxygen the only organisms that existed were *prokaryotic* (Table 1.1), comprising only simple cells lacking a nucleus defined by a membrane (see Chapter 2). It is probable that these first *prokaryotes* were heat-loving organisms, called Archaea, which translates as 'ancient ones'. The descendants of these earliest organisms survive today in extremely hot environments that are hostile to most other forms of life.

surroundings, a legacy from the pre-biotic 'soup' of chemicals that was left on the cooling Earth after its genesis. These would have provided them with energy and the building blocks for making cells. But as these natural resources were exhausted, a key event was the *evolution* of the process called *photosynthesis*, whereby sunlight is harnessed to provide an alternative external source of energy (see Chapters 2 and 8).

The study of very old rock formations in Australia has suggested that this event must have occurred more than 3600 million years ago, for by that time there were present on the planet simple organisms consisting of single cells or chains of cells that resemble the blue-green *Cyanobacteria* (*blue-green algae*) that grow in shallow, stagnant water or as a greenish

**Fig. 1.2** It is the usual experience that inanimate things, left to themselves, like this VW Beetle, eventually reach a state of disorder. In contrast, living things, like the plants of oilseed rape, are able to create order out of disorder, assembling atoms and molecules to form tissues and bodies of great complexity and sophistication. *Photograph by David S. Ingram.*

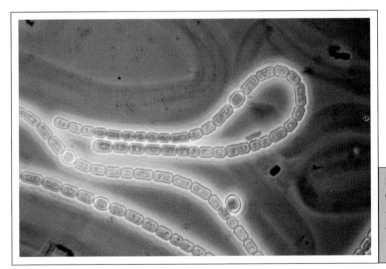

**Fig. 1.3** A chain of cells of the cyanobacterium (blue-green) *Anabaena. Light microscope photograph by Patrick Echlin, Multi-imaging Centre, University of Cambridge.*

*Bacteria*, which appeared soon after, are also prokaryotes, and some of these and some of the Archaea were, and still are, capable of photosynthesis.

The presence of oxygen in the atmosphere also led about 1500 million years ago to the gradual evolution, by natural selection, of *eukaryotic* cells (see Fig. 2.4), which had a clearly defined, membrane-bound nucleus, complex chromosomes and membrane-bound *organelles*. The latter are subcellular structures with specialised functions. They include *mitochondria*, where respiration occurs and, in plants, *chloroplasts* (Fig. 1.4), where photosynthesis occurs. Eukaryotic cells provided the building blocks for the evolution of all complex organisms, from seaweeds and shrimps to oak trees and orangutans. It is probable that organelles such as chloroplasts and mitochondria, which possess their own genetic information in addition to that contained in the nucleus, first evolved as a result of the incorporation of free-living prokaryotes into the evolving eukaryotic cells.

At first the earliest photosynthetic organisms lived just below the surface of the oceans, but with time the mineral resources of these open waters were depleted by the teeming life within them and organisms began to develop more abundantly near the shores, where

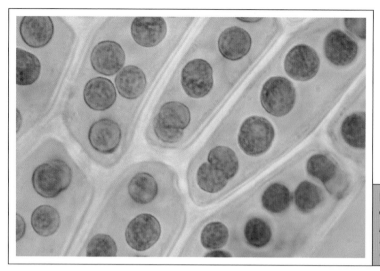

**Fig. 1.4** Chloroplasts in the cells of a moss leaf. *Light microscope photograph by Patrick Echlin, Multi-imaging Centre, University of Cambridge.*

the waters were enriched by minerals released by wave action or carried from the land by rivers and streams. The more varied and somewhat harsher environment of the shore gradually led to the evolutionary selection, some 650 million years ago, of complex, multicellular photosynthetic organisms with their tissues differentiated to form structures which anchored them to the rocks, flattened structures with a large surface-area-to-volume ratio which facilitated the collection of light and air for photosynthesis and with primitive conducting systems enabling them to carry the products of photosynthesis from the surface structures to the submerged parts of the plant. The descendants of these organisms are the large seaweeds such as *Laminaria* spp. (kelps) of present-day coastlines.

## COLONISATION OF THE LAND

The next critical stage in the evolution of the great diversity of complex plants that were eventually to colonise almost the entire Earth was the move from water to land, which occurred some 510 million years ago. The evolution of plants on the sea shore and the colonisation of the land did not mean, however, that plant life ceased to exist in open waters. Microscopic plants still occupy the surface waters of oceans, lakes and rivers in vast numbers as the largely unicellular, photosynthetic algae collectively called the *phyto-*

*plankton.* Indeed, these organisms are responsible for almost 50% of the Earth's photosynthetic productivity, and are a major sink for carbon dioxide ($CO_2$) as well as providing the base of the food chain for most of the life in the oceans and in fresh water.

The colonisation of the land was made possible by the evolution of a number of structural characteristics, in response to the selection pressures of life out of water, that are common to all land plants, even today (see Chapters 2 and 3). The terrestrial environment is rich in carbon dioxide and oxygen, and these diffuse more freely in air than in water, so are more readily available on land. Light is also abundant, undimmed by the filtering effect of water, although the infrared component imposes a significant heat load on the organs that absorb it. Finally, mineral ions are readily available in the soil. The remaining requirement is water, and this, together with the heat load from sunlight, is the key limiting factor to life on land.

The most successful land plants, therefore, had evolved roots to provide anchorage and to collect water from the soil. Flattened structures, the leaves, collected the energy of sunlight and with their large surface-area-to-volume ratio were able to maximise this process while providing the shortest possible pathways for the uptake of carbon dioxide for photosynthesis and oxygen for respiration. The evolution of stems provided a means of holding the leaves in the most advantageous position for the collection of sunlight. A conducting system comprised of two parts, the *xylem* and *phloem*, provided, respectively, a means of transporting water to the leaves and of conducting sugars and growth substances from the leaves to other parts of the plant. The evolution of *secondary thickening*, in which the xylem of the stem, branches and roots proliferates to form wood, enabled the development of shrubby plants and trees.

Two further problems, connected with the leaves, required solutions. Leaves lose water very rapidly and are also subject to a significant heat load from the sunlight. The most successful evolutionary response to these selection pressures was a thin, cellular outer skin, the *epidermis*, covered by a waterproof waxy coating, the *cuticle*, on the aerial parts of the plant. Such tissues, by preventing the loss of water, create two further problems, however: the cooling effect of the evaporation of water as the latent heat of vaporisation is lost, and a barrier is formed to the exchange of carbon dioxide and oxygen for photosynthesis and respiration, respectively. The most successful solution to both problems was the evolution of pores in the cuticle and epidermis, each surrounded by two specialised cells with the capacity of changing shape and thereby opening and closing the pores. These *stomata* enabled land plants to regulate the loss of water and the uptake of gases (see Chapter 8).

Land plants must grow continuously in order to compete with other plants for the available light, and to mine increasing volumes of soil in the relentless search for nutrients and water. Growth must not occur at random over the whole plant body, however, but must be localised in specific regions of cell division and differentiation. These regions are called *meristems*, with the principle ones, the apical meristems, being located at the tips of the stems and roots. Subsidiary meristems are involved in, for example, the development and expansion of leaves and in secondary thickening.

The final significant evolutionary steps in the colonisation of the land concerned the development of reproductive structures that were resistant to desiccation and other environmental stresses. The sexual phases (*gametophytes*) of the life cycle of the first land plants were dependent on water, the female egg cells being fertilised by swimming male gametes. The result was the formation of *sporophyte* generations producing relatively vulnerable *spores*, minute propagules not differentiated internally. Later, more successful plants evolved in which the two phases of the life cycle were brought together and the reproductive cells held in protective, multicellular structures.

The male gametes of these plants, the *pollen grains*, were non-motile and desiccation-resistant, being transferred from plant to plant by wind and later by other agents such as insects. Instead of spores, *seeds* were produced as a result of sexual reproduction. These were structures in which the fertilised embryo was protected from environmental stresses and herbivores by layers of cells which constituted the seed coat. In one of the major groups of seed plants, the Gymnosperms (Table 1.1) the seeds were unprotected. Indeed, the name in translation means naked seeds. The present-day descendants of these plants are the conifers and their relatives. In the second group, the Angiosperms (Table 1.1) the seeds were enclosed in protective layers which constituted the fruit. The present-day descendants of these are the flowering plants. The developing embryos were also provided with a supply of stored food, which meant they could survive unfavourable climatic conditions in a dormant state, germinating and becoming established rapidly as soon as conditions improved.

The seed plants of the present day include the cycads, the gingkos, both of which retain motile male gametes, the conifers, and the angiosperms, or flowering plants (Table 1.1). It is these that dominate the world's flora and which have given rise to the majority of plants that are grown in gardens. They are thus the main subject of the rest of this book.

The early stages of the colonisation of the land by plants involved a great diversity of primitive forms which, although successful at the time, lacked one or more of the suite of characteristics that ultimately made the seed plants the dominant group. The descendants of many of these primitive plants, although less diverse and less successful than their ancestors, still grow on Earth today. Their ability to compete with the seed plants, except in highly specialised habitats, is very limited. They will, however, be familiar to gardeners as *liverworts*, *mosses*, *horsetails* and *ferns* (see Tables 1.1 and 1.2; Figs 1.5, 1.6 and 1.7).

## COMMUNITIES AND THE DIVERSITY OF LIFE FORMS

Once the first land plants had become established, they spread to occupy almost the entire surface of the

**Table 1.1** The main groups of photosynthetic organisms mentioned in Chapter 1.

| Superkingdom | **PROKARYOTAE (cells without a clearly defined nucleus)** | |
|---|---|---|
| **Kingdom** | **Monera** | |
| | Phylum | Cyanobacteria ('blue-green algae') |
| **Superkingdom** | EUKARYOTAE (cells with a membrane-bound nucleus) | |
| **Kingdom** | **Protoctista** | |
| | Phytoplankton green, red and brown 'algae' | |
| | **Plantae** | |
| | Bryophytes | |
| | Phylum | Hepatophyta ('liverworts') |
| | | Bryophyta ('mosses') |
| | Vascular plants without seeds | |
| | Phylum | Sphenophyta ('horsetails')* |
| | | Filicinophyta ('ferns')* |
| | Vascular plants with seeds | |
| | Phylum | Cycadophyta ('cycads')** |
| | | Ginkgophyta ('ginkgo')** |
| | | Coniferophyta ('conifers')** |
| | | Anthophyta ('angiosperms', flowering plants): |
| | | Class    Dicotyledons (two cotyledons in the seed) |
| | | Monocotyledons (a single cotyledon in the seed) |

\* Often referred to together as the Pteridophytes.
\*\* Often referred to together as the Gymnosperms.

Earth. Ultimately, extensive communities of plants and also animals, which had been evolving on land alongside the plants, came into being, their characteristics being largely determined by climate. These *biomes* were the equivalent of today's deserts, tundras, savannahs, rainforests and temperate grasslands, forests and woodlands. How these communities evolved is still only poorly understood. So far, knowledge is largely limited to that arising from studies of the evolution of the individual species that inhabit them, but this can provide no more than a glimpse of the infinitely more complex range of processes and interactions that must have been involved.

The biomes of the present day are made up of smaller communities called ecological systems, or *ecosystems*. These in turn are made up of assemblages of the plants, animals and other living organisms that occupy them, together with their non-living environment. Ecosystems are extremely stable assemblages, although the individuals that occupy them have a defined life span, ranging from a few hours or days in the case of some microorganisms to hundreds or even more than a thousand years in the case of some forest trees.

In the case of plants, the shape, size and form of the individual species are complementary, so that each has access to the light, water, carbon dioxide and the mineral nutrients required for growth and reproduction. Similarly, flowering, seed set and germination are so timed as to enable each species to grow and reproduce at the time of year most favourable for itself and for the other species, such as insect pollinators, with which it has co-evolved. None of this complementarity owes its origins to altruism, of course, but has resulted from competition between species during the evolution of the ecosystem. Indeed, all organisms present in the ecosystem compete for resources, and every organism, no matter how large or small, provides a food source for another organism. By this means, the energy captured by plants from sunlight is passed on in a regulated way throughout the entire ecosystem before being dissipated. Energy from the sun must enter the ecosystem constantly, but all nutrients are cycled through living organisms, eventually being returned to the soil, decomposed by bacteria and fungi, and recycled.

The diversity of biomes, ecosystems and specialised *habitats* (shaded, exposed, arid, saline, aquatic, cold

**Table 1.2** The characteristics of 'primitive' land plants.

| Group and characteristics | Garden habitats |
| --- | --- |
| **Algae** (Fig. 1.3)<br>An informal grouping of simple organisms ranging from unicellular blue-greens and green plankton to macroscopic colonies or chains of green cells and more complex brown-green and red-green seaweeds. The blue-greens are prokaryotes and the rest eukaryotes. Some blue-greens fix atmospheric nitrogen. | Blue-greens sometimes form a slime on wet lawns. Damp or wet places and in ponds. |
| **Lichens** (Lichenised fungi)<br>Symbiotic associations of green algae and certain fungi in the Ascomycotina. | Exposed surfaces such as rocks, walls and tree trunks, where neither partner could survive alone. |
| **Liverworts** (Hepatophyta) (Fig. 1.5)<br>Simple, small green plants without a cuticle, stomata or lignified xylem. Exhibit alternation of generations in which the haploid gametophyte is the dominant phase with the diploid sporophyte attached to and dependent on it. Male gametes are motile, requiring water for fertilisation. Reproduction is by spores and sometimes by production of multicellular *gemmae*. There are two types of gametophyte: in *thallose liverworts* it comprises a flattish, lobed plate of cells; and in *leafy liverworts* it is in the form of a small plantlet with rows of thin flattened leaves on either side of a thin stem and a third row of reduced leaves on the underside; both types have simple rhizoids for attachment. | Thallose types grow in moist, open sites such as shaded paths, and as pot weeds. Leafy types grow among other crowded plants, especially grasses. |
| **Mosses** (Bryophyta) (Fig. 1.6)<br>Simple, small green plants lacking a cuticle, stomata, lignified xylem and roots. May have specialised cells for conducting water and transporting sugars. Exhibit alternation of generations with the gametophyte being dominant, as in liverworts. Gametophytes are usually in the form of small leafy plantlets with simple rhizoids for attachment. Male gametes motile, requiring water for fertilisation. Reproduction by spores. | Moist, shady places or in boggy ground. Many can survive long periods of desiccation, making it possible for some species to grow on roofs, walls and paths and in lawns. |
| **Peridophytes: Ferns** (Filicophyta) and **Horsetails** (Sphenophyta) (Fig. 1.7)<br>Small to large green plants, sometimes small trees, with a stout rhizome and roots. Possess a cuticle with stomata, lignified xylem and phloem. Exhibit alternation of generations with a relatively large, free-living diploid sporophyte and relatively small, vulnerable, heart-shaped gametophyte (the *prothallus*) only a few cells thick. Motile male gametes require water for fertilisation. Diploid sporophyte grows out of prothallus after fertilisation. Sporophyte produces wind-blown spores.<br><br>   At present, the largest pteridophytes are the tree ferns, but in the geological past, as in the Carboniferous, pteridophyte trees grew to great size and formed vast forests. The seed habit evolved in some ferns. Probable that the vulnerability of the gametophyte to desiccation led to the decline of the pteridophytes as the seed plants evolved and out-competed them. | Moist, shady places, especially under trees. Gametophytes require wet conditions but sporophytes can often tolerate some drought. Horsetails, having a stout rhizome, can survive in most conditions. |

**Fig. 1.5** The thallose liverwort *Marchantia*. Note the flattened, green gametophytes with male (shaped like umbrellas) and female (shaped like the ribs of an umbrella) structures growing up from them. Male and female structures are found on separate plants. *Photograph by David S. Ingram.*

**Fig. 1.6** The leafy, tuft-forming gametophytes of the common garden moss *Bryum* with horny, flask-shaped sporophytes growing *in situ* on long stalks from the tips of the gametophytes, where sexual reproduction occurred earlier in the life cycle. *Photograph by Chris Prior, Royal Horticultural Society.*

and montane) within them has led to the evolution among the seed plants of a great diversity of life forms. These are summarised in Tables 1.3 and 1.4, for they have provided horticulturists and plant breeders with the raw material for developing the wide range of plants that grow in gardens today. The subject of garden ecosystems will be considered in Chapter 19.

The transfer of energy and the cycling of nutrients in an ecosystem involves complex sequences of events, with every organism playing a specific part

and with the components of the environment having their role too. It follows that such a stable, well-oiled machine will lose its stability and enter a period of rapid change if any component is changed. Changes in climate, for example, may have dramatic and unpredictable effects. Similarly, the loss of species, in some cases even a single species, may initiate change. The speed of change will depend very much on the nature and size of the perturbation.

The appearance of humankind on the planet did not occur until about 2 million years ago. To begin

**Fig. 1.7** Fern sporophytes, including tree ferns, growing at the Royal Botanic Garden Edinburgh. *Photograph by David S. Ingram.*

with our ancestors lived as hunter-gatherers, constantly moving from place to place and using the diversity of plant and animal life for food and other purposes wherever they found it. The cultivation of crops in agriculture and horticulture was a very recent development, first occurring only some 12 000 years ago. This was a most significant event, however, for it made it possible for humans to live in settlements: villages, towns and eventually cities. Time brought sophistication of agricultural and horticultural technology, an appreciation of the importance of cultivating plants for aesthetic reasons as well as to supply food and raw materials, and an ever-deepening understanding of the scientific principles that underlie plant evolution, growth, development, reproduction and classification, the subject of most of the rest of this book.

Time also brought population growth, industry and commerce. The resulting relentless expansion of cities and transport systems, over-exploitation of the natural environment, and increasing levels of pollution, especially of the atmosphere and oceans, have led to global environmental changes. These include *global warming*, ozone depletion and the catastrophic erosion of biodiversity. They represent the greatest challenges humankind has ever had to face. They are discussed from the horticultural perspective in Chapter 18.

**Table 1.3** The diversity of basic types of seed plants.

| Type | In the wild | In the garden (with examples) |
|---|---|---|
| **Annuals**<br>Usually herbaceous plants that complete their life cycle of germination, growth, flowering and seed set during one growing season, sometimes in only a few weeks. | Opportunistic primary colonisers of disturbed ground or, in arid regions, able to germinate from dormant seeds, flower and set seed again very rapidly following rain at the end of a period of drought. | Weeds in newly cultivated soil (groundsel, *Senecio vulgaris*).<br>    May be used to provide 'instant flowers' (poppies, *Papaver* spp.) and rapidly maturing vegetables from sown seed (lettuce, *Lactuca sativa*).<br><br>*(Continued)* |

**Table 1.3** (*Continued*)

| | | |
|---|---|---|
| **Biennials** Short-lived herbaceous plants that flower and set seed in the second season following germination. Usually require a trigger, such as low temperature (vernalisation) during winter to induce the formation of flowers. Sometimes have woody flower stalks. | Opportunistic early colonisers of disturbed or bare ground whose life cycle is regulated with precision by the seasons to ensure that flowering occurs at the most favourable time for pollination and seed set. | Weeds in newly cultivated soil (common foxglove, *Digitalis purpurea*). May be used to provide flowers rapidly, usually early in the growing season (wallflower, *Erysimum cheiri*). Spring vegetables, from seed sown during the summer or autumn of previous year (broccoli, *Brassica oleracea* Italica group). Some biennial vegetables are grown as annuals (carrot, *Daucus carota*). |
| **Herbaceous perennials** Plants that continue to grow for more than two seasons. Usually non-woody, but may have woody flower stalks. May require an environmental trigger such as low temperature (vernalisation) to induce flowering. May be evergreen, never dying back, or may die down during unfavourable periods such as summer drought or winter cold, regrowing from a perennial root stock. | Long-term, highly competitive components of grasslands and other open habitats, including mountain and alpine. Comprise the bulk of the herb layer in scrub, woodlands and forests. | Persistent aggressive weeds in permanent plantings (willowherbs, *Epilobium* spp.) including lawns (dandelion, *Taraxacum officinale*). Permanent plantings in a diversity of situations including: lawns (grasses, Poaceae), alpine gardens (*Lewisia* spp.), herbaceous and mixed borders (phlox, *Phlox* spp.), underplantings among trees and shrubs (germanders, *Teucrium* spp.) and herb gardens (oregano, *Origanum* spp.). Permanent sources of vegetables (good king Henry, *Chenopodium bonus-henricus*) or fruit (strawberry, *Fragaria* spp.). |
| **Shrubs** Woody perennial plants with many branches arising at or close to ground level and lacking an obvious trunk. May be deciduous, losing their leaves during unfavourable periods such as winter cold or summer drought; or evergreen, retaining leaves throughout the year, allowing photosynthesis to occur immediately following the alleviation of unfavourable conditions such as the shade from deciduous trees, winter cold or summer drought. Older evergreen leaves are eventually shed. | Principal and dominant component of scrublands (e.g. Mediterranean-type region or tundra), or alpine regions, or grasslands (e.g. savannah) or as significant understorey components of woodlands and forests of every type in temperate, sub-tropical and tropical regions. | Provide height, shape, form and flowers in every part of the garden, but especially valuable in mixed borders (*Weigela* spp.), shrubberies (mock orange, *Philadelphus* spp.), among trees (*Rhododendron* spp.), as specimens in lawns (*Cornus kousa*), in specialised plantings (*Rosa* spp.), herb gardens (rosemary, *Rosmarinus officinalis*) and alpine gardens (dwarf conifers and willows, *Salix* spp.). Permanent sources of fruit in season (currants, *Ribes* spp.). |
| **Trees** Woody perennial plants with a single main stem (or sometimes more than one main stem), usually branching well above the ground to form an elevated crown. The distinction between large shrubs and small trees is not clear cut. As with shrubs (see above) may be deciduous or evergreen. May be very long lived (oaks, *Quercus* spp.) or relatively short lived (birches, *Betula* spp.). | Dominant and overshadowing component of woodlands and forests in all regions of the world, or may occur as isolated specimens in grasslands such as savannahs or other open situations. | Provide height, shape, form, shade and sometimes flowers in all parts of the garden. There is a great diversity of types including long-lived specimens for large gardens (beeches, *Fagus* spp.), smaller, less long-lived specimens for smaller gardens (*Acer* spp.) and flowering types (*Prunus* spp.). Provide permanent supplies of fruit in season (apples, *Malus* spp.; cherries and plums, *Prunus* spp.). Seedlings may occur as weeds (sycamore, *Acer pseudoplatanus*). |

**Table 1.4** Examples of the diversity of adaptations of the basic types of seed plants. Such adaptations have evolved in response to selection by particular environmental conditions, sometimes extreme. Many adaptations have often evolved in parallel in many different and unrelated species and groups of plants. The great diversity of flower and fruit types are not dealt with here but are referred to in the text, principally in Chapters 2 and 3.

| In the wild | In the garden (with examples) |
|---|---|
| **Alpines and dwarf plants**<br>Long-lived herbaceous or frequently woody perennials reduced in size and with other adaptations to withstand the often short growing season, strong winds, free-draining soils and periods of exposure to extremes of heat, cold and desiccation that typify mountainous habitats and northern climates. | Valuable in rock or alpine gardens and containers (dwarf conifers, *Juniperus* spp.; dwarf broad-leaved shrubs, *Salix herbacea*; rosette- and mound-forming herbaceous perennials, *Saxifraga* spp.). |
| **Aquatic and bog plants**<br>Usually herbaceous perennials adapted to grow in bogs and at the margins of lakes or streams (*Iris* spp.), or as rooted plants with floating leaves in still water (water-lilies, *Nymphaea* spp.) or moving water (water crowfoot, *Ranunculus aquatilis*), or as free-floating plants in still water (duckweed, *Lemna minor*).<br><br>Adaptations include stomata only on the upper surface of floating leaves, air-conducting (aerenchyma) cells in submerged plants and mechanisms to trap insects (see Insectivorous plants). | May be used to provide form, flowers and oxygen in gardens with still or moving water (see left for examples).<br><br>May occur as weeds in some water gardens (*Lemna minor*, *Ranunculus aquatilis*). |
| **Aromatic plants**<br>Herbaceous and shrubby forms, sometimes trees, especially from relatively dry climates (e.g. Mediterranean-type), that produce aromatic oils and other chemicals (see Chapter 12), which may deter herbivores, parasites and pathogens. | Culinary herbs and spices (mints, *Mentha* spp.; thymes, *Thymus* spp.).<br><br>May also be used to provide scent as well as form and flowers in non-culinary plantings (*Eucalyptus* spp., *Cistus* spp.). |
| **Climbers, vines and scramblers (Lianes)**<br>Plants adapted to gain height, thereby reaching sunlight, in shrublands, woodlands, forests and rocky ground by growing in, on or over other, usually woody, species or rocks. Mechanisms for attaching plants to other species include twining stems (honeysuckle, *Lonicera* spp.), leaf tendrils (peas, *Pisum* spp.), branch tendrils (vines, *Vitis* spp.), twisting petioles (*Clematis* spp.), hooked thorns (*Rubus* spp.; *Rosa* spp.) and adventitious roots arising from the stem (ivies, *Hedera* spp.). | Used in a diversity of situations such as in and through trees and hedges, against walls and fences and on trellises and pergolas to provide height, cover, form, flowers and fruit. See left for examples. May occur as weeds in permanent plantings (*Convolvulus* spp.). |
| **Epiphytes**<br>Usually herbaceous perennial plants adapted to gain height and thereby access to sunlight by growing on other species, usually on the branches of trees and large shrubs. Sometimes occur in temperate regions, but most frequent in tropical and sub-tropical rain and mist forests, where modified roots (orchids, Orchidaceae) or the whole plant surface (Spanish moss, *Tillandsia usneoides*) may absorb water from the atmosphere, or modified rosettes of leaves may collect rainwater in the 'tank' created by the leaf bases (bromeliads, Bromeliaceae). Minerals are in short supply, and in the absence of soil are absorbed direct from rainwater or from decomposing plant and animal remains trapped by the plants. | Main value in cultivation is as glasshouse plants (especially orchids, Orchidaceae) or as house plants (especially bromeliads, Bromeliaceae; e.g. urn plant, *Aechmea fasciata*).<br><br><br><br><br><br><br><br>(*Continued*) |

**Table 1.4** (*Continued*)

| | |
|---|---|
| **Geophytes**<br>Usually herbaceous, sometimes woody, species with a perennial, underground structure that remains dormant during unfavourable conditions such as shade from deciduous trees (spring bulbs and corms), extreme drought or cold (spring bulbs; *Dahlia* spp.; potatoes, *Solanum tuberosum*). | Used in permanent plantings to provide flowers in winter (*Cyclamen coum*), spring (*Narcissus*, *Tulipa* and *Crocus* spp.) and autumn (*Cyclamen hederifolium*). Or for summer plantings (*Dahlia* spp.).<br>    Or as vegetables (onion, *Allium cepa*; potatoes) or herbs (chives, *Allium schoenoprasum*; garlic, *A. tuberosum*).<br>    May occur as weeds (lesser celandine, *Ranunculus ficaria*; couch grass, *Elymus repens*). |
| **Hairy plants**<br>Mainly herbaceous perennials in which the epidermal cells of the leaf are modified to form hairs arising from the surface. These may be sparse (hollyhock, *Alcea rosea*) or dense (lamb's ears, *Stachys byzantina*). Depending on density, the hairs may create turbulence around the leaf, facilitating the uptake of $CO_2$, or extend the width of the still-air layer, thereby reducing transpiration, or may provide protection from excessive sunlight. | Plants with densely hairy leaves often appear silver or white and have value in mixed borders, especially in hot, dry conditions (lamb's ears, *Stachys byzantina*; lavender, *Lavendula* spp.). |
| **Halophytes**<br>Usually herbaceous or shrubby plants adapted to grow in soils or atmospheres with a high salt content, in deserts, around coasts and in salt marshes. High salt levels restrict water uptake, so adaptations include mechanisms to reduce water loss, including succulence, fleshy leaves, hairy or waxy leaves and sunken stomata. May also possess physiological mechanisms for excreting or excluding salt. | Coastal gardens (sea kale, *Crambe maritima*; sea lavender, *Limonium* spp.). Often have strong architectural shapes or appear white or silver, and may be valuable in creating special planting effects such as in 'white gardens'. |
| **Insectivorous plants**<br>Plants adapted to grow in positions low in nitrogenous minerals such as peat bogs (sundew, *Drosera* spp.) or as epiphytes on the branches of trees or shrubs (pitcher plants, *Nepenthes* spp.) by trapping insects, which are then digested by enzymes. Methods of trapping include: sticky leaves which roll (butterwort, *Pinguicula vulgaris*); glandular leaf hairs which secrete mucilage (sundews, *Drosera* spp.); leaves with toothed margins and hinged midribs, which snap shut (Venus flytrap, *Dionaea muscipula*); underwater bladders (bladderwort, *Utricularia* spp.); and water-filled pitchers (pitcher plant, *Nepenthes* spp.). | Usually only grown as specialised glasshouse plants (Venus flytrap, pitcher plant) or occasionally in bog gardens (sundews, butterwort, bladderwort). |
| **Parasitic plants**<br>Adapted to obtain a supply of carbohydrate from another species. May be completely dependent on the host and lack chlorophyll (holoparasites: toothworts, *Lathraea* spp.) or partly dependent on the host and having some chlorophyll (hemiparasites: yellow rattle, *Rhinanthus minor*). May grow beneath the soil, producing only flowers above ground (toothworts) or among other plants in meadows (yellow rattle) or as epiphytes (mistletoe, *Viscum album*). | May have decorative flowers (the holoparasite, *Lathryaea clandestina*), or may have decorative value in trees (mistletoe) or may be used in meadow plantings to add diversity (yellow rattle).<br>    Sometimes occur as weeds, especially in hot climates.<br>                                      (*Continued*) |

**Table 1.4** (*Continued*)

| | |
|---|---|
| **Resurrection plants**<br>Plants adapted to grow in arid regions. May dry out and remain dormant for long periods, being induced to grow again by the onset of rain (rose of Jericho, *Anastatica hierochuntica*). | Of little value but may sometimes be sold as curiosities for use as house plants. |
| **Rosette plants**<br>Plants in which the stem is reduced in length so that the leaves form a tight rosette which may be resistant to grazing (daisy, *Bellis perennis*), desiccation or cold winds (many alpines; *Primula* spp.). Stems often elongate to form flowers. | A great diversity of garden plants form rosettes, but especially valuable in rock and alpine gardens (*Lewisia* spp.).<br>    Also common as lawn weeds (daisy), where mowing is the equivalent of grazing. |
| **Sclerophylls**<br>Evergreen perennial plants, shrubs and trees with 'hard' leaves; i.e. thick cuticle and epidermis, with considerable internal support tissues, especially modified and lignified cells. Adapted to grow in arid regions and in Mediterranean-type climates. | Of great value as garden plants in a diversity of situations, especially in the densely populated south-east region of the UK, where lack of water makes it necessary to grow plants that are adapted to dry conditions (cherry laurel, *Prunus laurocerasus*; Portuguese laurel, *Prunus lusitania*; holly, *Ilex* spp.; many *Cistus* spp.; bay or true laurel, *Laurus nobilis*; holm oak, *Quercus ilex*; most conifers). |
| **Spiny plants**<br>Plants, especially shrubs and trees, adapted to resist damage from browsing and grazing mammals by virtue of particular organs being adapted as or reduced to spines, notably leaves (holly, *Ilex* spp.; members of the Cactaceae) and stem branches (many *Prunus* spp.; gorse, *Ulex* spp.). Often grow in arid or semi-arid environments. Where leaves are reduced to spines, as in the Cactaceae, photosynthesis usually occurs in the stem. | May be used in a variety of garden situations (*Prunus* spp., *Ilex* spp.) or as specialist glasshouse and house plants (Cactaceae). |
| **Succulents**<br>Plants from arid regions adapted to store water in modified cells in thickened fleshy leaves (houseleeks, *Sempervivum* spp.) and stems (many cacti, Cactaceae). | Valuable in rock gardens and on roofs (houseleeks; *Sedum* spp.), as house plants (*Kalanchoe* spp.) or as specialised house and glasshouse plants (cacti). |
| **Switch plants**<br>Shrubs and small trees adapted to dry regions, especially with a Mediterranean-type climate, by having reduced leaves, often lost during very dry periods, and photosynthetic, often winged stems with considerable internal support tissue comprising heavily thickened or lignified cells. | As with sclerophylls, of great value in gardens where water is in short supply (brooms, *Cytisus* spp.). |
| **Xerophytes**<br>Herbaceous plants, shrubs and trees adapted to grow in dry conditions (see above: Alpines and dwarf plants, Hairy plants, Resurrection plants, Rosette plants, Sclerophylls, Spiny plants, Succulents and Switch plants). | Of great value in gardens and situations where water is in short supply (see above). |

## CONCLUSION

At the heart of every biome and ecosystem is the great diversity of photosynthetic plants and microbes, for these are the only organisms capable of capturing the energy of sunlight and of generating the oxygen required to sustain life. They are also the source of the main food and other natural products that sustain human societies and, as every gardener knows, they are among the most beautiful living things on Earth. The erosion of biodiversity, global warming and other aspects of global environmental change should be of great concern to all. Acquiring an understanding of the scientific basis of plant classification, and how plants have evolved, grow and reproduce, not only makes gardening itself more pleasurable, and often more effective, but also engenders a sensitivity to the forces that threaten life on Earth and the knowledge required to mitigate them through individual and collective action (see Chapter 18). The chapters that follow provide a first step in the acquisition of such understanding.

## FURTHER READING

Attenborough, D. (1979) *Life on Earth*. BBC Books, London.

Attenborough, D. (1995) *The Private Life of Plants*. BBC Books, London.

Beerling, D. (2007) The Emerald Planet. Oxford University Press, Oxford.

Bernhardt, P. (1999) *The Rose's Kiss: a Natural History of Flowers*. Island Press/Shearwater Books, Washington DC.

Dawson, J. & Lucas, R. (2005) *The Nature of Plants – Habitats, Challenges and Adaptations*. Timber Press, Portland, OR.

Lewington, A. (2003) *Plants for People*. Eden Project Books, London.

Mabberley, D.J. (1997) *The Plant Book*, 2nd edn. Cambridge University Press, Cambridge.

Margulis, L. & Schwartz, K.V. (2000) *Five Kingdoms*, 3rd edn. W.H. Freeman and Company, New York.

Pollock, M. & Griffiths, M. (2005) *RHS Illustrated Dictionary of Gardening*. Dorling Kindersley, London.

Raven, P.H., Evert, R.F. & Eichhorn, S.E. (1999) *Biology of Plants*, 6th edn. Worth Publishers, New York.

Walters, S.M. (1993) *Wild and Garden Plants* (New Naturalist Series). Harper Collins Publishers, London.

Wilson, E.O. (1992) *The Diversity of Life*. Allen Lane, The Penguin Press, London.

Wilson, E.O. (2003) *The Future of Life*. Abacus, London.

# 2

# Know Your Plant: Structure and Function

## SUMMARY

In this chapter the structure of leaves, stems and roots is outlined. The processes that occur in these organs – photosynthesis, respiration, transpiration, water and mineral uptake and sugar, water and mineral transport – are described, with special emphasis on the intimate relationship between structure and function. The types and functions of plant hormones are summarised.

## INTRODUCTION

Plant structure is intimately associated with function, the two having evolved together to produce species that are able to grow in and adapt to a wide range of environmental conditions and to compete successfully with a diversity of other organisms. Very often, the result of such evolution is a compromise, for conflicting selection pressures can only rarely be reconciled completely. This is particularly apparent when the structure of the leaf is considered in relation to the range of functions it performs: notably photosynthesis, respiration and transpiration.

## ENERGY FLOW IN THE BIOSPHERE: PHOTOSYNTHESIS AND RESPIRATION

Photosynthesis, the process whereby plants fix the physical energy of sunlight into the chemical energy of sugars (Fig. 2.1), begins when a molecule of chlorophyll (Fig. 2.2) is excited by a quantum of light (a photon). Chlorophyll absorbs light in the blue and red regions of the visible spectrum (380–780 nm; where nm means nanometre) and reflects green light, which is why plants appear green. This happens deep in the chloroplast (Figs 1.4 and 2.3), where the electrons energised by sunlight are captured by *photosystems*, which are complexes of proteins and pigments located in the thylakoid membranes. This physical energy is then used to split molecules of water ($H_2O$). Electrons are transferred from the water to an electron acceptor called a quinone and oxygen ($O_2$) is released into the atmosphere as a by-product. The energy is then incorporated into energy-rich chemical molecules called ATP and NADPH.

This chemical energy is used in the main body (the *stroma*) of the chloroplasts to convert carbon dioxide ($CO_2$) from the atmosphere into sugars. The process is a cyclical one, sometimes called the Calvin cycle after Melvin Calvin who, together with co-workers, first discovered it. During the cycle, which takes place in the dark, the carbon dioxide combines with ribulose bisphosphate to form an unstable molecule containing six carbon atoms. It immediately breaks down to form two molecules of the three carbon compound glycerate 3-phosphate. Finally, ribulose bisphosphate is regenerated and the sugars glucose and fructose are produced.

The Calvin cycle, like all metabolic pathways, is driven by *enzymes*, proteins that act in minute

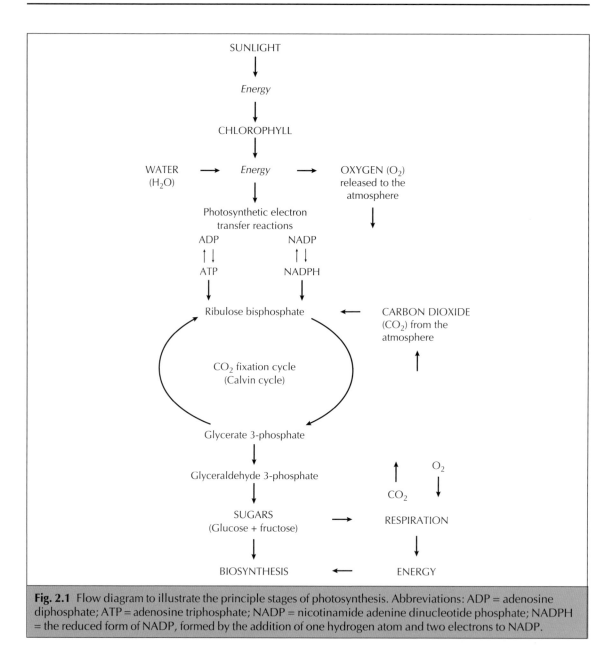

**Fig. 2.1** Flow diagram to illustrate the principle stages of photosynthesis. Abbreviations: ADP = adenosine diphosphate; ATP = adenosine triphosphate; NADP = nicotinamide adenine dinucleotide phosphate; NADPH = the reduced form of NADP, formed by the addition of one hydrogen atom and two electrons to NADP.

amounts in biological systems to promote chemical changes without being changed themselves. The key enzyme in the Calvin cycle, ribulose bisphosphate carboxylase, is probably the most abundant protein on Earth.

The sugars produced during photosynthesis may be combined with oxygen during *respiration* to release the stored energy. This takes place in *organelles* called *mitochondria* which occur in all cells of the plant body (Fig. 2.4). It is basically the same process as is used to release energy from carbohydrates in animals. The energy released is then used to drive the biosynthetic processes that lead to the formation of cellulose, fats, amino acids, proteins and the other molecules required by the plant for growth and reproduction.

Some 200 billion tonnes of carbon ($0.7 \times 10^{14}$ kg) from atmospheric carbon dioxide are fixed by

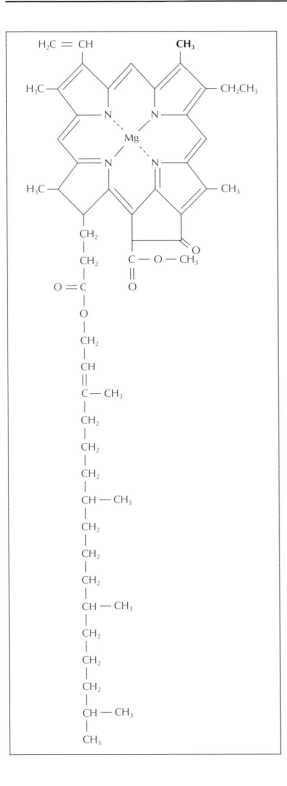

photosynthesis each year. Approximately half of this occurs on land, where the familiar leafy green plants predominate and approximately half is fixed from carbon dioxide in solution in the oceans, lakes and rivers, where the microscopic green plants called *algae* abound as part of the *plankton* (see Chapter 1). So, through photosynthesis, plants maintain the atmosphere by consuming carbon dioxide and releasing oxygen; and produce the energy that not only builds the plant body itself but also provides the primary source of food for almost all other living things.

## THE LEAF: INTO THE LABYRINTH

### The manufacturing centre

Although photosynthesis occurs in all the green parts of a plant, the principal site for this process is the leaf. Leaves come in many different shapes and sizes, but most are immediately identifiable as leaves because they have evolved with certain basic features in

**Fig. 2.2** (*left*) A molecule of the photosynthetic pigment *chlorophyll*. Chlorophyll is a complex molecule comprising a magnesium atom held within a ring of carbon atoms and possessing a long carbon–hydrogen tail. The latter anchors the molecule to the chloroplast membranes (see Fig. 2.3). Chlorophyll exists in various forms which differ from one another in structure and in light-absorbing properties. Chlorophyll **a**, illustrated here, occurs in the Cyanobacteria, and in all green plants, where it is the principal light-absorbing pigment. It absorbs light in the blue and to a lesser extent, the red parts of the spectrum. Plants also contain chlorophyll **b**, which differs from chlorophyll **a** in having a –CHO group instead of the –CH$_3$ printed in bold on the diagram. It absorbs light in slightly different parts of the spectrum from chlorophyll **a** and passes the energy so captured to chlorophyll **a** itself. It is thus an *accessory pigment* that serves to broaden the range of wavelengths of light that can be used in photosynthesis. A third form of the pigment, chlorophyll **c**, is found only in certain groups of algae.

**Fig. 2.3** An ultra-thin section of a plant *chloroplast.* Clearly visible are the *thylakoid* membranes, stacked at intervals to form more complex structures called *grana. Electron microscope photograph by Patrick Echlin, Multi-imaging Centre, University of Cambridge.*

common, not least the extent of their external and internal surfaces. The external surfaces maximise the collection of sunlight and the internal ones facilitate the movement of the gases carbon dioxide and oxygen between the air and the chloroplasts.

## Cells

Leaves are made up of *cells* (Fig. 2.4) and although these are usually differentiated to perform particular functions, the majority possess one or more large, membrane-bound, fluid-filled sacs called *vacuoles*, structures not normally found in animal cells. The two most important functions of vacuoles probably relate to photosynthesis, ensuring the most effective deployment of *cytoplasm*, the metabolically active material in the cell which contains the chloroplasts and the other organelles. First, the vacuoles force the cytoplasm towards the edges of the cells with the result that the diffusion paths for gases such as carbon dioxide into and out of the chloroplasts and mitochondria are as short as possible. This is a very important consideration since gases diffuse only very slowly through water, the medium in which the contents of the cytoplasm are dissolved or suspended.

Second, by restricting the cytoplasm to the edges of the cells, where it is most needed, the vacuoles ensure that the plant is able to achieve a large surface-area-to-volume ratio. This is a prerequisite for efficient photosynthesis, with a minimum investment of energy in making cytoplasm. Other, ancillary, functions of vacuoles may be to act as reservoirs for the storage of enzymes and metabolic products not immediately required by the cells, and to act as convenient stores for waste products. The latter are discarded when old leaves are shed, as in the autumn. The uptake of water into cells and vacuoles is described in Fig. 2.4.

## Layout of the leaf

The most obvious feature of the leaf (Fig. 2.5) is the relatively large, thin plate of tissue called the *lamina*, with its upper surface usually orientated towards the sun. The lamina is often attached to the stem by means of a stalk, the *petiole*, and this extends into the lamina as the *midrib*, branching to produce a network of *veins*. At the base of the petiole there may be further leafy outgrowths, called *stipules* (as in pea, *Pisum sativum*). In the angle between the petiole and the stem, called the *axil*, is a bud, the *axillary bud*, which may ultimately produce a shoot or a flower. There are many variations on this basic structure, some of which will be described below.

To appreciate the complex arrangement of the different cells and tissues within the leaf it is necessary to examine thin sections with a microscope, and to build up a three-dimensional picture (Fig. 2.6).

## Controlling gas and water exchange

An unprotected living structure with a large surface area relative to its volume, constantly exposed to the sun and the wind, will inevitably lose water very rapidly. This does not occur with leaves, because the entire surface is covered with a thin, waterproof layer, the *cuticle*, largely composed of *cutin*, a hydrophobic, fatty molecule called a polyester.

Beneath the cuticle lies the epidermis, which surrounds the entire leaf (Fig. 2.6). It comprises a single layer of cells, most of which are relatively unspecialised, without chloroplasts. Some, however, contain chloroplasts and are differentiated as pairs of

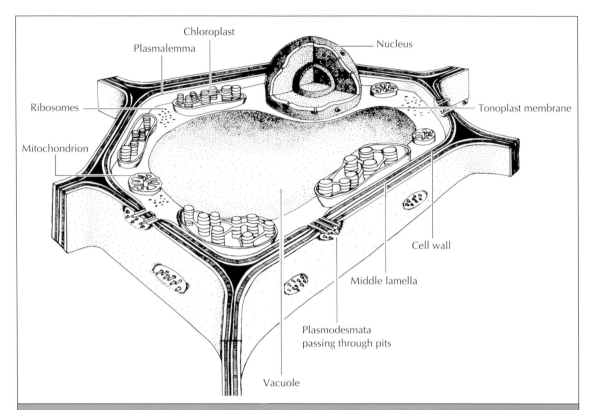

Chloroplast
Plasmalemma
Nucleus
Ribosomes
Tonoplast membrane
Mitochondrion
Cell wall
Middle lamella
Plasmodesmata
passing through pits
Vacuole

**Fig. 2.4** Diagram of a plant cell seen in section. Clearly visible within the *cytoplasm* are the following organelles: the *vacuole*, bounded by the *tonoplast* membrane; the *nucleus*, which contains the *chromosomes*; *chloroplasts*, the sites of photosynthesis; *mitochondria*, with their folded inner membranes, the sites of respiration; and *ribosomes*, the sites of protein synthesis. The cytoplasm is bounded by a membrane, the *plasmalemma*, and the whole cell is enclosed by the *cell wall*. The cell is linked to its neighbours by strands of cytoplasm, the *plasmodesmata*, which occur in groups and pass through holes in the cell wall called pits.

The two major cell membranes, the plasmalemma and the tonoplast, like all plant membranes, are made up of two layers of lipid (fatty) molecules with globular proteins that straddle the two layers occurring at intervals. These membranes are not impenetrable barriers: they block the movement of most *solutes* (dissolved substances), but allow the movement of water by simple *diffusion*. They are therefore said to be *differentially permeable membranes* (they are sometimes referred to as *semi-permeable membranes*) and the movement of water through them is called *osmosis*. Osmosis involves the flow of water, through a differentially permeable membrane, from a region of low solute concentration (and, therefore, high water concentration or high *water potential*) to a region of high solute concentration (and, therefore, low water concentration or low water potential). Water continues to move across the membrane until equilibrium is reached or until sufficient pressure is applied to the solution with low water potential to balance the pressure created by movement of water from the solution with high water potential.

In the case of the plant cell, water moves across the plasmalemma, into the cytoplasm and vacuoles, until the hydrostatic pressure exerted by the plasmalemma and the rigid cell wall prevents further movement. The cell is then said to be *turgid* and is stiff and hard, as a bicycle tyre is when the inner-tube is pumped full of air. When the cells are not turgid, the plant wilts. The hydrostatic pressure in a plant cell that is turgid is often referred to as the *turgor pressure*. The opposing, inwardly directed pressure exerted by the cell, that balances the turgor pressure, is called the *wall pressure*.

Hydrophobic molecules, such as oxygen, and small uncharged molecules, such as carbon dioxide, can also move across the cell membranes by simple diffusion. Other molecules such as nutrients, hormones, waste products and toxins that the cell must take up, secrete or excrete, as appropriate, are carried across the membranes by *transport proteins*. The transport process consumes energy, mainly from the energy-rich molecules of *ATP*.

The cell wall, which surrounds the cell, consists of a mesh or 'basketwork' of rigid *cellulose microfibrils* which have great physical strength. This microfibril mesh is permeated by a *matrix* of softer, somewhat sticky *hemicellulose* and *pectic polymers* and the whole structure is bound together by long sugar–protein molecules called *glycoproteins*. The walls of older cells may be impregnated with the aromatic polymer *lignin*, the principal component of woody tissue, or the fatty polymer *suberin*, the principal component of *cork*.

Cells are joined to one another by a *middle lamella*, a thin layer of pectic substances which are continuous with the pectic substances of the wall matrix. *Drawing by Chris Went.*

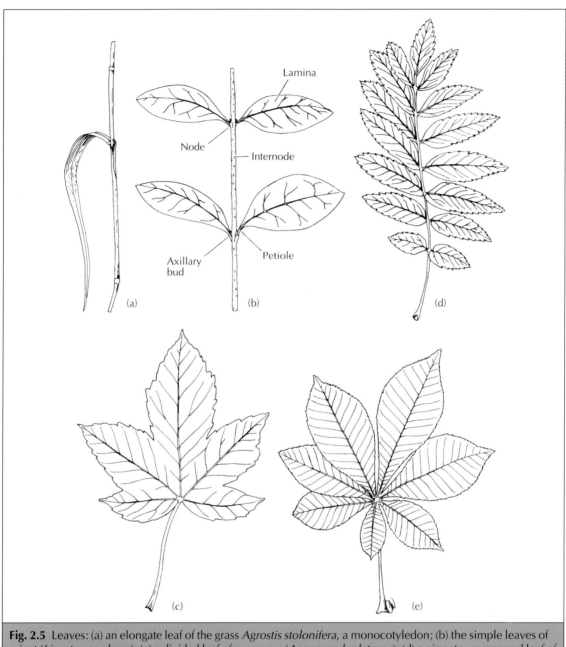

**Fig. 2.5** Leaves: (a) an elongate leaf of the grass *Agrostis stolonifera*, a monocotyledon; (b) the simple leaves of privet (*Ligustrum vulgare*); (c) a divided leaf of sycamore (*Acer pseudoplatanus*); (d) a pinnate, compound leaf of rowan (*Sorbus aucuparia*); (e) a palmate compound leaf of horsechestnut (*Aesculus hippocastanum*). Note that (b), (c), (d) and (e) are all dicotyledons. *Drawing by Chris Went.*

banana-shaped *guard cells*, which surround the specialised pores, the *stomata* (Fig. 2.6). The walls of the guard cells are thickened in such a way that the cells change shape as they take up or lose water, opening or closing the pores. The spacing and distribution of stomata and their role in controlling the exchange of carbon dioxide, oxygen and water vapour between the leaf and the atmosphere is described in Chapter 8.

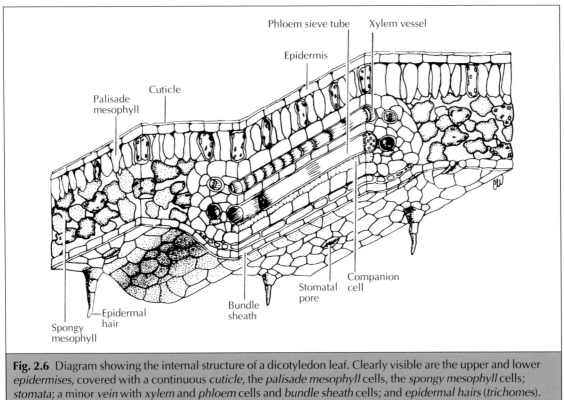

**Fig. 2.6** Diagram showing the internal structure of a dicotyledon leaf. Clearly visible are the upper and lower *epidermises*, covered with a continuous *cuticle*, the *palisade mesophyll* cells, the *spongy mesophyll* cells; *stomata*; a minor *vein* with *xylem* and *phloem* cells and *bundle sheath* cells; and *epidermal hairs* (*trichomes*). *Drawing by Chris Went.*

## Protection against harmful radiation

An important feature of the epidermis is that in most plants its cells contain pigments called flavanoids which filter out potentially damaging longer-wavelength ultraviolet (UV) light (320–400 nm). The cuticle of exposed desert plants, like *Yucca* spp., also contains chemicals that filter out shorter-wavelength ultraviolet light (280–320 nm). Other cells of the epidermis may be differentiated as hairs (*trichomes*), giving some leaves a downy appearance (e.g. dead-nettles, *Stachys* spp.), or may secrete waxes, deposited on the leaf surface as crystals, giving leaves a characteristic bloom (e.g. cabbage, *Brassica oleracea*), or aromatic oils (e.g. rosemary, *Rosmarinus officinalis*; see Chapter 12), giving plants a characteristic smell. Hairs, waxes and oils may all have specialised functions in adapting plants to particular extreme environments.

## Palisade tissues: the sites of photosynthesis

Beneath the upper epidermis is the *palisade tissue* of the leaf (Fig. 2.6), consisting of one or more layers of elongate cells containing numerous chloroplasts. This is where the energy of sunlight is collected and most of the photosynthetic processes occur. Plants that live in the shade, where light levels are low, usually have only one layer of palisade cells, but plants that grow in strong sunlight, where light penetrates deeper into the leaf, invest in two or even three layers of palisade cells to maximise the energy gain. The palisade cells are not attached to one another, but 'hang' from the upper epidermis. The moist walls of the cells thus have a very large surface area exposed to the internal air spaces of the leaf for the diffusion in and out of carbon dioxide and oxygen, respectively.

Beneath the palisade layer, providing support for the palisade cells, is the *spongy mesophyll* (Fig. 2.6). The cells are large and irregularly shaped, with relatively few chloroplasts. There are large air spaces between these cells, especially around the stomatal pores, where they form the sub-stomatal cavities.

Carbon dioxide represents only about 0.03% of the Earth's atmosphere, but this figure is now rising. Since plants have no moving parts to pump the gas into the leaf, it must diffuse along a very weak concentration gradient from its highest concentration in the outside air to the chloroplast molecules of the palisade cells, each stomatal pore connecting with about $100 \, mm^3$ of leaf tissue. The leaf's geometry, just described, is such as to maximise the steepness of this gradient and consequently the rate of flow of the gas. It will be shown in Chapter 8 that the spacing of the stomata is such that carbon dioxide enters the leaf at the maximum rate possible, within the constraints imposed by the need to conserve water. It then diffuses freely through the branching labyrinth of air spaces between the spongy and palisade mesophyll cells before entering the chloroplasts of the palisade by diffusing through the liquid phase of the cell walls, cytoplasm and chloroplast membranes (Fig. 2.6). The diffusion of carbon dioxide in liquid is about 1000 times slower than in air, which means that this last stage is the slowest part of the journey for each carbon dioxide molecule. However, the distance each molecule has to travel in liquid is very short, being about 1000 times less than the rest of the journey through the leaf. Moreover, the surface area of palisade cells available for absorption is some 15 to 40 times the area of the lower surface of the leaf. The net result is that when the stomata are open, carbon dioxide reaches the chloroplast at the incredibly high level of about 50–80% of its concentration in the outside atmosphere, and the rate of uptake of the gas by the chloroplasts roughly balances the rate at which it enters the stomata.

## The transport system

The leaf, like any good manufacturing unit, requires a reliable and efficient system of transport to bring in raw materials and to take away the manufactured products. This function is served by the *vascular bundles*, which are visible as the veins of the leaf, widest towards the leaf base, branching and narrow-

ing into the body and tip of the lamina so that the entire structure is permeated. As a rough generalisation, the main veins of monocotyledonous plants such as grasses (Poaceae) run roughly parallel with one another along the main axis of the leaf. In contrast, the veins of dicotyledonous plants such as *Acer* spp. are branched to a greater or lesser extent (Fig. 2.5). Each vascular bundle includes: *xylem* tissues, comprising *vessels* and fibres; *phloem* tissues, comprising *sieve tubes* and *companion cells*, and *cambium*, and is surrounded by a *bundle sheath* (Figs 2.6 and 2.7).

## Movement of water, minerals and hormones in the xylem

The xylem cells are elongate, have lost their living contents and have walls thickened with *cellulose* and spirals, rings or a network of the tough aromatic polymer *lignin*, the main component of wood. Together, these cells provide a rigid support for the thin plates of cells that constitute the rest of the leaf lamina. The xylem vessels have a second, equally important, function, which is to carry water to the leaf. This is essential to maintain the turgidity of the leaf cells, to provide an aqueous environment for the metabolic processes going on within those cells, and to carry dissolved mineral nutrients such as nitrates and phosphates, taken up from the soil by the roots. Some *plant hormones* (Table 2.1) manufactured in the roots, such as *cytokinins*, may also be carried in the xylem. The water in the xylem also contains dissolved citrate, an agent which helps to keep some mineral ions in solution. The pH of xylem fluid is slightly acid, at pH 5.5–5.7, which helps to prevent precipitation of dissolved minerals.

Xylem vessels, which may be up to 0.2 mm in diameter, lose their end walls during development. Linked end to end, they form continuous open tubes up to several metres in length. The xylem of the leaf is in continuity with that of the stem and root, and therefore provides an unbroken channel of communication from root to leaf (see Chapter 6). Moreover, the individual xylem vessels are linked to one another laterally through pores called *pits*, so sideways movement of water is also possible.

The conifers and their relatives (Gymnosperms), unlike the flowering plants (Angiosperms), do not

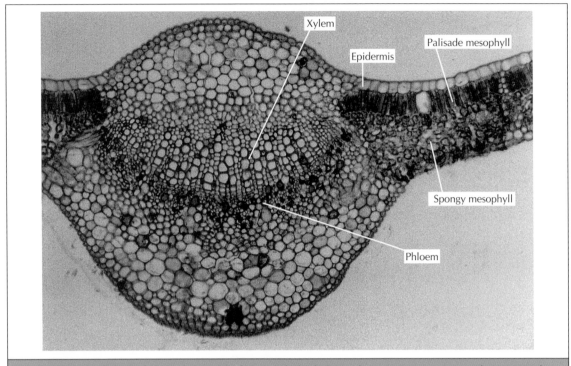

**Fig. 2.7** Section of the mid-rib and part of the lamina of a leaf of camphor (*Cinnamonum camphora*) viewed with a light microscope. Clearly visible are the *epidermises; palisade* and *spongy mesophylls; xylem; phloem* and *cambium* (between the xylem and phloem). The *bundle sheath* has not stained and is therefore indistinct. *Specimen prepared and photographed by B.G. Bowes, University of Glasgow.*

form vessels. Instead they possess less efficient water-transporting cells called *tracheids*. These too have lost their living contents, and are braced with cellulose and lignin. They are, however, spindle-shaped, with closed ends, and linked to one another only laterally through pits. They are also only a few millimetres long and very narrow, being only one tenth as wide as the Angiosperm vessels. They thus provide a less efficient water-transport system than that formed by vessels.

The liquid in the xylem moves upward through the plant to the leaves because it is under tension. This results from the evaporation of water from the stomata at the leaf surface, the process called *transpiration*, generating a pull force that is transmitted to the roots through the unbroken stream or column of water within the vessels or tracheids. Since water molecules, which are 'sticky', are bonded to one another more strongly than to the walls of the vessels and tracheids, this tension literally results in the water being pulled up through the plant and into the leaf by physical force. Indeed, the tensile strength of the water column is, amazingly, only ten times less than that of copper wire.

However, it is about a million times more difficult to pull water through living cells than through the dead xylem vessels. To facilitate the transport of water in the leaf, therefore, the distance it must travel in the mesophyll is reduced to a minimum by permeating this tissue with many fine veins so that water molecules never have to pass through more than six living cells after leaving a vein. To give some idea of the complexity of this branching system, it has been calculated that a single beech (*Fagus sylvatica*) leaf, for example, contains almost 4 metres of finely branched veins.

**Table 2.1** The main groups of plant hormones. Hormones are usually defined as organic compounds that cause a physiological response at very low concentrations. In most cases, they are synthesised in one part of a plant and transported to a distant site of action. The mode of transport is, therefore, important. They may also act at the site of synthesis. Only the five most important groups are considered here. Other naturally occurring compounds (e.g. brasicins, polyamines, jasmonic acid) have growth-regulatory activities at low concentrations, but are less well understood.

| Name | Main sites of synthesis | Transport | Some effects | Some practical uses of growth regulators in horticulture |
|---|---|---|---|---|
| **Auxins** (e.g. indole-3-acetic acid) (Chapters 3, 10, 11) | Leaf primordia, young leaves, developing seeds | Primarily by unidirectional polar movement in living cells Also bi-directional in phloem | Promote apical dominance, vascular tissue differentiation and fruit growth Induce formation of adventitious roots Stimulate ethylene synthesis Delay abscission | Promoting root initiation in cuttings and in micro-propagation systems (Chapter 10) Killing weeds (Chapters 10, 15, 16) |
| **Cytokinins** (e.g. zeatin) (Chapters 10, 11, 13) | Root tips | In xylem from roots to shoots | Promote cell division and lateral bud growth Induce shoot formation Delay senescence | Promoting shoot initiation in micropropagation systems (Chapter 10) |
| **Gibberellins** (e.g. gibberellic acid) (Chapters 9, 10, 11, 13) | Young tissues of shoot, developing seeds and probably roots | In phloem and xylem | Promote cell division, cell elongation and seed germination Promote flowering and bolting in long-day plants Cause reversion to juvenile form Prevent tuber formation in potato | Producing dwarfed plants with growth retardants ('anti-gibberellins') (Chapter 11) Promoting malting (Chapter 9) |

*(Continued)*

**Table 2.1** (*Continued*)

| | | | | |
|---|---|---|---|---|
| **Ethylene** (Chapters 8, 13, 17) | Most tissues in response to stress and in ripening or senescing tissues | Ethylene is a gas and moves by diffusion in and around the plant. May also be transported as precursor solutes | Promotes ripening in fruits and senescence in leaves and flowers. Induces abscission of leaves and fruits. Promotes flowering in bromeliads | Promoting flowering in bromeliads (especially pineapple) by ethylene-releasing compounds (Chapter 13). Promoting flowering in bulbs (Chapter 13). Retarding senescence in fruits and flowers by 'anti-ethylene' compounds (Chapter 17). Inhibiting flowering in sugar cane (Chapter 13) |
| **Abscisic acid** (Chapters 8, 10) | In root tips and leaves, especially in response to stress. Seeds | Mainly in phloem, but also in xylem | Closes stomata. Maintains dormancy in seed and possibly in buds. Increases tolerance to salinity, low temperature and drought. Controls deposition of reserve proteins during seed development | None at present but important in connection with water management (Chapter 8) |

## Transport of carbohydrates and hormones in the phloem

The xylem tissues of the leaf are located on the upper sides of the veins, linking with the centrally located xylem of the stem. The other major conducting tissue, the phloem, is located on the undersides of the veins (Figs 2.6 and 2.7), linking with the peripherally located phloem of the stem. The phloem tissues consist largely of vessels called *sieve cells* joined end to end to form sieve tubes, involved in the transport of sugars, and their associated companion cells.

The sugars glucose and fructose produced during photosynthesis are usually converted to sucrose for transport around the plant. Once formed, the sucrose passes out of the palisade cells into the *apoplast*, which is the non-living space within a plant comprising the cell walls and the intercellular spaces between them. From here it is taken up by the sieve cells of the phloem (Figs 2.6 and 2.7) for transport out of the leaf. The sieve cells are elongated cells, connected to one another by their end walls, which are perforated, hence the name. These cells are alive, with intact *plasmalemma* membranes, but their contents have liquified and they have lost their nuclei. The contents of each of the sieve cells are connected to one another at the ends by thin, membrane-bounded threads of liquid cytoplasm that pass through the holes in the end walls. Each sieve cell is also connected laterally to a companion cell, which has a nucleus and is packed with organelles and is therefore metabolically very active. It provides the energy required for the transport of sugars and other substances in the phloem.

Sucrose is pumped into the sieve cells and the companion cells by metabolic pumps located on the plasmalemmas and powered by high-energy molecules of ATP. These pump protons (hydrogen ions, or $H^+$ ions) out of the phloem cells into the cell walls and intercellular spaces and the sucrose is then carried back in with them on sucrose–proton transporter molecules. Often the area of plasmalemma available for this process is greatly increased in the companion cells by finger-like outgrowths of the wall which project into the cytoplasm and around which the plasmalemmas are folded. The sucrose-loading mechanism is so effective that the concentration within the phloem cells can reach 20–30%. Since the solution of sucrose outside the phloem cells is very weak, the resulting hydrostatic pressure generated in the phloem cells as a result of osmotic movement of water into them (see caption to Fig. 2.4) can be as high as 20–30 atmospheres (2–3 megapascals). For this reason the walls are thickened with hoops of extra cellulose, a polymer with great tensile strength, otherwise the cells would burst.

The phloem tissues of the leaf connect, via the stem, to all the other parts of the plant, where sucrose is off-loaded to provide the energy and wall-building material required for growth. Notable *sinks* for off-loaded sucrose are, of course, the root and shoot tips, where active growth is occurring, and storage organs such as roots, fruits and modified stems such as tubers and rhizomes. As sucrose is off-loaded into such tissues, the hydrostatic pressure in the sieve elements is reduced, creating hydrostatic pressure gradients in the phloem between the leaves and the rest of the plant. These gradients vary from 0.5–5.0 atmospheres (50–500 kilopascals) according to the rate of photosynthesis and the rate of off-loading into the sinks, and are sufficient to drive the contents of the sieve elements at speeds of up to 50 cm per hour.

It follows that any damage to a high-pressure system such as this will result in leakage of the vessel contents. Such leakage is especially apparent, for example, when the phloem is exposed to remove the sugar sap from sugar maple trees (*Acer saccharum*) in spring. In most plants, when injury to the phloem occurs, the perforations in the end walls of the sieve tubes are rapidly clogged with filaments of phloem protein and are then plugged permanently with a polysaccharide called callose, a substance widely deployed by plant cells to repair injuries.

In addition to transporting sugars, sieve tubes are also important in transporting other organic molecules about the plant, notably hormones such as *indole-3-acetic acid* (IAA) and the *gibberellins* (Table 2.1). Some minerals are also transported in the phloem as well as in the xylem. Because the loading of sucrose requires the extrusion of protons ($H^+$ ions), however, the phloem sap tends to be rather alkaline, at pH 8.0–8.5. This means that iron, calcium, manganese and copper precipitate out of solution and can only, therefore, be transported in the xylem. During periods of drought, when transpiration is not occurring and xylem transport is restricted, these elements soon become depleted in the leaf tissues (see Chapter 6).

When light levels are high and water and other nutrients are plentiful, photosynthesis may proceed so rapidly that the phloem transport system is unable to cope with the removal of all the sucrose produced. If this were not dealt with, sucrose would build up in the leaf tissues, disrupting the osmotic balance of the cells (see Fig. 2.4). Excess sucrose produced under such conditions is therefore rapidly converted to insoluble starch by starch synthetic enzymes, and starch grains are laid down within the chloroplasts and elsewhere in the palisade and spongy mesophyll cells. This stored carbohydrate may later be released for transport and use during times of low photosynthesis by reconversion of the starch to sucrose. A similar process of starch synthesis and breakdown also makes possible the storage of carbohydrates in specialised storage organs such as tubers, thickened roots or seeds (see below).

## Producing new xylem and phloem: the cambium

So far, no mention has been made of the thin layer of small, regularly shaped and metabolically very active cells in the vascular bundles of the leaf comprising the cambium. This tissue occurs between the xylem and phloem, and the cambium cells are capable of dividing to produce daughter cells on their upper and lower faces, which differentiate to form new xylem and phloem cells respectively. (See also Chapter 10, where the importance of the cambium in grafting is discussed.)

## The bundle sheath

Finally, in most plants the vascular bundles of the leaves are completely enclosed, except at the tips, by a prominent, single layer of cells comprising the bundle sheath (Figs 2.6 and 2.7). This effectively provides a living seal around the bundles, ensuring that air is not sucked into the xylem vessels from the surrounding intercellular spaces, causing breakage of the water columns and thereby disrupting water transport. The bundle sheath cells perform other functions too. They are important in transferring mineral ions from the non-living xylem vessels and tracheids to the surrounding living cells of the leaf, and like the companion cells of the phloem may have inward-pointing projections from the walls to increase the area of plasmalemma available for this process. Bundle sheath cells may also provide a

convenient repository for unwanted waste products until these can be lost from the plant when the leaves are shed in the autumn. Moreover, in some plants, the bundle sheath cells may be involved in a special kind of photosynthesis, called C-4 metabolism, which will be discussed in Chapters 8 and 13.

## Connecting with the stem

The leaf may be attached to the stem of the plant in a variety of ways. Most of the monocotyledons, such as the grasses (Poaceae), have relatively sword-shaped (lanceolate) leaves with parallel veins, attached to the plant by a broad base which often completely ensheathes the stem. This feature is particularly noticeable in the palms (Palmae), where the persistent, ensheathing leaf bases may constitute the bulk of what is loosely called the 'stem', serving both to support and protect the true stem tissues. In all cases, however, the parallel vascular bundles continue down into the stem and link there with the scattered vascular bundles that are characteristic of monocotyledons (see Fig. 2.11, below).

In most of the dicotyledons (Fig. 2.5), such as members of the rose family (Rosaceae), and some monocotyledons, such as members of the fig family (Ficaceae), the leaf lamina narrows at its base and a stalk, the petiole, of varying length in different species, attaches the leaf to the stem. Sometimes this contains only one, crescent-shaped vascular bundle, but in most cases there are several. These again link to the vascular bundles of the stem which, in dicotyledons, are arranged in a regular pattern (see Fig. 2.11, below).

The point of attachment of the petiole to the stem is sometimes slightly swollen, as in scarlet runner bean (*Phaseolus coccineus*), to form a structure called the *pulvinus*. Many of the cells located in the pulvinus respond to external stimuli, especially light intensity and light direction, by taking up or releasing water. This leads to changes in turgidity, causing the leaves of some plants such as *Oxalis* species to droop at night as the light levels fall, a so-called sleep movement, and to recover at dawn as the sun rises. Adaptations of this kind may originally have evolved in response to environmental extremes such as very cold night temperatures, where drooping leaves may afford some protection against frost (see Chapter 13).

Changes in the turgor of the cells of the pulvinus may also enable the leaves of some other plants like the scarlet runner bean to turn towards the sun, thereby maximising their ability to collect the energy of sunlight. The leaves of most plants, however, do not possess pulvini; leaf movement is still possible, but is much slower, being brought about by the differential rate of growth of the cells at the base of the stem in response to light coming from one direction only, as when a plant is partially shaded or grows against a wall. Such growth responses, called *phototropism*, are not restricted to the leaf but also occur in the stem so that the whole plant may appear bent towards the direction of its principal light source (see Chapter 8).

## Leaf fall

At the point where the base of the petiole meets the stem, a special layer of cells, the *abscission layer*, forms when the life of the leaf is over, to facilitate leaf fall. This occurs annually in deciduous species such as oak (*Quercus* spp.), but less frequently in evergreens such as holly (*Ilex* spp.). In this zone the xylem vessels may be narrower than normal or there may only be tracheids rather than vessels, to reduce the risk of air bubbles being drawn into the xylem vessels of the stem and blocking them if leaves are torn off by wind or browsing animals. When the time is reached for a senescent leaf to be lost, often in response to low temperatures and/or short days in temperate climates (see Chapter 13), a protective layer of corky cells forms at the base of the *abscission zone*. These are impregnated with the hydrophobic corky polymer *suberin* and are impermeable to water. The vessels of the vascular tissue are then plugged and the leaf falls. Immediately before this occurs, however, the plant breaks down the chlorophyll, starch and other useful substances in the leaf and transports them into the body of the stem. Only the waste products are left behind. Many of these are complex, highly coloured chemicals like the red, yellow and purple anthocyanins, and are only revealed in all their glory as the masking effect of the chlorophyll is removed, giving the autumn colour of so many of our garden trees and shrubs (see Chapters 12 and 13).

In some normally deciduous woody species, such as beech (*Fagus sylvatica*), an abscission zone does not form in juvenile plants and the dead leaves remain attached to the stem throughout the winter, perhaps providing the buds with a degree of protection from frost and damaging winds. This phenomenon is exploited by gardeners who plant beech hedges, maintaining juvenility by annual clipping (see Chapter 10).

## Leaf patterns: phyllotaxy

Finally, careful examination reveals that leaves are not attached to stems in a random way, but are arranged in a regular pattern which varies from species to species. In some dicotyledons they form a spiral along the stem, as in the strawberry tree (*Arbutus unedo*, Fig. 2.8), whereas in others, such as privet (*Ligustrum* spp., Fig. 2.5b), they may alternate with one another in opposite pairs or form whorls of

**Fig. 2.8** The *spiral phyllotaxy* of the strawberry tree (*Arbutus unedo*). *Photograph by David S. Ingram.*

three or more leaves. In sunflower (*Helianthus* spp.) the first leaves are in opposite pairs, but a spiral arrangement soon sets in as the plant grows. In most monocotyledons the leaves are in two opposite rows, although dicotyledonous patterns occur in some species. Sometimes the leaf petioles may be different lengths and the laminas different sizes according to the position of the leaf on the plant, an arrangement most easily seen in the maples (*Acer* spp.). These different patterns of leaf attachment, called *phyllotaxy*, have evolved to maximise the amount of light reaching the leaves by ensuring that each leaf shades its neighbours as little as possible. That different arrangements of leaves may be seen in different groups of plants shows that there may be a variety of structural solutions to this problem.

## Variations on a theme

The basic form of the leaf described above is a magnificent compromise that has evolved in response to the conflicting environmental and physiological constraints imposed on plants. However, such a compromise is only effective under conditions of an equable climate and adequate supplies of water, light, carbon dioxide and mineral nutrients. Any deviations from these optimal conditions reduce the effectiveness of the leaf as a photosynthetic organ, leading to the evolution of a diversity of variations on the basic form.

One of the commonest variations has been in the evolution of divided leaves, perhaps minimising wind and rain damage by posing less resistance, as in the maples (*Acer* spp.). In other species the leaf may have holes, as in the cheese plant (*Monstera deliciosa*), caused by the death of certain cells as the leaf develops, or drip-tips as in the weeping fig (*Ficus benjamina*), to facilitate run-off of water in tropical climates. In other cases compound leaves have evolved, again perhaps minimising wind and rain damage, and these may be pinnate as with ash (*Fraxinus excelsior*), or palmate as with horsechestnuts (*Aesculus* spp.).

Such compound structures may develop by programmed, differential division and growth of the cells of the young developing leaf, as with the divided leaves of dicotyledonous plants like ash. In such cases each leaflet of the compound leaf resembles a perfect simple leaf. In the palms, however, compound leaves develop in a completely different way. To begin with, the young leaf is intricately folded, with the folds packed together like a fan. As the leaf blade unfolds and expands, lines of cells along alternate folds die and break down so that the fingers of the older compound leaf separate from one another along the resulting lines of weakness.

The distribution of stomata also varies from species to species. Although usually found mainly on the lower surfaces of the leaves, they are restricted to the upper surfaces of the leaves of floating aquatic plants like the water-lily (*Nymphaea* spp.). In submerged aquatics, like Canadian pond weed (*Elodea canadensis*), both the cuticle and the stomata may be absent. Stomata may also be sunk in pits or grooves, as in pines (*Pinus* spp.), restricting the loss of water vapour under desiccating conditions by creating a moist microclimate with the grooves.

Other leaf modifications include: surface hairs, sometimes stinging or glandular (deadnettles, *Stachys* spp.; stinging nettles, *Urtica* spp.); surface scales (*Rhododendron* spp.); waxy blooms (*Brassica* spp.); thickened cuticle (cherry laurel, *Prunus laurocerasus*); reduction in size (brooms, *Cytisus* spp.); development of, or reduction to, spines (cacti, Cactaceae; Fig. 2.9); succulence (cacti); production of aromatic oils (lavenders, *Lavandula* spp.); formation of tendrils by the lamina (pea, *Pisum sativum*) or twining petiole (*Clematis* spp.); and formation of insect traps and protease enzymes (sundew, *Drosera* spp.). *Water stress* has been the principal selective force in the evolution of such modifications (see Chapter 8). They are summarised in Table 1.4.

Two final modifications not visible to the gardener, for they occur at the cellular level, are nevertheless of such importance that they deserve a special mention. These are C-4 photosynthesis and *crassulacean acid metabolism* (*CAM*). Both are modifications to the basic processes of photosynthesis, and may have evolved in response to the stress imposed by low levels of soil water and high temperatures. CAM and *C-4 plants* are described in detail in Chapter 8.

## THE STEM: REACH FOR THE SKY

The stem or shoot is the main axis of the plant, generating at its apex the leaves which it holds aloft to

**Fig. 2.9** *Echinocactus* sp. The spines of cacti are modified leaves, with the stem, often ridged or flattened, being modified for photosynthesis. *Photograph by Chris Prior, Royal Horticultural Society.*

collect the sun's energising light, and providing at its base a vital link to the roots and the water and mineral nutrients of the soil.

## The growing point

To find the growing point of the shoot, the so-called *apical meristem* (Fig. 2.10), it is necessary to strip away the immature leaves that surround and protect it from damage. It is usually dome-shaped, a millimetre or so in diameter, and very delicate. In flowering plants (Angiosperms) it usually comprises an outer layer, the *tunica*, which is one to several cells thick, enclosing an inner mass of cells, the *corpus*. The cells of the apex are packed with cytoplasm and are constantly dividing and differentiating to produce new tissues. Such centres of cell division and growth are called *meristems*, in this case the apical meristem.

As the apex grows, the cells of the tunica divide in one plane only, at right angles to the surface of the plant, which means that the tunica remains as a discrete layer. It ultimately gives rise to the epidermis of the shoot and the leaves. The cells of the corpus divide in a more irregular way and by expansion and division give rise to the internal tissues of the stem and leaves. In the stem apex of conifers and their relatives (Gymnosperms) the apex is not visibly differentiated into tunica and corpus, although the epidermal tissues of shoot and leaf still arise from the outer cells.

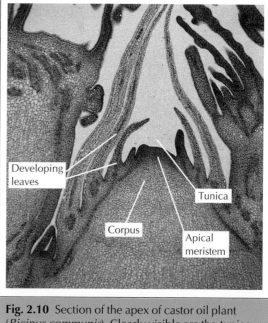

Developing leaves

Tunica

Corpus

Apical meristem

**Fig. 2.10** Section of the apex of castor oil plant (*Ricinus communis*). Clearly visible are the *tunica, corpus* and *leaf primordia* of the developing leaves. *Specimen prepared and photographed by B.G. Bowes, University of Glasgow.*

Nourished by nutrients and water carried in the vascular tissues from the mature leaves and the roots, and under the control of both the environment and internally produced hormones, the meristematic cells of the apex produce both the stem of the plant and the leaves.

## Forming new leaves

New leaves grow from small bumps of active cell division, called *leaf primordia*, on the sloping sides of the apical dome. These do not arise at random, their position relative to one another being determined genetically and regulated by the apex itself and the previously formed primordium. In this way a distinctive pattern develops, the phyllotaxy already referred to.

Inside the body of the shoot apex, just below each leaf primordium, the cells divide longitudinally to form a strand of cells, the *procambial strand*. As the apex grows upwards the procambial strands differentiate as distinct cambiums, which in turn produce the xylem and phloem of the vascular bundles of the shoot. The pattern of vascular tissues of the shoot is therefore determined by the position of the developing leaves they will service.

The leaves are finally shaped by a series of centres of cell division, subsidiary meristems, each with a fixed period of activity. First a meristem forms at the apex of the leaf primordium, forming a narrow peg about a millimetre in length. This will become the midrib. Next, cells down either side of this peg begin to divide, growing out to form thin plates of tissue which will become the lamina. Finally, a wave of cell divisions followed by differentiation passes along the whole length of the leaf to produce the internal tissues.

The final form of the leaf is determined by the precise position and duration of activity of the subsidiary meristems. For example, long, narrow leaves are produced when the meristems on the sides of the leaf peg have a very short period of activity, whereas large flat leaves are produced when the active period of the marginal meristems is much longer. Compound leaves form when the marginal meristem activity is intermittent rather than continuous. Small areas of meristematic activity, called meristemoids, located on the surfaces of the leaf, provide the finishing touches, such as the stomata.

The pattern of growth of leaves just described holds true for most dicotyledons and for some monocotyledons such as the grasses (Poaceae). In monocotyledons with tubular leaves, such as onions (*Allium* spp.), the meristems at the margin of the leaf are suppressed and the leaf arises from a new meristem that ensheaths the stem apex.

Most plants have buds which occur in the angle between the leaf and the stem, called the axil. These axillary buds often arise very early in leaf development, in the axils of the leaf primordia. Like the leaves, each possesses a procambial strand. The axillary buds do not usually grow out to form new shoots immediately, being held in a state of dormancy, perhaps by hormones secreted by the apex, until circumstances change, as when the apex is cut off by pruning (see Chapter 11). When axillary buds are not present, branch stems may form from buds that arise spontaneously from the stem tissues. These buds are said to be *adventitious*.

## A tower of strength

In the young stem of most dicotyledonous garden plants, such as sunflower (*Helianthus annuus*), the vascular bundles form a peripheral ring (Fig. 2.11) which extends through the whole stem, like the letters through a stick of seaside rock. The cells in the centre of the stem are large, irregular and thin-walled, with large vacuoles, and form a tissue called the *pith*. Sometimes the pith cells contain starch grains and are important for the storage of carbohydrate. External to the ring of vascular bundles and between the individual bundles, linking with the pith, are smaller, irregular cells comprising the *cortex*. These may be photosynthetic, especially in young plants. Beyond this is an epidermis, the cells of which may also contain chloroplasts and be photosynthetic. Scattered stomata may be present to facilitate gas movement in and out. And finally, like the leaves, the stem is ensheathed by a waterproof layer of cuticle.

The vascular tissues of most monocotyledons are arranged rather differently from those of dicotyledons, the bundles being scattered throughout a background tissue of large, irregular cells (Fig. 2.11). Occasionally a central pith is present, as in rushes (*Juncus* spp.).

Each vascular bundle of the stem (Fig. 2.12) comprises an outward facing layer of phloem-fibre cells, their walls thickened with lignin. These serve to strengthen the stem and to protect the phloem beneath. The phloem itself, as in the leaf, is composed mainly of sieve tubes and companion cells,

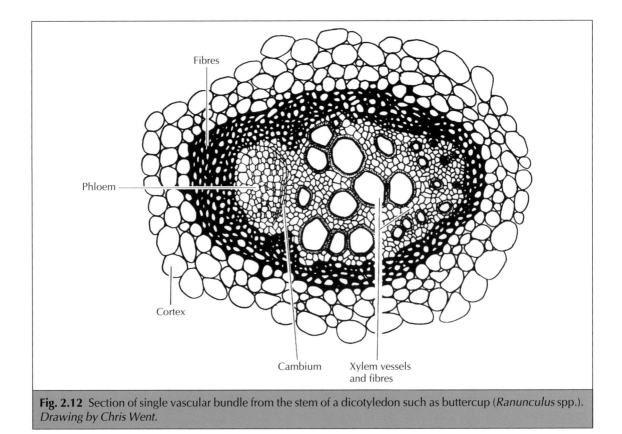

**Fig. 2.11** Arrangement of vascular bundles in the stems of (a) a monocotyledon and (b) a dicotyledon. *Drawing by Chris Went.*

involved in the transport of sugars and organic substances such as hormones (Table 2.1). The xylem tissue, composed of lignified water-conducting vessels, or tracheids, with lignified fibres between them, forms the inward-facing region of each bundle. It is

separated from the phloem by a cambium, the *fasicular cambium*, a tissue made up of files of small, box-shaped cells which divide continuously to produce phloem on the outer face and xylem on the inner. Eventually, as the stem matures, the cambiums of adjacent bundles link up by development of a so-called *interfasicular cambium* in the cells between the bundles. This has great significance in a process called *secondary thickening*, which in the maturing stem of trees and shrubs leads to the development of a mass of woody tissue (see below).

There are many variations on this basic arrangement of the vascular tissues. For example, in some plants phloem tissues occur both outside and inside the xylem. In many monocotyledons the xylem tissue may occupy the centre of each bundle, being completely surrounded by phloem.

In the young stem it is clear that the peripheral ring of vascular bundles provides not only an efficient transport system for water, sugars and other substances, but

**Fig. 2.12** Section of single vascular bundle from the stem of a dicotyledon such as buttercup (*Ranunculus* spp.). *Drawing by Chris Went.*

also strength and support for what would otherwise be a rather fragile structure. A hollow cylinder of strengthening tissue gives much greater resistance to mechanical damage caused by the side-to-side movement of the stem on a windy day than if the vascular tissues formed a solid core in the middle of the stem.

The simple vascular anatomy described above is more complex at stem *nodes*, the points at which the leaves, axillary buds and branches emerge, for the vascular supply to these organs sometimes necessitates complex patterns of branching of both the phloem and the xylem tissues: a veritable 'spaghetti junction' of the plant world. Moreover, there may be crosslinks between the individual bundles in the internodes, the portion of the stem separating the nodes. Finally, complex repositioning of vascular bundles also occurs in the *hypocotyl*, the portion of the stem which at its base links with the root, for in the root proper the xylem comprises a central core, with the phloem tissues dispersed around it (see below).

## Secondary thickening: the formation of wood and bark

As the stem of a dicotyledonous plant ages in woody plants, it undergoes a process called secondary thickening (Fig. 2.13), in which woody tissues are created to provide additional support, and an outer layer of bark is laid down to protect the phloem beneath.

First, the cambiums of adjacent vascular bundles link up to form a continuous cylinder of cambial cells, running the whole length of the stem. This forms a complete cylinder of xylem tissue on its internal face and a thinner layer of phloem tissue on its outer face. During the next season a further cylinder of xylem is produced, so that two annual rings of

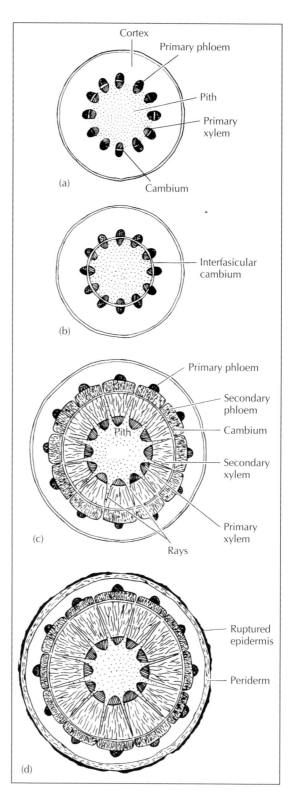

**Fig. 2.13** (*right*) Diagram illustrating the early stages of secondary thickening in the stem of a dicotyledon: (a) the unthickened stem; (b) development of the *interfasicular cambium*; (c) development of the secondary phloem and secondary xylem (the primary phloem and primary xylem are pushed to the periphery and centre of the stem, respectively) and the development of the *rays*; (d) development of the corky tissue of the *periderm* and rupture of the *epidermis*.

xylem are visible. A second layer of phloem is also formed, the layer from the previous year being disrupted to accommodate it. The epidermis is also disrupted, and replaced by a layer of corky cells, the *periderm*, produced by a completely new cylinder of cambium, the *cork cambium*, that arises in the cells of the outer layers of the stem. The corky layer is disrupted each year as the stem expands, and is replaced by a new layer of corky cells the following year.

In trees and shrubs this process is repeated annually until a large woody stem with prominent annual rings is produced. The central, older layers of xylem eventually become inactive, the vessels being plugged with resinous substances and the mass of tissue being protected from attacks by rot-causing fungi by the deposition of aromatic fungitoxic substances. This *heartwood* has a darker colour when seen in section than the outer, active zone of the stem, called the *sapwood*. Communication between the layers of xylem is maintained by slightly wavy radially arranged sheets of living cells running the length of the stem, called *rays*.

The evidence of annual rings of xylem may be clearly seen at the cut end of a tree after felling. The annual rings, cut tangentially, are what create the grain in planed wood. The sheets of ray tissues, also cut tangentially, create the 'figure', which cuts across the grain and is especially conspicuous in planed oak (*Quercus* spp.). The wood of each species has its own characteristic pattern of grain and figure. This is in part fixed genetically and in part determined by climate and rate of growth. Fast-growing trees like conifers (e.g. *Pinus* spp.) produce soft woods, whereas slower-growing species such as beeches (*Fagus* spp.) produce hard woods. The variation in the darkness or density of the annual rings results from the laying down of large water-conducting vessels during the spring, when new growth and new leaves are being produced, followed later in the season by the laying down of narrower vessels and fibres for strength.

The majority of monocotyledonous plants are herbaceous and do not undergo secondary thickening. However, in a few species, such as bamboos, secondary thickening does occur as a result of the activity of a layer of cells called a primary or secondary thickening meristem, which produces masses of woody cells. In other tree-like monocotyledons such as palms, woody tissue may be produced diffusely throughout the stem.

## Bark

Bark (Fig. 2.14) usually consists of disrupted phloem tissue, interspersed with layers of corky tissue. It thus provides a waterproof, protective covering for the woody stem. Gases such as oxygen and carbon dioxide can pass in and out, however, for at intervals there are large pores in the bark, called *lenticels*, loosely filled with large, thin-walled cells. Lenticels are especially prominent in the bark of elder (*Sambucus nigra*).

The structure of bark is distinctive for each species and depends upon the arrangement of cork cambiums which generate the corky cells and the amount of phloem and cork produced each year. For example, in deciduous oaks (*Quercus* spp., Fig. 2.14a) and other rough-barked trees the cork cambium is not a continuous cylinder, but instead comprises a series of plates re-formed each year as more phloem is produced, thereby creating separate plates of bark tissue made up of alternate layers of corky and phloem tissue. As the girth of the tree increases the outermost plates are pulled apart, giving the tree its fissured surface. In other species, such as cherries (*Prunus* spp.) the 'plates' of bark tissue may take the form of large sheets which split and peel away as the stem expands. In smooth-barked trees such as beeches the cork cambium remains as a continuous cylinder, increasing its girth each year and forming only a very thin layer of phloem (Fig. 2.14b).

## Stem modifications

As with the leaf, the selective forces of evolution have resulted in various modifications of the stem to cope with particular environmental conditions. These include: succulence (cacti; Fig. 2.9); conversion to thorns (*Crataegus* spp.); development of photosynthetic wings (brooms, *Cytisus* spp.) or blades called *cladodes* (butcher's broom, *Ruscus* spp.); reduced length to form a rosette (daisy, *Bellis perennis*); a twining, climbing habit (honeysuckles, *Lonicera* spp.); growth as underground organs of storage and perennation, such as rhizomes (*Iris* spp.), tubers (potato, *Solanum tuberosum*; Fig. 2.15) or corms (*Crocus* spp.). Water stress, competition for light and the need for organs of perennation have been the principal selective forces in the evolution of these modifications, which are summarised in Table 1.4.

(a)                                          (b)

**Fig. 2.14** Bark: (a) oak (*Quercus* spp.), a rough-barked tree in which the cork cambium comprises a series of plates, reformed each year, and pulled apart as the trunk expands; (b) beech (*Fagus* spp.), a smooth-barked tree in which the cork cambium forms a continuous cylinder that expands to accommodate the increasing girth of the tree. *Photographs by Chris Prior, Royal Horticultural Society.*

**Fig. 2.15** Potatoes are stems modified for storage, the 'eyes' being reduced leaves with buds in their axils. *Photograph by Chris Prior, Royal Horticultural Society.*

## THE ROOT: MINING FOR MINERALS AND WATER

### The structure and growth of the root

The most remarkable thing about roots is not their structure, but the fact that they grow and branch continually, forever exploring and exploiting new areas of soil for the minerals and water that are as essential for the health of the plant as the energy, carbon dioxide and oxygen harvested by the leaves. This growth occurs at the root apex, which is protected by a *root cap* and lubricated by mucilage (see Chapter 6).

The efficiency of the root as an absorbing organ depends on its absorbtive surface area relative to its volume, created by the root hairs (extensions of the epidermis that extend out into the soil) and the complex system of branches. It has been estimated that a 4-month-old rye plant (*Secale cereale*), for example, has a total area of over 600 square metres in contact with the soil. To put this figure into perspective, it is equivalent to the area of a lawn about 25 metres by 25 metres. How this vast structure functions and keeps the plant supplied with water and nutrients will be

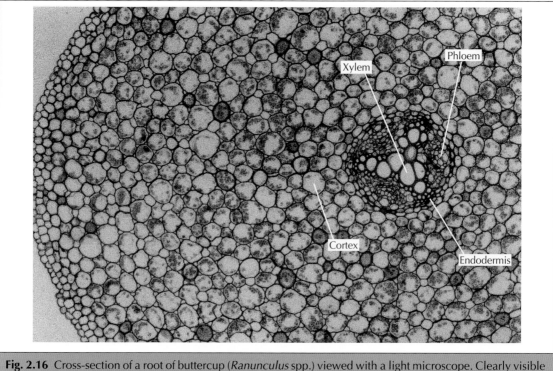

**Fig. 2.16** Cross-section of a root of buttercup (*Ranunculus* spp.) viewed with a light microscope. Clearly visible are the central core of vascular tissue (xylem and phloem), the *endodermis* and the wide *cortex*. *Specimen prepared and photographed by B.G. Bowes, University of Glasgow.*

described in Chapter 6. Here, the root will be considered as a structure for anchoring and supporting the plant, and as an organ that may have been modified during evolution and by selection and breeding to perform a variety of other functions.

The basic structure of the root body is shown in section in Figs 2.16 and 6.6. The xylem tissues, consisting of vessels and fibres, form a rod or central core with the phloem tissues deployed around them (the *stele*) and provide great mechanical strength to resist the pulling power caused by the wind or grazing animals.

The plant is even more firmly anchored by the extensive branch system of the root. Root branches arise behind the apex, deep in the central tissues. First the vascular connections are established and then the new root breaks through the outer layer of the cortex. Branch roots may branch again and again, ending in delicate, short-lived *feeder roots*, which are most actively involved in the uptake of nutrients

and water. They are replaced continuously as the root system explores and re-explores the soil in which the plant is fixed.

The basic pattern of branching established by the root system varies from species to species. In some garden plants, such as lupins (*Lupinus* spp.), there is a central *tap root* with side branches formed along its length. In others, such as Michaelmas daisies (*Aster* spp.), the root system has no central axis, consisting instead of a much-branched fibrous mass. Sometimes, as in maize (*Zea mays*), additional branch roots may arise from the stem base. These, called adventitious roots, provide support as well as enabling the plant to explore the surface layers of soil. Adventitious roots also arise at the base of stem or leaf cuttings, induced by the accumulation there of internal hormone levels triggered by excision or by the application of artificial rooting hormone (see Chapter 10).

As roots age they may undergo secondary thickening as the stem does. In dicotyledons this is initiated by the activation of a corrugated cylinder of dividing cells, the cambium, which develops between the mass of xylem tissue inside and the outer columns of phloem tissue. This cambium produces rings of new xylem vessels and fibres on its inner face, and rings of phloem on its outer face. Cells impregnated with the corky substance suberin may protect the outside of the thickened root, forming a layer of bark-like tissue. As in secondarily thickened stems, lenticels allow the movement of gases in and out. In most trees and in many grasses the corky layers are formed deep in the central root tissues, thus cutting off the supply of nutrients to the outer cortex, which dies and is in consequence sloughed off. The younger roots of trees and grasses may, as a result, be thicker than older roots.

## Storage roots and other modifications

Roots, like leaves and stems, may have become modified during evolutionary selection in a variety of ways to perform specific functions; the most obvious to the gardener is the over-winter storage of carbohydrates. Thus in turnip (*Brassica rapa*), for example, masses of large, thin-walled cells called *parenchyma*, packed with starch grains, are formed in the sec-

ondary xylem of the tap root as the plant ages. In swede (*Brassica napus*) both the root and the *hypocotyl*, the transition tissues that connect stem and root, swell in this way. In carrot (*Daucus carota*; Fig. 2.17), by contrast, carbohydrates are stored in masses of secondary phloem tissue in the root, the secondary xylem being limited to the central, yellowish core. Fibrous root systems may also swell by proliferation of parenchyma cells to form tuberous storage roots, as in *Dahlia* spp. In most cases the carbohydrate is stored as starch grains within the cytoplasm of the cells. In carrot, sugars are also present, giving them a sweet taste, and in Jerusalem artichoke (*Helianthus tuberosus*) the carbohydrate takes the form of a starch-like compound called inulin.

Roots may have become modified in a variety of other ways too. For example, the roots of some plants that grow in trees, such as certain tropical orchids, have an epidermis that is several cell layers thick. The outermost layers of cells are dead and form a structure, the *velamen*, that absorbs water from the saturated air. Some of the roots of crocus (*Crocus* spp.) and other species that produce corms are able to contract, drawing the newly formed corm, which grows on top of the parent corm, down into the soil. Other structural modifications are summarised in Table 1.4.

## Nitrogen fixation

A final group of modifications of roots, although not easily visible to the naked eye, are of vital importance to almost all plants. These are the *symbiotic* associations of roots with other organisms, mainly bacteria and fungi, to improve the supply of nutrients to the plant. The best known are the associations of plants in the legume family (Papilionaceae) such as beans (*Phaseolus* spp.), with nitrogen-fixing bacteria. In these associations the bacteria invade the root, causing nodules to form, and inside these the bacteria fix nitrogen gas from the atmosphere into nitrogenous salts (see Chapters 6 and 7). Nodules may easily be seen if the roots of bean or clover (*Trifolium* spp.) plants are gently dug up and washed free of soil (Fig. 2.18). The evidence of additional nitrogen being available around plants with nitrogen-fixing nodules can be seen in the darker green of grass in a lawn infested with clover. The bacteria are supplied with sugars by the plant. Some fungi are also capable of *nitrogen fixation*: for example, the fungus *Frankia*

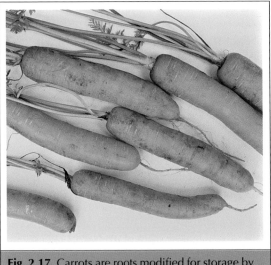

**Fig. 2.17** Carrots are roots modified for storage by proliferation of the phloem tissue. *Photograph by Chris Prior, Royal Horticultural Society.*

**Fig. 2.18** Nodules containing *nitrogen-fixing bacteria* on the roots of broad bean (*Vicia faba*). *Photograph by Debbie White, Royal Botanic Garden Edinburgh.*

forms nitrogen-fixing nodules on the roots of alder trees (*Alnus* spp.)

## Mycorrhizas

Less well known, but even more important for plants, are the associations between roots and beneficial fungi to form *mycorrhizas*. Mycorrhizal associations are almost ubiquitous in the plant kingdom, although a few families, notably the Brassicaceae, do not form them. Some say that the first plants to invade the land depended on mycorrhizas rather than roots to supply them with water and nutrients. There are various types of mycorrhizas, but the two most common are *endomycorrhizas*, formerly called the *vesicular-*

*arbuscular mycorrhizas* (*VA mycorrhizas*), formed by most herbaceous plants, and the *ectomycorrhizas*, formerly called sheathing mycorrhizas, formed by many trees. Fungi grow as thin, microscopic threads called hyphae (see Chapter 15). In endomycorrhizal associations these grow between the living cells of the root, forming at intervals large storage vesicles containing fatty lipids and specialised, finely subdivided branches called arbuscules. The latter penetrate the living cells of the root and form an interface for the exchange of nutrients. The hyphae then extend out into the soil, forming an extensive absorbtive network (Fig. 2.19).

In ectomycorrhizas the hyphae of the fungal partner form a thick mass around the outside of the feeder roots, causing them to thicken and branch abnormally (Fig. 2.20). The pattern of branching is characteristic of each association: for example, in the case of pines (*Pinus* spp.) the branches are dichotomous (y-shaped); whereas in beech (*Fagus* spp.) the branches arise at right angles from a central axis. The fungal sheath sends out branches on its inner face which form a network, called the *Hartig net*, between the cells of the root cortex (Fig. 2.21). It is here that nutrient exchange between plant and fungus occurs. The outer face of the sheath also sends out branching hyphae which form an absorbtive network in the soil. Ectomycorrhizal fungi often reproduce by forming toadstools in which spores, the minute equivalent of seeds, are formed. This is why characteristic toadstools are often associated with particular trees in woodlands and in the garden.

Research has shown that mycorrhizal fungi are especially involved in scavenging phosphates from the soil, for these minerals are largely insoluble, and to acquire sufficient for the plant's needs, especially in impoverished soils, a vast absorbing surface area is required. The fungi may also be involved in the uptake of other nutrients too, especially nitrates.

The mycorrhizal fungi extract sugars from the plant with which to sustain their own growth. The association is thus one that is truly beneficial to both partners, enabling them to grow together in habitats that would not be available to either of them growing alone.

Other specialised types of mycorrhiza are formed by heathers and other members of the Ericaceae and orchid family (Orchidaceae). In the latter the fungus provides the orchid with carbohydrates, released by breaking down organic matter in the soil.

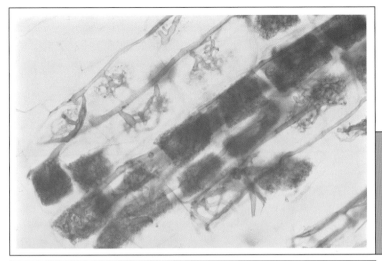

**Fig. 2.19** *Vesicular-arbuscular mycorrhiza* (*endomycorrhiza*): intercellular *hyphae* and arbuscules are visible in a squashed and stained root of cowpea (*Vigna* sp.). *Light microscope photograph supplied by P.A. Mason, Centre for Ecology and Hydrology, Edinburgh.*

**Fig. 2.20** *Ectomycorrhiza*: swollen and dichotomously branched infected roots of pine (*Pinus* sp.). *Photograph supplied by P.A. Mason, Centre for Ecology and Hydrology, Edinburgh.*

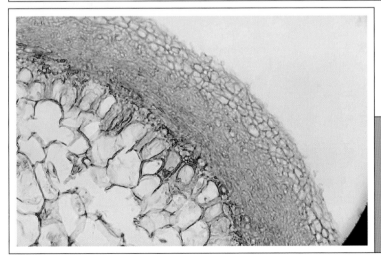

**Fig. 2.21** *Ectomycorrhiza*: section of an infected root of beech (*Fagus* sp.). A two layered fungal *sheath* can be seen around the outside of the root with the hyphae of the *Hartig net* growing between the cells of the cortex. *Specimen prepared and photographed by H.J. Hudson, University of Cambridge.*

## CONCLUSION

Knowledge of the structure and major functions of plant organs – leaves, stems and roots – and how they are interrelated is a prerequisite for exploration of the scientific principles and processes underlying the growth and management of plants in the garden, the subject of Chapters 4 to 20.

## FURTHER READING

Attenborough, D. (1995) *The Private Life of Plants*. BBC Books, London.

Bowes, B.G. (1997) *A Colour Atlas of Plant Structure*. Manson Publishing, London.

Capon, B. (1992) *Botany for Gardeners*. B.T. Batsford, London.

King, J. (1997) *Reaching for the Sun: How Plants Work*. Cambridge University Press, Cambridge.

Mabberley, D.J. (1997) *The Plant Book*, 2nd edn. Cambridge University Press, Cambridge.

Morton, O. (2007) *Eating The Sun – How Plants Power The Planet*. Fourth Estate, London.

Pollock, M. & Griffiths, M. (2005) *RHS Illustrated Dictionary of Gardening*. Dorling Kindersley, London.

Raven, P.H., Evert, R.F. & Eichhorn, S.E. (1999) *Biology of Plants*, 6th edn. Worth Publishers, New York.

Walters, S.M. (1993) *Wild and Garden Plants* (New Naturalist Series). Harper Collins Publishers, London.

# 3

# Reproduction: Securing the Future

## SUMMARY

In this chapter vegetative and sexual reproduction are compared. Next, the structures of cones, flowers, seeds and fruits are described. The composition of floral diagrams and floral formulae are outlined. The diversity of flower shapes and forms is explained in terms of co-evolution with pollinators. The diversity of fruit structures is related to seed dispersal.

## INTRODUCTION

Although some plants may live for years, or even centuries, all must reproduce to establish further generations or to spread to new habitats. Most plants are able to reproduce sexually by forming flowers and, following pollination, setting *seed* (see Chapter 9 and below). Some hybrids, however, have lost this ability and reproduce vegetatively. And some species are able to reproduce both sexually and vegetatively, thus increasing their reproductive flexibility.

## VEGETATIVE REPRODUCTION

Examples of vegetative reproduction include the formation of clumps of individuals from a proliferating crown, as with many herbaceous perennials such as lupin (*Lupinus* spp.). Similarly, some bulb-forming plants such as bluebell (*Hyacinthoides non-scripta*) may produce new bulbs from buds located between the bulb scales of the parent. Alternatively, branches may arch over and root, giving rise to new plants, as with bramble (*Rubus* spp.), or adventitious roots may

form from partially buried or broken branches, as in willow (*Salix* spp.). In some species modified stems called runners (*stolons*) grow along the ground, producing plantlets at the nodes, as in the case of strawberry (*Fragaria × ananassa*). In others, such as devil's backbone (*Kalanchoe daigremontiana*), plantlets may form at the edges of the leaves, subsequently falling to the ground and taking root. Finally, in some hybrid grasses, such as viviparous fescues (*Festuca* spp.), plantlets may replace the sexual structures of the flower. These and the many other examples of natural vegetative propagation depend upon the ability of plants to produce roots and shoots adventitiously from vegetative tissues, usually stems, but sometimes also leaves, roots or flower parts. Some plants, however, cannot do this, in particular many mature trees, for reasons that are not fully understood.

## SEXUAL REPRODUCTION

Although there is a great diversity of mechanisms for vegetative reproduction in the plant kingdom, the feature they have in common is that they result in progeny that are *clones* (see Chapters 5 and 10), genetically identical to the parent plant. Vegetative reproduction is thus harnessed by the gardener, as in taking cuttings or layering, to produce large numbers of uniform individuals (see Chapter 10 for details). Sexual reproduction, unlike vegetative reproduction, involves the mixing of the genetic material from two different individuals, the parents, and thus generates diversity (see Chapter 5). Genetic diversity among progeny is important, not least because it increases the possibility of survival of at least some individuals of a species where the environment is subject to change. It also creates the relatively rare possibility of new combinations of genetic information among progeny that may make a species more successful in an existing environment, perhaps, for example, because it can compete better with its neighbours or reproduce more effectively. Such a species is said to be more fit. The reproductive advantage so acquired is the basis of evolutionary change, the reproductively fittest individuals in each generation being at a competitive advantage over the less fit individuals of the same species. Plant breeders harness and speed up the evolutionary process by performing artificial crosses with selected individual plants, followed by further selection for desirable traits. Most of our garden plants were produced from wild plants in this way, a subject discussed in detail in Chapter 5.

## Cones and flowers

The process of sexual reproduction in seed plants involves the formation of male and female sexual structures, either in cones (Fig. 3.1), as in the *Gymnosperms* (conifers and their relatives), or in flowers, as in *Angiosperms* (the flowering plants; Fig. 3.2). In dicotyledons the flower bud is surrounded by protective modified leaves, called *sepals*, which are shed or which shrivel after the flower opens. Next come the *petals*, usually the most conspicuous part of the flower. In monocotyledons, where sepals are absent, these are called perianth segments (see Fig. 3.2). Petal and perianth segment numbers vary with family and may be separate, as in *Ranunculus* spp., or fused to one another, as in *Campanula* spp. The male structures, which come next, are the *anthers* in which the *microspores* (*pollen grains*) are produced. Numbers of anthers vary with family, as with petals. They are usually borne on stalks called *filaments*. Pollen grains have

**Fig. 3.1** Cones of *Abies* sp. *Photograph courtesy of the Royal Botanic Garden Edinburgh.*

thick, fatty walls and are often richly ornamented. They are carried by wind, insects and other arthropods, or occasionally other groups such as bats or birds, to the female organs, where pollination and fertilisation occur. In wind-pollinated plants such as grasses and conifers the flowers usually lack both colour and scent, but are borne in such a way on the plant as to maximise the likelihood of pollen being picked up by the wind and carried to a receptive female. Where arthropods or other animals are involved in pollination, *co-evolution* of the plant species and pollinators has usually resulted in the production of sweet nectar or scent in special glands at the base of the petals, or particular flower colours, to attract the appropriate pollinating species (see Chapter 12). White flowers, which appear to lack colour, reflect ultraviolet (UV) light in different ways and may therefore appear as different colours to the compound eyes of insects that can detect wavelengths invisible to the human eye (see Chapter 12). Scent is especially important as an attractant in the case of night-flying pollinators such as moths or bats. The shape of the flower – radially symmetrical (actinomorphic) or bilaterally symmetrical (dimorphic), with separate or fused petals – is also fine-tuned to the pollinator with which the species co-evolved. Finally, in some groups such as members of the Euphorbiaceae, the petals are insignificant and a ring of modified, coloured leaves (bracts) around the flowers serves to attract pollinators.

The female structures (the *ovules*) of Gymnosperms are unprotected by other tissues and the pollen reaches them direct, immediately followed by fertilisation. In the Angiosperms the ovules are enclosed by an *ovary*, which may be above the ring of petals (superior, as in *Campanula* spp.) or below it (inferior, as in *Rosa* spp.). The ovary possesses a *stigma* at one end, supported by a *style*, for reception of the pollen (Fig. 3.2). The pollen grains therefore germinate on the stigma surface to produce long pollen tubes which grow through the surface and down through the tissues of the style, eventually reaching the ovaries, where fertilisation occurs, a process described in detail in Chapter 9.

Male and female organs in the Gymnosperms are usually produced in separate, structurally different cones. It is easy to distinguish the soft, pollen-forming male ones from the much larger, woody, female cones on pine trees (*Pinus* spp.). The male cones fall off the trees as soon as the pollen has been released, whereas the female cones continue to grow and mature over a long period. In the Angiosperms the male and female organs may be formed in different flowers, as in hazel (*Corylus* spp.), or in the same flower, as in buttercup (*Ranunculus* spp.). However, most plants have inbuilt incompatibility mechanisms to prevent self pollination, which would lead to inbreeding (see Chapter 5), or pollination by an unrelated species, which would produce a nonsense hybrid. Normally this relies on the recognition by the cells of the stigma of specific proteins on the surface of the pollen grains. Following recognition of alien pollen, fertilisation is aborted (see Chapter 5).

Plants bear their flowers in different ways, perhaps in part to facilitate pollination by wind or the insect and other pollinators with which they have co-evolved. Often they are borne singly, as in poppies (*Papaver* spp.); sometimes they are borne all the way up the stem, as in foxglove (*Digitalis* spp.), to form a

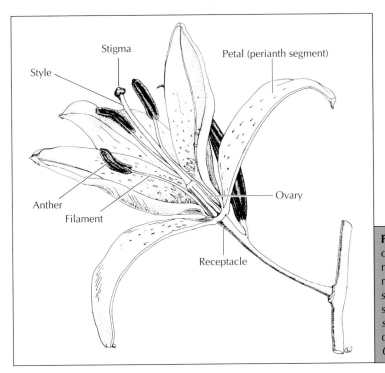

Style

Stigma

Petal (perianth segment)

Anther

Filament

Ovary

Receptacle

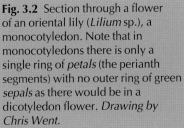

**Fig. 3.2** Section through a flower of an oriental lily (*Lilium* sp.), a monocotyledon. Note that in monocotyledons there is only a single ring of *petals* (the perianth segments) with no outer ring of green *sepals* as there would be in a dicotyledon flower. *Drawing by Chris Went.*

**Fig. 3.3** The compound flower of *Echinacea pallida*, a member of the Asteraceae. Note the outer ray florets and the inner mass of tube-like florets. *Photograph by Chris Prior, Royal Horticultural Society.*

*spike*, or with each flower on the spike having a short stalk, forming a *raceme*, as in *Delphinium* spp.; or they may be borne on a branched structure to form a *panicle*, as in roses (*Rosa* spp.). In onions (*Allium* spp.) all the flower-bearing branches arise from a single apex, forming a dense head of flowers called an *umbel*, and in *Hydrangea* spp. these branches are subdivided to form a flat head of flowers called a *corymb*. In hazel (*Corylus* spp.) the male flowers form a *catkin*. In the daisy family (Asteraceae) all the flowers arise from a single platform, the *receptacle*, in such a dense mass as to form a *composite flower* that resembles a single flower (Fig. 3.3). Very often, as in Fig. 3.3, the inner and outer flowers may be very different, the latter being expanded to give the appearance of petals and the former being very small and tube-like.

## Alternation of generations

The genetic information needed to co-ordinate the development of a new plant is encoded in the *genes* within the *DNA* of the cells, packaged in elongate structures, the *chromosomes*, contained within the cell nuclei (see Chapter 5). The number of chromosomes per cell is fixed for each species, but varies among species. Chromosomes normally occur as almost identical pairs. Indeed, they are identical for genes defining all the major characteristics of the plant, but may vary for minor characteristics like rate of growth, height or flower colour (see Chapter 5). Each cell of a plant therefore contains two almost, but not quite, identical sets of genetic information and is said to be *diploid*. When a cell divides in the shoot or root or apex to produce two daughter cells, the chromosomes also divide to produce identical pairs of daughter chromosomes. This division is termed *mitosis*. The process of sexual reproduction, in contrast, first involves a specialised form of nuclear division called *meiosis*, which results in a halving of the genetic material of the parent to produce specialised *haploid* cells, the microspores and ovules already described. Fertilisation of an ovule by a microspore restores the genetic complement and the resulting *zygote* is therefore diploid. This process is dealt with in detail in Chapter 5.

Following fertilisation a diploid seed develops from the zygote, and this is capable of germination to produce a new diploid plant which contains genetic materials from each of the parents. This alternation of haploid and diploid generations in seed plants has evolved over millions of years from a situation seen in more primitive plants, such as the ferns (see Chapter 1), in which there is an alternation between separate, free-living sexual (*gametophyte*) and spore-bearing (*sporophyte*) generations.

Because of its minute size and dependence on water for growth and fertilisation, the gametophyte generation of the fern is, as described in Chapter 1, very vulnerable and is thus the Achilles' heel of the life cycle. In seed plants, however, the risk of damage to the gametophyte generation has been reduced by its total incorporation into the sporophyte generation until after fertilisation has occurred. Thus, although there is an alternation of sporophyte and gametophyte generations, this is not apparent to the casual observer. The new sporophyte resulting from fertilisation can receive food from the parent sporophyte and can be dispersed in a complex unit called the seed. This contains the minute new sporophyte (the *embryo*), together with stored food materials to fuel germination and establishment of the seedling, a process described in Chapter 9. The whole structure is protected by a seed coat (see below and Chapter 9).

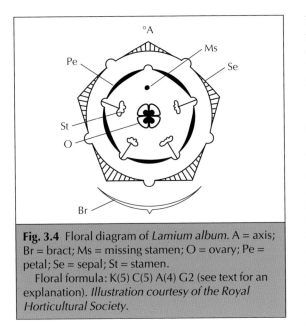

**Fig. 3.4** Floral diagram of *Lamium album*. A = axis; Br = bract; Ms = missing stamen; O = ovary; Pe = petal; Se = sepal; St = stamen.
Floral formula: K(5) C(5) A(4) G2 (see text for an explanation). *Illustration courtesy of the Royal Horticultural Society.*

## Floral diagrams and formulae

The disposition of the parts of the flower is usually displayed in formal descriptions (as in *Floras*) using a floral diagram (Fig. 3.4). This provides a condensed and simplified view of the flower from above, with the individual parts displayed in one plane. The organs shown, starting from the centre, are: ovary; stamens; petals or perianth segments (the corolla); the modified leaves that enclose the bud (the calyx); and any reduced leaves that may be attached to the flower stalk (bracts and bracteoles). The ovary is usually shown in cross-section to reveal the arrangement of the carpels (see below). The anthers are drawn to show where they open to release the pollen. An absent organ important to the symmetry of the flower is indicated by a dot or asterisk. If one part of the flower is attached to another, the link is shown by a line. The separation of petals and sepals is also made clear.

A floral formula (Fig. 3.4) may be used to supplement the floral diagram, in which **K** = calyx, **C** = corolla, **A** = androecium (male structures) and **G** = gynoecium (female structures). Figures are added to each letter showing the number of each part in a single flower. If the number exceeds 12, the symbol ∞ is used. If a flower part is absent, this is indicated by zero (0). Where whorls of sepals or petals are linked, this is indicated by square brackets. A bar above or below the number for the gynoecium indicates whether the ovary is superior or inferior. The floral formula does not indicate the overall shape and form of the flower. This is usually rectified by including in the formal description a half flower section.

## Seeds and fruits

The embryos of Angiosperms may have single *cotyledons*, the seed leaves that enclose the embryo, as in the monocotyledons (e.g. the Poaceae, grasses and their relatives) or two or more cotyledons, with the embryo between them, as in the dicotyledons (broad-leaved flowering plants). The embryos and cotyledons are in turn surrounded by the *integuments*, which become the seed coat (*testa*) (see Chapter 9). Finally, the seeds are contained within *carpels*, the walls of which thicken to become the fruit. Carpels evolved from leaf-like structures folded over and fused along the edges.

Most of these structures may be seen very easily with the naked eye in the pea (*Pisum sativum*). If a mature pod, the fruit (Fig. 3.5), is split open along its length the ancestral leaf-like form of the single carpel is obvious. The 'beak' end is the remains of the stigma with its short style. The remains of the flower, now shrivelled, may still be attached to the stalk end. Attached along one margin of the pod are the individual seeds, the 'peas', surrounded by a tough coat, the testa, derived originally from the integuments. This may be split to reveal the contents. Before doing so, however, the point where the pea was attached to the pod should be examined with a lens. The remains of the *micropyle*, the minute slot-like hole through which fertilisation by the pollen tube actually occurred, will be visible. If the testa is next removed, the two fleshy cotyledons, the seed leaves, which enclose the embryo, will be revealed. The embryo itself, close to the point of attachment to the pod, will be seen to have an embryonic shoot, the *plumule*, and an embryonic root, the *radicle*. Following germination these will grow to produce a new plant.

## Other fruits

There are many other forms of seeds and fruits, some more and some less complex than the pea. All have

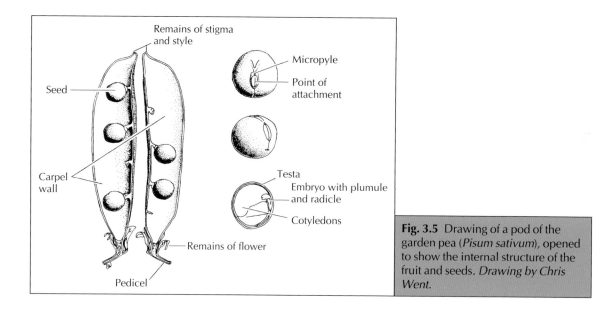

**Fig. 3.5** Drawing of a pod of the garden pea (*Pisum sativum*), opened to show the internal structure of the fruit and seeds. *Drawing by Chris Went.*

evolved to provide protection for the embryo and to facilitate the dispersal of the seeds. In the weed species dandelion (*Taraxacum officinale*), for example, each carpel contains only one seed, and hardens to give protection, but also develops a parachute of hairs at one end that enables it to be dispersed by wind. The paired carpels of *Acer* spp., each with a single seed, also become hard and develop membranous 'wings', again to aid wind dispersal. In wild progenitors of beans (*Vicia* and *Phaseolus* spp.) and in their relatives such as brooms (*Cytisus* spp.) the carpel walls dry and harden in such a way as to develop internal tensions. On a warm day, as the fruit ripens, these tensions cause the pods to twist and split, throwing the seeds considerable distances.

In other species the carpel develops into a fleshy structure, often brightly coloured and sweet, encouraging dispersal by animals and birds. These eat and digest the fruits, but the seeds, which are resistant to digestive enzymes, pass straight through the gut and may be deposited in the faeces many miles from the place of their formation. In plum (*Prunus* spp.; Fig. 3.6), for example, a simple seed is surrounded by a three-layered *pericarp* or fruit wall, giving the stone, the flesh and skin. In tomato (*Lycopersicon esculentum*) there may be two, three or more fleshy carpels comprising the fruit, each containing many seeds attached to a central thickened region called the *placenta*.

**Fig. 3.6** Plums are *true fruits* with a three-layered fruit wall, the *pericarp*, consisting of skin, flesh and stone. The *seed*, which has two *cotyledons* and is surrounded by a brown, papery *testa*, is contained within the stone. *Photograph by Chris Prior, Royal Horticultural Society.*

Finally in this brief list of examples, there are fruits in which the carpels harden to provide protection to the seed, but the structure on which the flower developed, the receptacle, swells to become fleshy and attractive to animals and birds. In strawberry

**Fig. 3.7** Strawberries are *false fruits* in which the red flesh is an expanded, fleshy *receptacle*, with the remains of the flower clearly visible around the base. The true fruits, which are small with a hard brown *pericarp*, are scattered over the surface of the receptacle. *Photograph by Chris Prior, Royal Horticultural Society.*

(*Fragaria* × *ananassa*; Fig. 3.7), for example, the receptacle grows to form the familiar red pyramid structure, a *false fruit*, with the tiny woody true fruits (*achenes*), each containing a single seed, dotted all over the surface. In apple (*Malus* spp.) the fleshy receptacle grows up and completely surrounds the fruit. If an apple is cut in half the core, comprising the seeds, each with a brown testa, contained within about five horny carpels, are clearly visible. It is this core that is the *true fruit*, while the surrounding receptacle, with the remains of the flower emerging through the end opposite the stalk, is the false fruit. It functions in the same way as a fleshy true fruit, however, being attractive to animals and birds and thereby aiding seed dispersal.

Over the centuries gardeners and plant breeders have selected plants with particular seed or fruit characteristics that bring either visual benefits, like the Chinese lantern (*Physalis alkekengi*), or are a good source of food, such as tomatoes and eating apples. This selection has exaggerated particular parts of the seed and fruit, making it relatively easy to make out the component parts.

## CONCLUSION

Flowers and fruits provide much of the colour and seasonal interest in a garden. They are also sophist-icated structures vital for reproduction, variation and dispersal of species (Chapters 5 and 9). They have evolved in response to environmental forces and have co-evolved with pollinators and dispersal agents to produce a great natural diversity of structures, shapes, forms and colours. This diversity constitutes the raw material used by plant breeders who, by making crosses and selecting desirable hybrids (Chapter 5), have produced the vast array of decorative plants now available to the gardener.

## FURTHER READING

Attenborough, D. (1995) *The Private Life of Plants*. BBC Books, London.

Bernhardt, P. (1999) *The Rose's Kiss: a Natural History of Flowers*. Island Press/Shearwater Books, Washington DC.

Bowes, B.G. (1997) *A Colour Atlas of Plant Structure*. Manson Publishing, London.

Capon, B. (1992) *Botany for Gardeners*. B.T. Batsford, London.

Mabberley, D.J. (1997) *The Plant Book*, 2nd edn. Cambridge University Press, Cambridge.

Pollock, M. & Griffiths, M. (2005) *RHS Illustrated Dictionary of Gardening*. Dorling Kindersley, London.

Raven, P.H., Evert, R.F. & Eichhorn, S.E. (1999) *Biology of Plants*, 6th edn. Worth Publishers, New York.

Walters, S.M. (1993) *Wild and Garden Plants* (New Naturalist Series). HarperCollins Publishers, London.

# 4

# Naming Plants

## SUMMARY

This chapter first deals with the identification and naming of plants and the meaning and structure of the names. The basic principles of plant taxonomy are discussed next, with special emphasis being given to cultivated plant taxonomy. The reasons why plant names are sometimes changed by taxonomists are then outlined. Finally, the quest for stability of botanical and cultural plant names is described in the context of recent developments in taxonomic methods, notably the use of data derived from DNA sequences and the use of computers to handle and make connections among and between large data sets.

## INTRODUCTION

The naming of plants begins with the simple spontaneity that we associate with the naming of other everyday objects, using so-called common names, but as soon as any precision is required this quickly leads to a need for some understanding of the biological principles of reproduction, heredity and evolution. Furthermore, to understand the complexity of variation observed among plants, a framework of apparently arcane rules has been established: the language of botanical nomenclature. But with the basic knowledge of the principles that underpin this language, a nurseryman's list of plants becomes as informative to the gardener as a musical score is to a musician.

How is a plant specimen analysed to yield its name? The name is not an intrinsic property of the plant itself; the person naming, or in other words identifying, the plant in Fig. 4.1 will expect a name to have been coined already. This will have been done by a *taxonomist*, a person who studies the classification and naming of plants, and has already studied similar plants and defined their limits of variation, as well as the scope and usage of their names. Once a name has been deduced for the plant from the available evidence, our identifier can find out more about the plant, such as where it grows wild, how to cultivate it, its possible uses and the names of related plants. Discovering the name of a plant is no idle activity: it is the key to what has already been written about it.

## HOW TO IDENTIFY A PLANT

There are a various approaches to naming plants, but each requires as much information as possible to be gathered about the specimen. The extent to which the gathered material is representative of the whole plant and its growth and reproductive characteristics will determine the route or routes taken; as, of course, will the expertise and information resources of the identifier. All the options are easier if the specimens are as nearly complete as possible. Details of leaves, leaf arrangement, stem and bark type, presence of latex (a milky white fluid), root form, flowers and fruits are all important; missing parts may result in

**Fig. 4.1** Identifying plants.
*Photograph courtesy of the Royal Horticultural Society.*

incomplete or inaccurate identification. Associated information such as details of the origin and hardiness of the specimen are of equal significance.

How might someone attempt to identify a pelargonium, found upon a window-sill, such as that illustrated in Fig. 4.2? Tens of thousands of pelargoniums have been named and this might suggest that a hopeless, or at least arduous, task lay ahead; but the strategy which follows is regularly used by botanists to tackle such problems.

1. By using the appropriate specialist literature, such as *The European Garden Flora* [Walters et al. (eds), 1984–2000] or other similar works, which give as complete a coverage of plants known in cultivation as can be expected of a reference work. Regional botanical accounts, called *Floras*, are useful when something is known about the plant's origin. Monographs provide a complete and structured account of entire plant groups; they are often the foundation of our understanding of the classification for the group, too. Articles in periodicals such as *The Plantsman*, published by the Royal Horticultural Society (RHS), provide up-to-date information and detailed accounts of less diverse plant groups.

2. By making comparisons with illustrations or specimens. These may be in books or on the Internet, or named plants in botanic gardens and collections

**Fig. 4.2** A plant of *Pelargonium crispum* 'Variegatum'.

such as those held by members of the National Council for the Conservation of Plants and Gardens, or even pressed specimens in a *herbarium*. A herbarium is a collection of preserved, usually pressed, and named plants, such as those at the Natural History Museum, the Royal Botanic Garden Edinburgh, the Royal Botanic Gardens, Kew, and the Royal Horticultural Society Garden, Wisley.

3. By asking a botanical or horticultural specialist at, for example, the institutions listed above.

The literature option usually involves the use of *botanical keys* where the specimen's features are matched to each of a series of paired statements. The best keys are always dichotomous, with an either/or choice until a satisfactory solution has been reached. In most cases, keys are followed by descriptions and illustrations that should be used to check that this step-by-step approach has led to the right identification. It is easy to go wrong, especially with large groups of plants. It may even be prudent to follow up further references. The identity of a specimen successfully keyed in *The European Garden Flora*, a well-structured and comprehensive starting point, may be verified by looking up the illustrations cited. If it is a tree or shrub cultivated outdoors in the British Isles, the standard work, *Trees and Shrubs Hardy in the British Isles* (Bean, 1970–1988), might then be consulted. These four volumes and supplement contain very good descriptions and facts, but unfortunately no keys and very few illustrations. For bulbous plants there are many books with descriptions and illustrations; similarly for cacti and succulent species.

## THE MEANING AND STRUCTURE OF NAMES

Prior to 1753 plants were known by phrase names, which were difficult to distinguish from simple descriptions. For example, the plant known as *Achillea ptarmica* (sneezewort) once carried the phrase name *Achillea foliis lanceolatis acuminatis argute serratis*, which simply means the *Achillea* with lanceolate leaves which taper to a sharp point and have sharply pointed, saw-toothed edges. It was the Swedish botanist Carl von Linné who formalised in the eighteenth century the *binomial* system of

nomenclature consisting of two basic terms, the *genus* and *species*, that is still in use today.

Latin is the language of nomenclature because it was the accepted international language of science until the twentieth century. Indeed, even von Linné's name was Latinised to Linnaeus. Although it may make plant names seem more obscure today to the layperson, the legacy of the Latin binomial system has been a consistent and familiar language enabling us to understand the precise meaning of plant names from 1753 onwards. The remarkable book *Botanical Latin* (Stearn, 1992a) has greatly helped to make these names intelligible, and to enable the system to be continued in a less classically trained age. It makes fascinating reading.

When a plant is discovered and thought to be new to science, it is given a Latin name and formally described with a Latin diagnosis according to a set of agreed international rules, the *International Code of Botanical Nomenclature* (ICBN). This requires that a specimen be made and chosen to serve as a reference point, the type specimen of the name. This specimen will fix the use of the name unless altered through the formal procedures of the ICBN.

Just as the word knife is a generic term for cutting implements, *Pelargonium* is the generic name (genus) for a group of plants held together by common characteristics. Following this thread, *Pelargonium crispum*, the name of the plant in Fig. 4.2, is more specific, as table knife is for knives.

Genus and species refer to two ranks in a hierarchy of relationship, reflecting different degrees of variation (see Table 4.1). *Pelargonium crispum* belongs to the family Geraniaceae which encompasses the other genera *Geranium* (the cranesbills or hardy garden geraniums) and *Erodium* (the storksbills). At the other end of the scale small degrees of variation, such as in flower colour or leaf variegation, might be recognised as botanical varieties or, if raised in cultivation, as cultivars, for example *Pelargonium crispum* 'Variegatum', the cultivar of *P. crispum* illustrated in Fig. 4.2.

It is now recommended that all botanical names of Latin form are written in italics, or underlined where italics are unavailable (Fig. 4.3). Names of species and the lower botanical ranks are not written with an initial capital letter, even when they are named after a person or place. *Cultivars* are always written with initial capital letters and placed in single quotation marks. Since 1959, in order to emphasise the

**Table 4.1** Classification of a variegated pelargonium, showing the hierarchy of categories and their names.

| Category | Scientific name | Vernacular name |
|----------|-----------------|-----------------|
| Class | Angiospermae | Angiosperms; flowering plants |
| Subclass | Dicotyledoneae | Dicotyledons, Dicots |
| Order | Geraniales | – |
| Family | Geraniaceae | – |
| Subfamily | Geranioideae | – |
| Tribe* | – | – |
| Genus | *Pelargonium* | The pelargoniums (confusingly also called geraniums) |
| Subgenus* | – | – |
| Species | *crispum* | – |
| Subspecies* | – | – |
| Variety | *crispum* | – |
| Forma* | – | – |
| Cultivar | 'Variegatum' | – |

* Note that not all ranks are used in all cases.

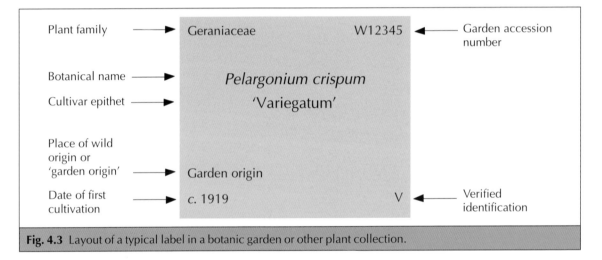

**Fig. 4.3** Layout of a typical label in a botanic garden or other plant collection.

different treatment of botanical and horticultural names, it has not been permissible to coin all-new cultivar names in Latin form, such as words with the endings *-us* or *-um*. 'Variegatum', having been coined around 1919, long before the 1959 ruling, is an acceptable epithet, despite its Latin form.

## TAXONOMY: ORDER IN DIVERSITY

### The botanical framework

The evolution of plants implies a common ancestry and, since a natural classification based on an understanding of relationships will have the most predictive value – and thus be the most useful overall – this is what botanists strive to understand and reflect in their system of naming. However, in the absence of a fossil record, much of this is based on speculation. Characters, such as spines or flower colour, have often evolved several times in different groups. Such *convergent evolution* has confused the past assessment of relationships.

Understanding the time scale of ancestry will often indicate how closely plant groups are related, but because some groups evolve faster than others the situation is not straightforward. Some groups have evolved very little in a long time: *Metasequoia glyptostroboides* (dawn redwood) was first described from fossil deposits before the living tree was discovered and *Ginkgo biloba* (the maidenhair tree) is frequently referred to as a 'living fossil'. Others have shown much more rapid evolution, for example the large *Echium* species in the Canary Islands, where different species have evolved and become adapted to the different ecological niches on the different islands.

The taxonomic botanist looks for *discontinuous variation* to characterise the units to which names are applied. These discontinuities are brought about by three *isolating mechanisms*: reproductive, in which plants may flower at different times; ecological, in which plants become adapted to separated environments, as in the case of the wood and water avens, *Geum urbanum* in hedgerows and *Geum rivale* in streamside locations respectively; or distributional, in which plants become stranded on geographically different mountain tops, valleys or islands, with changing environmental conditions.

*Hybridisation* breaks down discontinuities arising from these isolating mechanisms. It gives rise to new complexities of variation for the taxonomist to unravel, a process which is often fraught with difficulties. Hybridisation occurs frequently and spontaneously in gardens, where isolating mechanisms are removed, not simply because plants from different areas are thrust together; it is also deliberately used by plant breeders in their quest for new and better garden plants.

There is an inherent conflict in trying to divide this complex and often little-understood variation of plants into the neat, hierarchical units of botanical names. Little wonder that errors of judgement are made; as with plotting the outline of a plough in the stars, there are no absolutely definitive rules to dictate what is included in a group or what is excluded.

## New developments

The palette of characters available to the taxonomist has expanded greatly in recent years, and with them a new perspective on how relationships may be deduced. Most recently, the ability to read the DNA sequences of plants has transformed our approach to taxonomic science. The resulting new knowledge and understanding must be reflected in the classification, and therefore in the naming, if the basic principles of the system are to hold true.

Two general approaches to analysis have arisen to handle the large amounts of information generated in taxonomic studies. The first, *phenetics*, brings together a wide range of characteristics of each plant in a broad comparison of similarity. A very refined branch of statistics is required to handle the diversity of measurement data, such as leaf length. More recently, *cladistics* has dominated as an analytical approach. In this system, each character is represented by two states: primitive and derived. All characters used in the analysis are brought together in a branching diagram like a family tree, such that the number of branches is minimised, while still concurring with the evidence of each character. The relationships of three plants, with one character, are easy to deduce, but the number of possible combinations very rapidly increases as plants and characters are added. A fast computer is just as essential a tool for modern cladistic studies of DNA sequence information as it is for any phenetic analysis.

Even the most sopisticated and detailed taxonomic analysis is open to interpretation, and a diversity of opinions may arise in the assignment of names to the natural groups of plants, called *taxa* (singular, taxon*)*. Thus *Ulmus angustifolia* (Goodyer's elm) of some authors is recognised by others as *Ulmus minor* subspecies *angustifolia*: the same taxon, but a different name.

## TAXONOMY OF CULTIVATED PLANTS

Botanical names alone have long been used to label plants of cultivated origin. Just as hybrids occur naturally and have been named as such, for example the oak *Quercus × rosacea*, a hybrid of the wild species of oak *Quercus robur* and *Quercus petraea*, the products of deliberate hybridisation may be named in the same way. The distinction between natural and deliberate hybridisation is blurred. Russian comfrey (*Symphytum × uplandicum*), for example, is a natural hybrid, but is derived from an introduced plant from South-west Asia (*Symphytum asperum*) and the native *Symphytum officinale*, and gives rise to some very garden-worthy variants. The involvement of humans is inextricably bound up in the origin of such plants. The use of a botanical epithet, however, only identifies the two parent taxa; it does not identify the *clone* (a genetically identical group; see Chapter 10), or even the clone of either parent. The name applies to all independent crosses of representatives of the two parent taxa. The system is therefore of only limited value in identifying individual plants of merit to the gardener; an indicator of potential value, but infinitely varying between and beyond the characteristics of the parents.

The *cultivar* is the basic unit of cultivated plant taxonomy, which borrows the framework of the ICBN as far as possible, usually to the genus, though often to the species. A cultivar must possess the attributes of distinctiveness, uniformity and stability to merit recognition, and its characteristics must be retained through propagation. The extremity of form of many cultivated plants cannot usually be transferred by seed from one sexual generation to the next, and so vegetative propagation is often required (see Chapter 10); hence most cultivars are clonal selections.

Botanical names make reference to a single specimen as a representative of the group, called the type specimen. Other plants are given the same name on the basis of shared characteristics with those of the type. Cultivated plant names, however, are ascribed in a different way, usually to identify an unvarying clone, an unusual and often extreme form, and have evolved to suit the practical purposes to which plants are put in the service of humanity. A *standard*, the horticultural equivalent of a botanical type, differs from the latter in that it exhibits the precise characteristics of the cultivar and is not just one example plant drawn from a pool of continuous variation. Only relatively recently has an attempt been made to impose a structured system of classification upon the diversity of plants of cultivated origin, where clonal selection is only part of the story. The utilitarian demands of cultivated plants make classification a complex affair; the system must accommodate not only individual plants, but sometimes one or more of the following.

1. A perpetuated single stage of a plant's growth cycle, such as the juvenile foliage of some conifer cultivars.
2. A self-sustaining part of a plant, such as the lateral growths of some conifers which are grown as prostrate plants, as with *Abies amabilis* 'Spreading Star'.
3. A characteristically virus-infected plant, such as *Abutilon* 'Souvenir de Bonn'.

4. A clonal selection of aberrant growth on a plant, such as the witches' broom, *Picea abies* 'Pygmaea'.
5. An assemblage of seed-derived individuals raised from within a population, sharing the same distinctive characteristic, individuals which did not share the defining characteristic having been 'weeded out'.
6. A line produced by repeated and exclusive self-fertilisation.
7. A genetically modified plant (see Chapter 5).
8. A graft *chimera*, where the plant is made up of cells originating from both of two united plants (see Chapter 5), but not a simple graft, where scion and rootstock are separate cultivars.

The *F1* hybrid (see Chapter 5) is a familiar type of cultivar: two different parents, of different species, are cross-pollinated to give rise to the cultivar, which is often sterile. As the cultivar does not breed true, it can only be maintained by repeated crossing of the parents. F1 hybrids are common because they tend to be vigorous plants.

## Distinctiveness

Although essential to the definition of a cultivar, distinctiveness may relate to rather obscure attributes: for instance, long season of fruiting or resistance to stem collapse (often called lodging), an attribute important for grain crops, as well as the more familiar attributes of colour, shape and scent. As variation is usually eliminated within a cultivar, distinctiveness may be assessed to the finest degree. This may make cultivars far more difficult to distinguish from one another than botanical taxa. In common with the definition of a botanical taxon, however, the description of distinctiveness can only relate attributes to those of other known taxa. Thus 'flowers purple' may adequately distinguish a cultivar today, but tomorrow's new plant with 'glossier purple flowers' blurs the distinction.

## Uniformity

Non-uniform plants fall outside the definition of the named cultivar, even if wholly derived from it, and uniformity of a cultivar need not necessarily be spontaneous. Seed-raised cultivars are often only maintained through selection of plants that match the cultivar's physical description, others being discarded. *Sports*, non-typical individuals or parts of individuals (such as a white flower on an otherwise red-flowered carnation) from otherwise uniform clones, may not be assigned the same name.

## Stability

To merit naming, the characteristics that make a cultivar worthy of attention must also be stable in the selected individual. A variegation may be interesting at the time, but the plant exhibiting it will not be worthy of a new name unless it can be sustained. Attributes maintained by the gardener's attentions through the life of the plant do not count: a pleached lime is taxonomically not different from its free-grown counterpart, whereas the fastigiate beech, *Fagus sylvatica* 'Dawyck', maintains its form naturally.

## WHY PLANTS CHANGE THEIR NAMES

The reasons for name changes fall into three categories: taxonomic, nomenclatural and misidentification.

## Taxonomic changes

Taxonomic changes are usually the result of advances in botanical knowledge, such as when a species is found to have been classified in the wrong genus, a genus placed in the wrong family or what was considered to be two genera becomes recognised as one. For example, analysis of morphological and DNA characteristics of *Cimicifuga* and *Actaea* in the 1990s showed *Cimicifuga* to be nested within the bounds of *Actaea* and therefore only one genus name was needed. *Actaea* as the oldest published of the two generic names took priority and was retained, with species such as *Cimicifuga racemosa* and *Cimicifuga simplex* now known as *Actaea racemosa* and *Actaea simplex*.

Some changes reflect taxonomic opinion. It is often difficult to draw a line between matters of fact and opinion when making taxonomic judgements. Some botanists prefer not to recognise the rank of

subspecies, elevating these taxa to species and sections of genera to genera. Such people are splitters, part of a more radical tradition which has often been favoured by Russian and eastern European botanists. The user is left with a choice of classifications and names, with no botanical rules to help choose which to use.

The improvement of classifications may be judged by their benefit to the widest range of users of the plants, including plant breeders and phytochemists. Name changes will be inevitable as we move in this direction.

## Nomenclatural

Nomenclatural changes arise where plant names are contrary to the ICBN. They may only be avoided by the action of international committees which continually decide cases to avoid unfortunate name changes. In recent years, it has been shown that the annual chrysanthemum (a species native to Southwest Morocco) is a very different genus from the shrubby chrysanthemums of the Canary Islands, and very different from the European cornfield weed species and the Japanese ones so well known to horticulturists. The ICBN required the annual chrysanthemum to be called *Ismene* according to the strictly fair rule of priority by earlier publication: an important nomenclatural principle, but one which is not readily appreciated by most users of plant names. Thus the Canary Island ones were called *Argyranthemum*, the weedy ones remained *Chrysanthemum* and the Japanese cultivated ones called *Dendranthema*.

After an outcry from horticulturists, the international committee on names decided that the type should be changed, so that the Japanese species retain the use of the name *Chrysanthemum*. The weedy species, including our native *Chrysanthemum segetum*, become *Xanthophthalmum*, certainly upsetting European weed scientists. Committee members involved take all views on board in an effort to obtain nomenclatural stability, though clearly not everyone's needs can be satisfied all of the time.

## Misidentification

Often species are introduced into cultivation under the wrong name: they have simply been misidentified at some stage. Names such as *Corydalis decipiens* and *Achillea taygetea* have become well established after such early confusion. The latter name is likely to have been inaccurately applied to *Achillea clypeolata*, which was itself not well known at the time; and generations of hybridisation have separated the garden plant still further from the rare, isolated plant of the Greek Mount Taygetos, which would have little place outside the alpine house. The take-up of a name in horticulture may remain unquestioned for a considerable period of time, and some upset may be caused when the true identification is revealed.

To help with the stability of names and resolve problems, the RHS has set up an Advisory Panel on Nomenclature and Taxonomy. This panel meets regularly and has played a very considerable role in stabilising and correcting plant names in the RHS Horticultural Database and its associated publications, particularly the *RHS Plant Finder*. If discrepancies are found in nomenclature and classification, it is recommended that the names in use in the *RHS Plant Finder* are followed.

## THE QUEST FOR STABILITY AND LINKING INFORMATION SYSTEMS FOR THE FUTURE

Gardeners are the most vocal of those seeking stability of plant names because their needs are often the least demanding of accurate nomenclature. They require a straightforward and distinct label for a plant, and expect to gain no more than an indication of its characteristic value in the garden. Whereas most other users of plant names have stronger reasons for a name to convey some underlying meaning, the desire for stability strikes a chord, to a greater or lesser extent, with all. The problem has been in deciding upon a single, functional and universally acceptable set of names.

It is not only gardeners who wish to be able to cross-reference botanical and horticultural literature. Botanists often turn to the rich heritage of horticulture for conservation purposes. For example some species, such as the cedar of Lebanon (*Cedrus libani*), that are relatively genetically uniform in the wild, show considerable genetic variation in cultivation. Plants like this may have been taken from long-

vanished, diverse, wild populations, making them an invaluable genetic resource today.

Trade in plants is an area of increasing legislative activity, and future government regulation is only predictable in that it will probably expand, leading to increasing interest in names to define those plants affected. Breeders protecting their intellectual property rights through Plant Breeders' Rights (PBR) or Plant Patents, together with the restriction of international trade in endangered species, the computerisation of stock-control systems and consumer attachment to brand names all push towards standardisation of plant nomenclature. PBR are becoming familiar across a wide range of ornamental and food plants, protecting a breeder's investment in careful selection and manipulation of the plants. Without such protection, a competitor is free to propagate and distribute the same plant, which all too easily undermines the profitability of the original breeding work.

## Stability of botanical plant names

*Index Kewensis* is a comprehensive guide to plant names, other than those of specific horticultural application, and is their first place of publication. It has been used by many as a standard reference to plant nomenclature, but is not the reference to accepted nomenclature that it is so often taken to be. *Index Kewensis* (now part of the larger International Plant Names Index initiative, www.ipni.org) does, however, help to prevent the appearance of de-stabilising, earlier-published names for well-known plants which the avid taxonomist might rescue from the obscurity of a little-read scientific journal.

*Index Kewensis* and other such publications follow after the event, tracking the publication of names as they appear. A proposal for registration of botanical names has been made, which would prevent their appearance in obscure journals, there to lie as a time-bomb to de-stabilise nomenclature in the future. However, this proposal for compulsory registration of new names has been resisted by many botanists, who are wary of its potential effect upon their intellectual freedom.

Since 1991, the Names in Current Use (NCU) project has provided a platform of stability for genus names. This has helped avoid a worrying tendency for botanists to discover obscure references which invalidate or take precedence over commonly accepted genus names. The NCU vascular plant list was based largely upon the standard work entitled *Vascular Plant Families and Genera* (Brummitt, 1992), published by the Royal Botanic Gardens, Kew.

A new approach to stability has appeared with the advent of the Internet. It is now possible to link information without necessarily adopting a standard naming system: thus the same information about plants might be viewed by different users, using different names. Major information systems such as the International Plant Names Index (IPNI; www.ipni.org), Species 2000 (www.species2000.org) and the International Legume Database and Information Service (ILDIS; www.ildis.org) have already been established as a first foray into this sphere. The RHS Horticultural Database, currently offering an independent perspective on naming, will link into other such systems in the future.

## Stability of cultivated plant names

Stability in the naming of cultivated plants (cultivars) is affected by many of the same factors influencing botanical nomenclature: priority, changing taxonomic perceptions, common usage and orthography (spelling). However, the influence of market forces brings new challenges to the naming of cultivars.

Voluntary registration of cultivated plant names began in 1955 as an attempt to avoid the unnecessary duplication of cultivated epithets, and has been implemented successfully for most of the major cultivated plant groups, although not adopted as universally by those who name plants as might have been hoped. A list of International Cultivar Registration Authorities (ICRA) may be found in the *International Code of Nomenclature for Cultivated Plants* (1995), and an updated version is available online (www.ishs.org/icra). Among popular groups of plants there is considerable pressure to allow re-use of certain names with fashionable associations or other marketable qualities. The name 'Wedding Day' may help sell a cut flower, but the temptation to re-use the name time and time again can only cause confusion.

The trend towards use of coded cultivar names adds further confusion; by adoption of a meaningless combination of letters and numbers as the 'cultivar name', the seller is able to attach other names to the plant without infringing any legal or other naming requirement. This allows nursery industries to protect their intellectual property rights over new plants without 'wasting' good, marketable names before time. These other names are known as trade designations or selling names, and should be distinctively styled in print: *Choisya* **Sundance** ('Lich') is a well-known example. Roses, with their particularly rich horticultural heritage, labour under the greatest proliferation of coded cultivar names. Because roses are often named after well-known personalities of the time, a rose breeder cannot risk applying the name before the plant is ready to be sold in large numbers, as the namesake may lose popularity.

Since 1987 *The Plant Finder* (later known as the *RHS Plant Finder*) has offered a standard and readily available reference to the names of plants and their annual availability in Britain, usefully bridging the divide between cultivated and botanical nomenclature, with some emphasis on adopting names of practical value to gardeners.

Despite the need to ensure precision in naming, cultivated plants are no more amenable to our rules than their wild counterparts. As a cultivar becomes more established in cultivation, and passes through successive generations, variation may creep in. At what point should new names be applied to describe this creeping variation? For the sake of stability in such cases, the cultivar-group name has been introduced. Thus *Achillea* 'The Pearl', a seed-raised cultivar known since before 1900, has become so variable that a cultivar-group name (*Achillea* The Pearl Group) has been coined, and selections made therein (e.g. *Achillea* The Pearl Group 'Boule de Neige').

Vernacular or common, descriptive names have a rich tradition of application to cultivated plants, but create the same difficulty as that applied to wild plants: imprecision. Nevertheless, the cultivar names we employ today are largely founded upon descriptive names, and a careful transition has been made to reduce confusion. Descriptive names relate principally to the place of origin, the raiser's name, or obvious characteristics of the plant. For example, Fortune's double yellow rose is now treated as *Rosa* × *odorata* 'Pseudindica' or *R.* 'Fortune's Double Yellow'. The Cultivated Code now seeks to prevent misunderstanding through the avoidance of obviously descriptive terms associated with common names for cultivated plants, and the terms 'form' and 'variety', which are reserved with stricter meaning by the Botanical Code.

The Cultivated Code has been introduced to help stabilise names of plants in cultivation. While providing a sound, voluntary code of practice for those giving and using such names, it must give leeway to developments in the industry and the new ways in which plants are produced, marketed and sold. Horticultural taxonomists are needed to unravel this complexity for the sake of gardeners, who show an increasing interest in the wonderful diversity of plants available to them.

## CONCLUSION

It may be concluded that since the introduction of the binomial system of plant nomenclature of Carl von Linné in 1753, increasing knowledge and understanding of wild and cultivated plant taxonomy and evolution has led to relatively sophisticated and stable systems for the naming, identification and classification of plants. The introduction of the *International Code of Botanical Nomenclature*, the *International Code of Nomenclature for Cultivated Plants* and the *RHS Plant Finder* have been especially important in the recent quest for stability of names. Recent trends in plant taxonomy, especially the use and analysis of DNA sequences, are increasing the precision and sophistication of plant nomenclature, and therefore its stability, but even the most sophisticated and detailed taxonomic analysis is open to interpretation, and a diversity of opinions will always arise over the assignment of names to particular taxa.

## FURTHER READING

Australian Geranium Society (1978, 1985) *A Check List and Register of Pelargonium Cultivar Names, Part 1 A–B; Part 2 C–F*. Australian Geranium Society, Sydney.

Bean, W.J. (1970–1988) *Trees and Shrubs Hardy in the British Isles*, 8th edn, G. Taylor and D.L. Clarke (eds),
supplement edited by D.L. Clarke, 4 volumes and supplement. John Murray, London.

Brickell, C.D. (ed.) (2002) *The Royal Horticultural Society Encyclopedia of Plants and Flowers*. Dorling Kindersley, London.

Brickell, C.D. (ed.) (2003) *The Royal Horticultural Society A–Z Encyclopedia of Garden Plants*. Dorling Kindersley, London.

Bridson, D. & Forman, L. (eds) (1992) *The Herbarium Handbook*, revised edn. Royal Botanic Gardens, Kew.

Brummitt, R.K. (1992) *Vascular Plant Families and Genera*. Royal Botanic Gardens, Kew.

Cullen, J. et al. (1986–2000) *The European Garden Flora*, 6 volumes. Cambridge University Press, Cambridge.

Greuter, W., Barrie, F.R., Burdet, H.M., Chaloner, W.G., Demoulin, V., Hawksworth, D.L., Jørgensen, P.M., Nicolson, D.H., Silva, P.C., Trehane, P. & McNeill, J. (eds) (1994) *International Code of Botanical Nomenclature (Tokyo Code)*. Koeltz Scientific Books, Königstein, [*Regnum Vegetable* **131**].

Griffiths, M. (1994) *Index of Garden Plants*. Macmillan Press, London.

Huxley, A. (ed.) (1997) *The New RHS Dictionary of Gardening*, 4 volumes. Macmillan Reference, London.

Kelly, J. (ed.) (1995) *The Hillier Gardener's Guide to Trees and Shrubs*. David & Charles, Newton Abbot.

Lord, W.A. (ed.) (published annually) *RHS Plant Finder*. Dorling Kindersley, London.

Mabberley, D.J. (1997) *The Plant-Book*, 2nd edn. Cambridge University Press, Cambridge.

Miller, D. (1996) *Pelargoniums: a Gardener's Guide to the Species and their Cultivars and Hybrids*. B.T. Batsford, London.

Phillips, R. & Rix, E.M. (1989) *Shrubs*. Pan, London.

Royal Horticultural Society (2001) *The RHS Colour Chart*. The Royal Horticultural Society, London.

Stearn, W.T. (1992a) *Botanical Latin*, 4th edn. David & Charles, Newton Abbot.

Stearn, W.T. (1992b) *Stearn's Dictionary of Plant Names for Gardeners*. Cassell Publishers, London.

Trehane, P., Brickell, C.D., Baum, B.R., Hetterscheid, W.L.A., Leslie, A.C., McNeill, J., Spongberg, S.A. & Vrugtman, F. (eds) (1995) *International Code of Nomenclature for Cultivated Plants – 1995*. Quarterjack Publishing, Wimborne.

Vaughan, J.G. & Geissler, C.A. (1997) *The New Oxford Book of Food Plants*. Oxford University Press, Oxford.

Walters, S.M. et al. (eds) (1984–2000) *The European Garden Flora*. Cambridge University Press, Cambridge.

<h1>5</h1>

# Selecting and Breeding Plants

## SUMMARY

In this chapter the structure, replication and functioning of plant DNA, genes and chromosomes are described. Next the generation of genetic variation, by recombination events, mutation, the activity of transpons and the formation of chimeras, is considered. The concepts and importance of homo- and heterozygosity, and dominant and recessive alleles, are analysed. The use of different breeding systems to utilise genetic variation in the production of improved plant cultivars is demonstrated. And finally, recombinant DNA technology is described and its potential value to the horticultural plant breeder assessed.

## INTRODUCTION: ADAPTATION AND DESIGN

The majority of plants cultivated in our gardens originate from wild plants, adapted to survive and compete in a natural environment, that have been selected and bred to meet human requirements for decoration, food, fuel or products such as chemicals, medicines, timber and fibre. In ornamentals the gardener looks for prolific or delayed flowering, extended and/or repeat flowering periods, and particular foliage, flower colours and shapes. In vegetable varieties the requirements are for prolific fruit and seed set, good seed retention at maturity, large root size, bolting resistance, flavour and colour quality, and earliness or lateness of maturity. Fruit varieties may be required to set a lot of fruit but with a low seed content; also the size, colour and shape of fruits are important. In crops grown for food or raw materials, the harvested product as it occurs in the wild plant may possess qualities, such as the presence of toxic chemicals, that make it unsuitable for its intended use and these must be removed by the plant breeder.

About 10 000–12 000 years ago the earliest farmers began to make major changes to the landscape by burning and clearing forests and heath to create fields, and then cultivating the cleared ground. They also began selecting and domesticating plants. The earliest records of what is undoubtedly domesticated plant material come from sites in the Near East (wheats, *Triticum* spp., flax, *Linum* spp. and pulses) and the New World (beans, *Phaseolus vulgaris*, and *Curcubita* spp.), dated to 6000–8000 BC. These crops were developed by cultivating wild plants and choosing, year-on-year, seed for the next season from those individual plants that performed well, according to the criteria of the people who grew and used them. These seeds were then planted in the following season. Unbeknown to the farmers, they were selecting for particular genes, forms of genes and gene combinations that together provided the characteristics they required in their crops.

Plant domestication was probably not restricted to crops grown for food and raw materials. Certainly the ancient Greeks and Romans understood the concept of gardens for pleasure rather than utility, and the earlier cultures of Egypt and Babylon also seem to have cultivated gardens, sometimes for vegetables but also for pleasure, and probably as a status symbol. Thus we can imagine that early gardeners were also selecting plants for improved flowers, foliage and form.

Initially, the spread of new plant types was relatively slow, but the advent of extensive exploration and trade by sea facilitated the discovery and movement of plants that had economic or social value. Those who probably made the greatest contribution to the distribution of plants around the world were the 'plant hunters'. Beginning in the sixteenth century, they collected enormous numbers of plants from remote regions of the globe to feed the curiosity of professional and amateur botanists and gardeners, and the economic interests of plant breeders, nursery owners and seed producers. Thus the trade in novel plants for food, raw materials, flowers and foliage distributed exotic species and new plant types all over the world.

Plant breeding entered a new phase in the early 1900s with the birth of modern genetic science. It was a gardener, Gregor Mendel, studying the inheritance of traits in the common garden pea (*Pisum sativum*), who first established the fundamentals of genetics in the 1850s. Mendel was abbot of the monastery at Brunn in Austria, now Brno in the Czech Republic, and an amateur scientist. Through his experiments in the monastery garden he uncovered the principles of heredity, although he knew nothing of genes, DNA or any of the other key biological components involved in transferring hereditary information from one generation of plants to the next. Precisely what Mendel understood from his discoveries is still debated, but his major contribution was to recognise that there are 'units' that carry inherited information from parents to offspring. These units, and the information they carry, do not blend in the offspring but are passed from generation to generation relatively unchanged (but see later). Mendel realised that it is the interaction between these units of inheritance in the offspring that account for the similarities and differences between members of different generations.

The history of genetics is a fascinating biological detective story. But in order to understand how genes work, how they account for plant characteristics and how genetics has been, and still is, used to produce new plant types, it is easier to begin from a practical rather than an historical perspective.

## GENES

Plants depend on enzymes (see Chapter 2) to carry out the processes that are essential for their life cycles. Assembling the appropriate enzymes in the correct amounts and locations enables a cell to manage all the chemical reactions it needs to operate efficiently at a particular developmental stage or under particular internal or external conditions. The plant requires a system to store and control the use of all the information that is needed to assemble this complex biological system. This is the function of the genes.

The genes of most organisms are made from the molecule *DNA* (*deoxyribonucleic acid*). Some viruses, however, use a closely related molecule, *RNA* (*ribonucleic acid*), to encode genetic information. RNA also has a role in all organisms in the reading of DNA-based genes. As DNA and RNA function in the same way, in terms of encoding genetic information, the fact that they are chemically slightly different is unimportant here.

DNA is made from four chemical building blocks, called bases. Each base has two parts: a sugar component, which is the same in all four bases, linked to a *nucleotide*. There are four nucleotides in the DNA molecule, giving four different bases. The initial letter of the nucleotides – adenine, cytosine, guanine and thymine – are commonly used to identify the bases. Any base may be joined to any other by linking them together through the sugar component of the DNA (Fig. 5.1). In this way bases may be assembled into strings of tremendous length, reflecting the large quantity of information they have to store. Each DNA molecule is in the form of a spiral helix, a significant configuration as will become apparent below. For example, if unravelled, the entire DNA content of a wheat cell would stretch to nearly 1.7 metres, carrying about 100 000 genes.

The way in which DNA encodes genetic information is both elegant and simple. The four nucleotides function like an alphabet of four letters, ACGT, from which words and sentences may be constructed.

Two sorts of words may be formed. The first are simple three-letter combinations, called *codons*, which represent amino acids (see Chapter 2). A protein consists of a string of amino acids, anything from about one hundred to several thousand. There

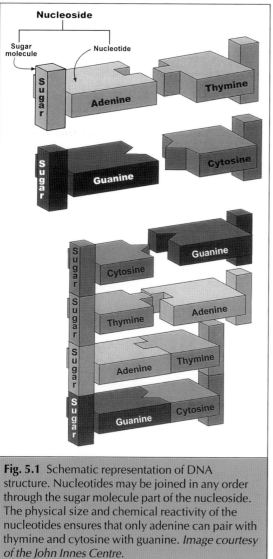

**Fig. 5.1** Schematic representation of DNA structure. Nucleotides may be joined in any order through the sugar molecule part of the nucleoside. The physical size and chemical reactivity of the nucleotides ensures that only adenine can pair with thymine and cytosine with guanine. *Image courtesy of the John Innes Centre.*

The second sort of word is longer and more complex. These sequences of bases convey instructions about how, when and where the coding regions are to be read. All of a plant's cells contain the same genetic information. So for a plant to be able to produce a variety of different cells, tissues and organs and to respond to its internal and external environment it must be able to regulate the activity of individual genes or groups of genes. Regulation is achieved by specific base sequences that control gene activity; they effectively tell plant cells, during development, under what conditions and in which tissues a gene is to be switched on or off. They also regulate the level of gene activity by controlling how frequently a gene is read by the cell. Because of their function in regulating gene activity these sequences are called regulatory or control sequences, to distinguish them from the protein-coding sequences.

The DNA is read by a cluster of several enzymes that physically move along the DNA strand and, under the direction of the regulatory sequences, make RNA copies of the coding regions of genes. This copying process is called transcription. As RNA and DNA function in the same way, the cell is in effect making multiple, identical copies of the information content of the gene. These RNA copies are then distributed to another type of enzyme cluster, which reads along the RNA molecule, recognises each group of three nucleotides and matches the appropriate amino acid to each codon. These enzymes are translating the genetic code into proteins (Fig. 5.2).

There are a few genes that do not make proteins but instead produce some of the key chemical components in the transcription/translation apparatus.

## Transcription factors

Transcription factors have a very important role in gene regulation. As their name suggests, they are involved in the process of copying the DNA to RNA prior to protein synthesis. How transcription factors work, and their importance in biology, are revealed from research on the control of flower structure.

The basic structure of any flower is four concentric rings of floral organs: an outer ring of sepals, a ring of petals, a ring of stamens and, in the centre of the flower, the carpels (see Chapter 3). There are many common flower types where this standard pattern is disrupted. Double flowers, where stamens or sepals

are 20 naturally occurring amino acids and each is represented by at least one, sometimes more, distinct codon(s). By reading along a DNA molecule a cell can assemble a sequence of amino acids according to the sequence of bases in the DNA. Different sequences and amounts of amino acids produce different proteins with different chemical and biological activities. There are also codons that act as punctuation in the genetic code; one codon, which also encodes the amino acid methionine, identifies the start of a gene and there are several codons that can mark the end of a gene.

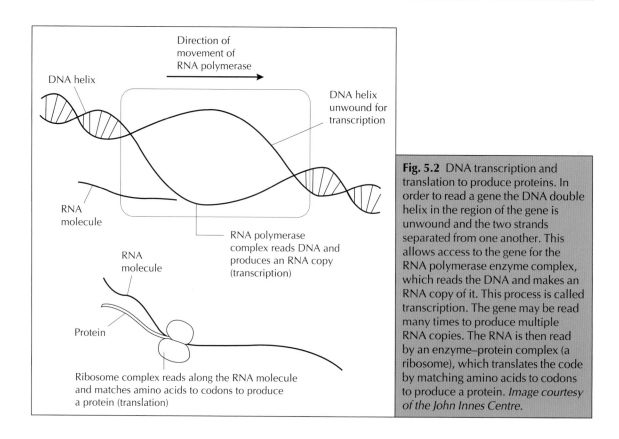

Direction of
movement of
RNA polymerase

DNA helix

DNA helix
unwound for
transcription

RNA
molecule

RNA
molecule

RNA polymerase
complex reads DNA and
produces an RNA copy
(transcription)

Protein

Ribosome complex reads along the RNA molecule
and matches amino acids to codons to produce
a protein (translation)

**Fig. 5.2** DNA transcription and translation to produce proteins. In order to read a gene the DNA double helix in the region of the gene is unwound and the two strands separated from one another. This allows access to the gene for the RNA polymerase enzyme complex, which reads the DNA and makes an RNA copy of it. This process is called transcription. The gene may be read many times to produce multiple RNA copies. The RNA is then read by an enzyme–protein complex (a ribosome), which translates the code by matching amino acids to codons to produce a protein. *Image courtesy of the John Innes Centre.*

are replaced by extra petals, are common. The flowers of single and double roses, and Shirley poppies, clearly show reduced numbers of stamens and, in the double flowers, increased numbers of petals. Other flowers have different floral-organ substitutions. For example, in the viridiflora rose (*Rosa × odorata* 'Viridiflora') all the floral organs are replaced by sepals. These changes are caused by gene mutations (see below). Studying mutations that cause novel flower structures has made it possible to understand that three key genes determine the organisation of the organs within the flower. The model of how they operate is called the ABC model, so for convenience we will use A, B and C to label the genes.

A, B and C are genes that produce proteins that control the activity of other genes; they are transcription factors. The products of A, B and C interact to produce their effect, which is to establish an invisible map that defines the four different zones in the group of dividing cells at the shoot tip, or floral meristem, that will produce sepals, petals, anthers

and carpels. The genes are active very early in flower bud development before the primordia, which develop into the flower organs, become visible on the surface of the floral meristem. Gene A is active at a low level throughout the meristem, but is most active in the meristem's outer zone. Gene C is active in the middle of the meristem. Let us imagine that the product of gene A colours the zone where it is active red, and that the product of gene C is blue. Gene B is active in a zone that overlaps the inner edge of zone A and the outer edge of zone C. Let us imagine that its product is yellow.

The activity of these genes, and the presence of their products, divides the meristem into four zones of imaginary colours. Where only the product of gene A is present (red) the developing primordia will form sepals; where only the product of gene C is present (blue) the developing primordia will become carpels; the products of A and B together (red + yellow = orange) induce petal formation; and the products of C and B together (blue + yellow = green) induce stamen formation (Fig. 5.3a).

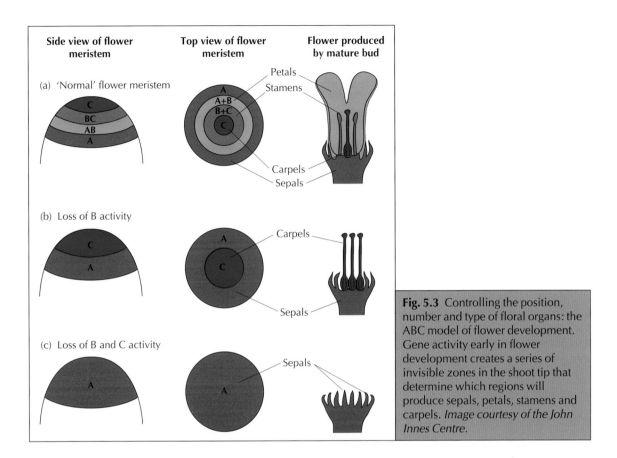

**Fig. 5.3** Controlling the position, number and type of floral organs: the ABC model of flower development. Gene activity early in flower development creates a series of invisible zones in the shoot tip that determine which regions will produce sepals, petals, stamens and carpels. *Image courtesy of the John Innes Centre.*

A change in the activity of one or more of these genes will result in a change in the pattern of organs in the flower. For example, a mutation that renders gene B inactive will result in flowers that have no petals or stamens but extra rings of sepals and carpels (Fig. 5.3b). Loss of B and C function will result in flowers with only sepals because the product of gene A is present in all the cells of the meristem, even though the level is so low as normally to be of no significance in determining flower structure when genes B and/or C are present (Fig 5.3c).

In species that normally have flowers with bilateral symmetry (see Chapter 3), such as snapdragon (*Antirrhinum majus*), toadflax (*Linaria vulgaris*) and African violet (*Saintpaulia*) hybrids, mutations are sometimes found that result in flowers with radial symmetry (Fig. 5.4a, b, c). Remarkably, such dramatic changes in flower shape can be caused by a change in only a single gene. This gene is active only in the region that will become the uppermost (dorsal) part of the flower, and the protein it produces interacts with other genes to determine the ultimate structure of the flower. The localised activity of a gene, and its product, establishes a polarity (which way is 'up') in the developing bud. This allows organs in different positions relative to 'up' to develop differently.

## Chromosomes

The DNA and genes in a plant cell are largely contained in the nucleus, but a small proportion of the cell's total DNA is found in the *chloroplasts* and *mitochondria*.

The nuclear DNA of any cell does not consist of one continuous fibre, but of several shorter pieces, the *chromosomes*. These are not broken fragments, but defined pieces whose ends are marked by specific DNA base sequences. During most of the cell's life the nuclear DNA exists as fibres, visible only under powerful electron microscopes at magnifications

**Fig. 5.4** Common species exhibiting flowers with radial and bilateral symmetry. (a) *Antirrhinum majus* (wild type, *left*, with bilateral symmetry and mutant with radial symmetry). (b) *Linaria vulgaris* (wild type, *top*, with bilateral symmetry and mutant with radial symmetry, *below*). (c) *Saintpaulia* hybrid (varieties 'Kazuko', *top*, with bilateral symmetry and 'Nada' with radial symmetry, *below*). *Photographs courtesy of the John Innes Centre.*

greater than 30 000 times. As cells prepare for cell division, however, each DNA fibre undergoes a complex 'packing' procedure. All along its length, pairs of loops of the fibre are wound around small particles of protein. The partly wound molecule is then wound again to create a spiral, which in turn is wound into another spiral. The result is that, immediately prior to cell division, the DNA becomes condensed into compact packets, the chromosomes, which are usually visible at magnifications of 200–400 times under a standard laboratory light microscope. Following division the DNA is unpacked.

## DNA replication

It has already been shown that bases are linked in long strings, through the sugar molecule that is common to every base, and that any base can be joined to any other base. Bases may also pair with other bases, but their physical size and chemistry allows pairing only of adenine with thymine and cytosine with guanine (Fig. 5.1). The DNA molecule exists as the famous double helix, which is arranged like a spiral ladder, with the paired bases forming the rungs and the long, sugar-linked, strands the supports. It is possible to create two new copies of a double helix by breaking the pairing between the bases, to produce two separate strands. Pairing new bases with the bases exposed on the single strands then produces two new double strands. This is called *semi-conservative replication*, because it uses an existing string of bases to create a new double strand. This method of copying information is extremely effective, and is backed up by sophisticated error-checking and repair systems.

## GENOMES

The complete genetic information content of an organism is called its *genome*. This includes all the genes, but may also include DNA that does not encode genetic information or control the reading of the code.

There are some important DNA sequences that define the middle (*centromere*) and ends (*telomeres*) of the chromosomes. These have a very significant role in the biology of the plant because they are important in cell division. There are other regions of the genome, however, where the DNA does not have an obvious function; it does not encode information, and is not associated with control or the centromere or telomere. This DNA is typically an assortment of sequences, some of which can be identified as fragments of sequence from plant viruses, plant genes and transposable elements (see below). It has the appearance of a genetic junk yard and for this reason is often described as junk DNA. The amounts of this type of DNA vary enormously among different species, but it typically occurs as long stretches of junk among clusters of active genes.

The shorthand description of this DNA as junk is potentially misleading as it suggests that it is useless. However, biological systems rarely retain materials that are entirely without purpose because natural selection operates against inefficient systems. Although its function is unknown, scientists have proposed that junk DNA may have a role as packing material that contributes to chromosome structure and size. Another explanation may be that these large stretches of non-coding, and therefore non-critical, material reduce the proportion of the genome that contains active genes (critical information). In the event of a virus or transposable element (see below) inserting itself into the genome, the probability that the insertion will occur in an active gene is reduced. Thus junk DNA may reduce the likelihood of certain types of damage occurring to active genes.

The size of the plant genome, in terms of total DNA content, varies dramatically between plants, with no particular relationship between genome size and plant size. For example, the insignificant weed *Arabidopsis thaliana* (thale cress) contains half the DNA of the horsechestnut tree (*Aesculus hippocastanum*), while the broad bean (*Vicia faba*) contains 200 times the amount of DNA found in *A. thaliana*. Nor does genome size reflect a difference in the number of genes; plants contain between about 25 000 and 100 000 genes, which in itself is only a four-fold difference.

One other factor has a major effect on the total DNA content of the genome, namely the *ploidy* level of the plant. The cells of most flowering plant species contain two complete copies of the genome, one copy from the male parent and the other from the female parent; these cells are called *diploids*. As we will see later, the gametes (egg and pollen) contain only one copy of the genome and are called *haploids*. Some plants, however, are polyploids. During their evolution these have undergone a process called *polyploidisation*, resulting in multiple copies of the genome. The most common forms are *triploids* (three copies of the genome), *tetraploids* (four copies of the genome) and *hexaploids* (six copies of the genome). The mechanisms by which polyploids arise are complex, but are basically the result of errors in cell division. Treating plants with chemicals that interfere with cell division, and so increasing the likelihood of polyploid formation, is a way of inducing polyploidisation deliberately. Colchicine, a naturally occurring toxic compound found in the autumn crocus (*Colchicum autumnale*), is used for this purpose. Polyploidy creates its own problems but plants that are stable polyploids may be both physically larger and genetically more robust than their diploid relatives. For this reason breeders have used polyploidisation to try to produce new varieties, for example of *Antirrhinum* spp., lettuce (*Lactuca sativa*) and tomato (*Lycopersicon esculentum*). Polyploidy is often associated with larger flowers, but may also be associated with later flowering. Many cultivated plants are natural polyploids, for example potato (*Solanum tuberosum*), cyclamen (*Cyclamen persicum*) and daffodil (*Narcissus* cultivars 'Cheerfulness' and 'Golden Dawn').

## Gamete formation

Most plant cells contain two complete copies of the genome. In the developing flower bud, at the time of egg and pollen formation, the cells that are destined to become the *gametes*, or *germ cells*, undergo a particular type of cell division called *meiosis*. This is a reduction division that ensures that, as a cell divides, only one copy of the genome is passed on to each of the daughter cells. Consequently, these cells have only one, rather than two, copies of the genome. At fertilisation the haploid pollen and egg cells fuse to produce a fertilised egg or *zygote*, which now contains two copies of the genome, one from each parent. The diploid fertilised egg then undergoes a form of cell division called *mitosis*. This common form of cell division, by which the plant produces the majority of its cells during its growth, replicates the entire complement of genetic information in the parent cell and passes on identical copies of both copies of the genome to each of the daughter

cells at division. Eventually, repeated division of the zygote produces a new plant consisting of diploid cells in which the genomes contributed by both parents are present.

## Recombination

Meiosis has a second function. It ensures that the genetic information from the two parental genomes is mixed, before gamete formation, through a process called *recombination*. During the early stages of meiosis equivalent chromosomes from each of the parent genomes pair up, so that throughout their entire length the two genomes lie side by side. The cell then cuts and rejoins the DNA in such a way that equivalent stretches of the genomes are switched. This is recombination and results in two 'new' genomes, that are still complete sets of genetic information, but are now made up from a mixture of the genomes that the plant received from its parents. Immediately following recombination, reduction division occurs so that the new genomes are distributed into separate germ cells. The points in the genome at which the cuts and splices occur, and the number of events, varies from cell to cell, so the germ cells produced by any one plant will contain a wide range of recombined genomes. Recombination ensures that only in very rare cases, those where no recombination occurred, will germ cells contain a genome that is identical to that of one or other of their 'grandparents'.

## Allelic variation

Why does recombination occur and why is it important to mix fragments of parental genomes? It has already been shown that genes can exist in different forms, the simplest example being where an active gene is damaged and rendered inactive on one chromosome of a pair, but not the other. The active and inactive forms of the gene are *alleles* and the difference in the trait controlled by the two alleles is called *allelic variation*. However, damage to a gene may not render it completely inactive but increase or decrease its activity or the activity of its product. In a situation where several gene products interact to control a trait, the products of different alleles may interact in subtly different ways and so produce equally subtle effects on the trait. Recombination produces new combinations of alleles in each generation and, through these new combinations, potentially more competitive individuals in a wild environment, or more attractive/productive individuals in a managed one. Both plant hunters and plant breeders are, in different ways, searching for new sources of allelic variation.

## MUTATION

*Mutation* is a natural process and describes any change to the genetic code. Mutants are familiar to gardeners as *sports*. Mutations result from the occasional copying errors that may occur during the normally accurate process of DNA replication. They may be induced by natural or artificially generated radiation, a naturally occurring or artificially applied chemical mutagen or as a result of the activities of viruses and transposable elements.

Mutations may occur anywhere in the genome. They range from a single base change in one codon of a gene to the loss or duplication of large pieces of genetic information or even whole chromosomes, representing a significant percentage of the genome. The larger the change, the more likely it is that it will so damage a plant's metabolism that it will be lethal. However, large duplications and deletions of sections of genomes are not uncommon in polyploids where they are reasonably well tolerated, as genes lost by deletion may well be duplicated in other copies of the genome.

Breeders artificially subject seeds to radiation and chemical treatments to increase the frequency of mutations in their breeding material. The process is random, so the breeder cannot determine which genes will be mutated and hence which characteristics will be changed. However, if conducted on a large-enough scale using many thousands of plants, it is usually possible to find interesting changes in traits that may be used in subsequent breeding programmes. This approach has been used with some success, for example, in breeding programmes for fruit colour changes in apples, compact growth in apples and pears, and flower colour changes in ornamentals, including *Alstromeria*, *Chrysanthemum*, *Dahlia* and *Streptocarpus* spp.

## Transposons

In the 1960s and 1970s the groundbreaking work of Barbara McLintock on unstable genetic traits in maize led to an understanding of what *transposons* are and how they function. They are naturally occurring fragments of DNA that are able to move around the genome by cutting themselves out of the DNA and splicing themselves in again elsewhere. They usually consist of a small piece of DNA encoding a few genes with a special sequence of bases at each end. Typically, the genes on the transposon encode the enzymes that actually carry out the cutting and splicing. Transposon insertion is random, so a transposon may by chance insert itself into a gene, and so disrupts its activity. If the transposon subsequently moves, it may cut itself out of the gene so precisely that the gene is reassembled intact and will function correctly again. The process is not always this precise, however, in which case the gene remains permanently damaged. The activity of some types of transposons causes higher than normal frequencies of breaks in chromosomes and thus may lead to losses of large fragments of the genome. Some plant viruses are also able to insert themselves into the genome and in so doing may disrupt gene activity. All of these activities are described as mutation.

Transposons may affect the activity of any gene, but are best known to gardeners through their influence on flower colour. Affected plants have enjoyed some popularity in recent years, with the fashion among breeders to produce flower varieties with randomly striped and spotted petals. However, the beautiful random red stripes and spots on the petals of one of our oldest rose varieties, *Rosa mundi* (*Rosa gallica* 'Versicolor'), and many other old roses, are also the result of transposon activity. The ancestor of *Rosa mundi* was a red rose carrying a transposon. At some point this transposon happened to insert itself into one of the genes involved in production of red pigment, damaging the gene and preventing the production of the pigment. This event must have happened in a germ line or meristem cell, as it has been possible to propagate the resulting sport, a 'white' rose without pigment in the petals. However, the transposon continues to move in these white roses, and each time it does so it may restore the structure and activity of the pigment gene. As transposons are active within individual cells, those cells where the transposon moves, and any daughter cells they produce, are capable of producing red pigment. If this happens during flower bud development in one of the cells that is involved in producing a petal, then when the flower opens a red spot will mark the presence of that cell and any daughter cells it produces thereafter. If the transposon moves early in flower bud development, the affected cell will produce many daughter cells, resulting in a large red stripe or spot on the opened petal. If the transposon moves late in bud development the cell will divide only a few times before petal formation is complete, and consequently only a small red spot will be evident. If the transposon is highly active, moving frequently, there is a high probability that it will move in several cells in each petal during the period of bud development, and several spots and stripes are likely to occur on each mature petal. A less active transposon will move less frequently, and the probability of it moving in more than one or two cells per petal during bud development will be small, so only a few spots and stripes are likely to occur on each petal (Fig. 5.5a). The snapdragon and pelargonium (*Pelargonium* spp.) are other common garden flowers in which transposon activity may be seen as striping and spotting of the petals (Fig. 5.5b, c).

The far-sighted and accurate explanation of the functioning of transposons by Barbara McLintock was initially treated with some incredulity, but now, having been accepted, has laid the foundation for studies that have culminated in our now being able to use transposons to search for other genes. This technique is called transposon-tagging and relies on the fact that transposons sometimes move in germ line cells (see below), which means that the mutations they cause are inherited. The disruption of a gene by a transposon may be seen as a change in a visual plant trait, as with the colour changes described above. A relatively simple genetic test may be used to demonstrate whether the mutation is caused by insertion of a transposon. If it is, the transposon can be isolated, since the DNA sequences of the transposons found in cultivated plants are known, and thus the transposon may be used to obtain the gene that it inserted itself into. Once isolated from the plant the transposon may be cut out of the gene and the original gene reconstructed for use in further studies on the genetic control of the character it controls.

Transposons occur in many of the plants studied so far, although in some cases they seem to be inactive, their presence being detected not by their biological activity but by their characteristic DNA sequences. They are fascinating genetic phenomena that are of interest to gardeners because of their inherent instability, generating an endless supply of unique patterns in flower and foliage colours.

## Somatic and germ line mutations

So far, it has been assumed that genetic changes are inherited. However, only a few cells in the plant will ever have the opportunity to pass on genetic material, together with any mutations that have occurred during their lifetimes, to the next sexual generation

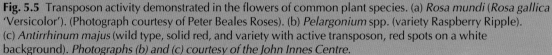

**Fig. 5.5** Transposon activity demonstrated in the flowers of common plant species. (a) *Rosa mundi* (*Rosa gallica* 'Versicolor'). (Photograph courtesy of Peter Beales Roses). (b) *Pelargonium* spp. (variety Raspberry Ripple). (c) *Antirrhinum majus* (wild type, solid red, and variety with active transposon, red spots on a white background). *Photographs (b) and (c) courtesy of the John Innes Centre.*

of plants. These are the germ line cells, whose route through growth and development will eventually result in their producing eggs or pollen. The majority of cells in the plant are somatic cells. They will not produce germ cells, and their genetic material, with any mutations it carries, will not normally be inherited by the next generation. So while mutations occur in both somatic and germ cells, somatic mutations are not transmitted from one sexual generation to the next, except when the plant is propagated vegetatively.

## Chimeras

Somatic mutations are, however, transmitted within a plant to any daughter cells produced by the cell carrying the mutation. Plant *chimeras* provide some good examples of somatic mutations and provide an interesting insight into how the internal anatomy of plants is organised.

Named after the chimera of Greek mythology, which was part lion, part goat and part snake, plant chimeras are made up of genetically distinct groups

or layers of cells. They may be one of two basic types. The first, graft chimeras, are familiar to many gardeners, perhaps the best known being +*Laburnocytisus adamii*, the result of grafting purple broom (*Cytisus purpureus*) and laburnum (*Laburnum* spp.). The first recorded graft chimera, however, was the 'Bizzaria' orange, produced in 1644 when a seedling orange (*Citrus* × *aurantium*) scion was grafted onto a citron (*Citrus medica*) stock. The 'Bizzaria' orange tree produced leaves, flowers and fruits of both citron and orange, as well as some fruits that were a blend of citron and orange. Similarly, +*Laburnocytisus* is distinct from either parent but frequently produces branches that have reverted to one or other parental type.

These chimeras may develop and be maintained because flowering plants have an internal anatomy that is layered (see Chapter 2). The cells in the growing tips, the meristems, of monocotyledons are organised as an inner core and an outer layer. In dicotyledons there is a third layer which is sandwiched between the outer layer and inner core (Fig. 5.6). The outer layer becomes the epidermis of the plant, the internal core forms all the internal

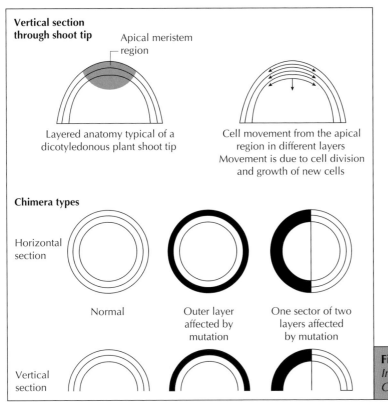

**Fig. 5.6** Chimera structure in plants. *Image courtesy of the John Innes Centre.*

tissues and the intermediate layer, when present, produces a layer of cells immediately below the epidermis. This layering makes it possible for one partner in a graft to 'grow over' the inner layer(s) of the other partner, resulting in a plant made up of cell layers from two different species. For example, +*Laburnocytisus* has outer cell layers of *C. purpureus* and an inner core of cells of *Laburnum* spp.

A second, less extreme type of chimera results when a mutation occurs in a dividing cell in one layer and this mutant cell type is then propagated through that cell layer by cell division. The precise position of the cell in the meristem will determine not only which layer(s), but also how much of the layer(s), is affected. Such chimeras are extremely common, but only apparent when the mutation is in a gene affecting an easily visible trait such as flower or foliage colour.

Studying chimeras has made it possible to understand a great deal about plant growth and development. For example, it has been discovered that the cells produced by the outer layer of the meristem cover the entire outer surface of a leaf or petal and contribute all of the cells of the leaf and petal edges. The cells produced by the meristem core make up most of the rest of the leaf and petal.

Variegation in plants may be caused by chimeras. For example, if the cells in the outer layer of the meristem carry a mutation that prevents them from making chlorophyll, the resulting shoot will have leaves with a pale margin and a green centre. The leaf margin is made up only from the chlorophyll-less cells of the outer layer. The leaf centre is also covered with a colourless 'skin' of cells from the outer layer, but the effect of the mutation is not seen because the green cells underneath show through. Variegated varieties of *Hosta undulata*, *Pelargonium* × *hortorum* and *Ficus benjamina* are caused by this type of chimera (Fig. 5.7b). If the situation is reversed, the cells in the meristem core being unable to make chlorophyll but the cells of the outer layers being normal, the leaves will have a pale centre with a green margin. *Euonymus japonicus* 'Aureus' (Fig. 5.7a) and *Elaeagnus pungens* 'Maculata' are this type of chimera. *E. japonicus* is particularly interesting as it exists in two common forms that are the chimeric reverse of one another: the leaves of one are white with a green margin, the other green with a white margin. Usually the chlorophyll-less regions of chimeras are not white, but appear cream, golden or

yellow, as in *Euonymus* spp. (Fig. 5.7a). This is because leaf and stem cells usually produce coloured pigments which are masked by the green chlorophyll, but visible in tissues that contain no chlorophyll. Although the layered structure of the plant meristem is not so well defined in conifers, they also produce chimeras, as in *Chamaecyparis lawsoniana* 'Albovariegata' and *Chamaecyparis thyoides* 'Variegata' (Fig. 5.7c).

The pinwheel African violet (*Saintpaulia* spp.) is a chimera in which the cells of the outer meristem layers are unable to make petal pigment, resulting in petals with colourless (white) margins and a pigmented centre (Fig. 5.7d).

The pattern and extent of variegation in leaves and flowers of chimeras is determined by how many of the meristem layers, and how much of each layer, is affected by a mutation. Interesting effects arise if a few cells in one sector of a meristem layer are affected by a mutation. In this case a streak or sector may arise that runs down one side of a plant, including all leaves, the stem and side shoots derived from cells in that sector (Fig. 5.6). The demarcation between layers may not be strictly maintained during cell division, so there may be a limited and variable amount of 'migration' of cells from one layer into another, which can add further variation to the patterns produced in chimeras.

The mutations which generate a chimera only affect the cells in one or two layers, so the chimera will not be inherited. This is because the germ cells (egg and pollen) usually originate from only one of the cell layers and, therefore, either carry the mutation or do not. Chimeras demonstrate an important distinction between two types of mutation: first, germ line mutation, in which the change may be inherited because it has occurred in a cell that will eventually give rise to an egg or pollen cell; and second, somatic mutation, in which the change occurs in a somatic cell and therefore is very unlikely to be inherited via a seed generation.

However, it may be possible to maintain and propagate somatic mutations if the mutation arises in a piece of plant tissue that can be propagated vegetatively by taking cuttings or by grafting. The tissues used for such propagation must include all these cell layers from the donor plant, and the new shoots that develop must also include all these cell layers, otherwise the original chimeric effect will not be present in the propagated plants (see also Chapter 10).

**Fig. 5.7** Chimeras. (a) *Euonymus japonicus* 'Aureus' showing a chimeric structure in which the outer cell layers of the plant are green (due to chlorophyll) but the cells in the inner core of the plant are unable to make chlorophyll and are therefore colourless. This results in a colourless (golden) leaf centre bordered by green. (b) A variegated variety of *Ficus benjamina* showing a chimeric structure which is the reverse of that described above. (c) A branch tip from the Gymnosperm *Chamaecyparis thyoides* 'Variegata'. The meristem structure and layering of the plant's anatomy is not so regular as in Angiosperm species. Consequently, the green/white patterning, due to the inability of some cells to make chlorophyll, is random. (d) 'Pinwheel' African violet (*Saintpaulia*) showing a chimeric structure in the flower. The cells in the outer layers of the plant are unable to make petal pigments so that the petals have a pigment-less (white) outer margin and pigmented centre. *Photographs courtesy of the John Innes Centre.*

New mutations and chimeras may be uncovered in vegetative propagation systems that depend on inducing somatic cells, for example in the leaf, to undergo cell division and form new meristems and hence new plants. If a cell carrying a somatic mutation is the founding cell, or one of the founding cells, of a new meristem, then the mutation will be inherited by all, or a proportion of, the cells in the plant(s) produced from that meristem.

## HETERO- AND HOMOZYGOSITY

Plant cells contain two copies of the genome. In most cases, cross-species hybrids being one exception, these genomes will be equivalent in having the same genes, controlling the same characteristics, in the same locations along the DNA molecule. However, the alleles present at each gene location (*genetic locus*) may be different in the two genomes. The more closely related two plants are the more likely it is that the majority of the alleles in their two genomes will be the same; the more distant the relationship, the more likely it is that there will be differences.

Plants in which the two genomes carry identical alleles at all gene loci are described as *homozygous* (homo = the same). Genetically homozygous plants will be *true breeding* as the recombination events that occur between genomes, at gamete formation, will simply shuffle two identical genomes containing identical sets of alleles. All the eggs and pollen will carry identical genomes and self-fertilisation will generate offspring that in every case are genetically identical to one another and their parents, although in practice minor variations may occur. When these plants produce gametes, these too will be identical and the next generation of plants will be identical to their parents and their grandparents, and so on, although again, in practice, minor variations may occur.

Homozygosity may also be achieved with plant populations. By allowing only plants within a particular population to fertilise one another, and then selecting offspring that are of a uniform type at each generation, a plant breeder can establish a population of homozygous plants that are genetically nearly identical to one another. Any hybridisation between plants in this population will generate offspring that are very similar to their parents because no new allele combinations are being produced when gametes are produced or when fertilisation takes place.

By contrast, if a plant is *heterozygous*, having two parental genomes that are very different, then at gamete formation a large number of new allele combinations will be generated by recombination between the genomes. *Self-fertilisation* of such a plant may not result in offspring that are very different from one another or the parents, but in subsequent generations the genetic heterozygosity will become apparent as variation in traits between related plants. Similarly, hybridisation between plants in a heterozygous population will result in many new combinations of alleles, potentially generating new variation for traits, and thus maintaining the heterozygosity of the population in the next generation.

# Dominant and recessive alleles

One of the key observations made by Gregor Mendel in his experiments with peas was that when two different inbred (homozygous) varieties were hybridised with one another, the offspring, the first filial generation (*F1*), were all alike. They had a mixture of traits, some from one parent and some from the other. If he self-pollinated these plants their offspring, the second filial generation (*F2*), did not always breed true to the parent type, but sometimes exhibited characteristics from their grandparents that had not been obvious in their parents.

This observation led Mendel to understand that alleles are not equal. Typically, pairwise comparison of alleles for a particular gene will reveal that some alleles are *dominant* over other, so-called *recessive*, alleles. There are several physiological explanations for dominance and recessivity, but at the genetic level the effect is relatively straightforward. The presence of a dominant allele will mask the presence of a recessive allele. Hence, only in an individual where, in both genomes, the locus for a particular gene has the same recessive allele will the effect of that allele be seen in the plant. Such a plant is referred to as homozygous recessive for that gene. In a situation where the plant is heterozygous at the gene locus, that is, has a dominant allele in the locus on one genome and a recessive allele in the equivalent locus in the other genome, then the dominant allele will mask the effect of the recessive allele. Such heterozygous plants are usually identical, in terms of the character expressed by the gene, to plants that are homozygous for the dominant allele, having the dominant form of the allele at the gene locus in both genomes.

This illustrates an important concept in genetics. The genetic make-up of a plant, the *genotype*, is a major contributor to the appearance, the *phenotype*, of the plant. However, different genotypes can give rise to similar phenotypes. For example, if the allele for red flower colour is dominant over the allele for white flower colour then the homozygous dominant plants, in which both genomes have the allele for red flower colour at the gene locus for flower colour, and heterozygous plants, in which one genome has the allele for red flower colour and the other the allele for white flower colour, will have red flowers. The effect of the white flower colour allele will only be seen in homozygous recessive plants, in which both genomes have the allele for white flower colour at the gene locus for flower colour.

The dominant/recessive relationship is not always this straightforward, however, and allele relationships may be complex. Alleles may show semi-dominance or co-dominance, meaning that the dominant and recessive forms are not completely dominant or recessive with respect to one another. Consequently, the phenotype controlled by alleles with incomplete dominance is not either/or, but a blend of the dominant and recessive character.

So far we have dealt with the relationship between genes and traits as a simple situation where one gene controls one trait. Many characters are of this type: for example, a mutation in a single gene in *Rosa mundi*, caused by transposon inactivation, is sufficient to switch off pigment production in the petals. It is also possible, however, for a plant trait to be controlled by more than one gene. For example, where the colour of a flower's petals results from a mixture of pigments, then a change in the activity of any of the genes involved in producing the pigments could result in a change in flower colour.

Other genes interact in more subtle ways. For example, the protein products from several genes may be required to assemble particular enzymes, enzyme complexes or structural elements of cells. Many different aspects of cell and plant development can affect some traits, such as yield, and consequently a large number of different genes contribute to this characteristic of the plant. Intuitively one would expect that the more genes that are involved in determining a particular trait, the less likely it is that a change in any one gene would have a major effect on that character. However, when there are many genes controlling a character, it is usual to find that a few genes, so-called major genes, have relatively major effects, while the remainder have only minor effects on the trait.

Genes that have major effects are typically little affected by the environment, so the phenotype is closely related to the genotype. However, traits controlled by genes with minor effects, or by several genes, may well be more susceptible to the effects of the environment. Thus, variation in plant phenotype may have both a genetic and an environmental component. For example, the activities of some genes associated with flowering are affected by day length (see Chapter 13).

## BREEDING SYSTEMS

Most plant species are naturally heterozygous. Homozygous species do occur naturally, however, indicating that both of these genetic/reproductive strategies are competitive in the wild. Whether a plant is naturally hetero- or homozygous is determined mainly by the mating system of the species, as it is this that largely controls the movement of new alleles into and out of the breeding population.

Heterozygous species are typically outbreeders and tend to be perennials. They have mating systems that promote outcrossing and inhibit self-fertilisation and inbreeding. Enforced inbreeding in these species often results in a dramatic loss of vigour, size and fertility over one or a few generations. Such *inbreeding depression* results from the accumulation, in inbred plants, of recessive deleterious genes whose presence is masked in heterozygous plants. A plant population that contains several alleles for any particular gene locus, distributed among individual plants in the population, may have an adaptive advantage in a wild environment. For example, different alleles for seed dormancy could ensure that seeds from that plant population germinate over an extended period, thereby increasing the probability of some germination occurring during a period that is followed by favourable growth conditions. Heterozygosity within individual plants ensures that their progeny represent a variety of new allele combinations, resulting in a wide range of genetic and phenotypic variation for plant characters. This gives the species and individuals an opportunity to respond to changing conditions. The downside of this strategy is that the increased frequency of new gene combinations increases the probability of deleterious combinations occurring, and thus the frequency of less-fit individuals in the offspring of parents. To some extent this is offset by the perennial habit whereby the parent plants produce seed over several years, thereby compensating for the higher proportion of unsuccessful progeny produced in any one year.

Homozygous species are inbreeders. They are adapted largely to self-pollinate and usually do not suffer from inbreeding depression even when inbred over many generations. These species contain a pool of alleles that are highly adapted to local conditions, making them very competitive providing that the environment remains stable. However, mechanisms to ensure self-pollination restrict the flow of new alleles into the population from outside and reduce the frequency of new allele combinations. This reduces genetic variation in the offspring making the population more vulnerable to changing conditions.

In the wild, outbreeding is the basic state and inbreeding has evolved from it, probably on several separate occasions. The breeding of crops and ornamentals from wild species has led to a general reduction in heterozygosity and increased homozygosity in individuals and population. This is because breeders, and before them farmers, growers and gardeners, have selected for uniformity in their plants by rejecting off types.

Homozygosity has the advantage that characteristics are fixed. The plants thus breed true, generation after generation. This greatly increases the ease of selecting for improved plant types and of producing seed that will breed true in the next generation. However, this is not always possible (because inbreeding depression sometimes occurs), or indeed desirable or necessary.

There are five basic plant-breeding schemes, and these largely reflect the constraints imposed on the breeder by the mating systems of the plants in question. Thus (1) inbreeding annual plants, which do not suffer from inbreeding depression, are bred as inbred, pure (homozygous) lines; (2) outbreeding annuals, which suffer inbreeding depression if forcibly inbred, are bred as open-pollinated populations; (3) homozygous or heterozygous inbred or outbred lines may be bred as F1 hybrids; (4) perennials, usually outbreeders, are bred as open-pollinated populations or (5) if they can readily be propagated vegetatively, as clones.

## Breeding inbred lines

Hybridisation of two different homozygous lines will result in a uniform F1 population, but self-pollination of these plants will give a very variable population in the F2 generation. Subsequent rounds of self-pollination will reduce the variation seen between the offspring of any one parent, and by the fifth or sixth generation each *selfed* line will effectively be homozygous. During the early stages of the breeding cycle the breeder normally selects only for characters

which are highly heritable and/or easy to assess. Characters with low heritability or that are difficult to assess are generally not selected until the plants are more or less homozygous, and the characters have been fixed.

The pedigree system is the classic method of conducting an inbred-line breeding programme. Suitable parents are hybridised and the resulting seed collected and sown to produce an F1 population of plants. These are allowed to self-pollinate and the resulting seed sown to give the F2 generation. Each of the F2 plants will produce enough seed to sow and produce a family of closely related individuals. Breeders usually begin to select for best types at this stage, and seed from selected plants, in different lines, is used to produce the next generation of families. Growth and selection of families continues for several more generations, until homozygosity is reached in the F5 to F8 generations. At this point traits will be fixed and a few of the best plants will then be selected and retained for larger-scale trials and evaluation.

## Breeding open-pollinated populations

Breeding by this method is typically used with plants that have mating systems that encourage out-crossing and inhibit self-pollination. Such plants usually suffer from inbreeding depression if forced to inbreed, making it impossible to establish homozygosity and so fix the required alleles and the characteristics they determine. In open-pollinated plants, breeders must select at the population level for the alleles that they require and against those they do not, while maintaining a relatively high level of heterozygosity. In this way they can enrich a closed population with desired alleles and traits. Individual plants from this closed population will not produce seeds that breed true to type, but at the population level the variety will have the improved characters selected by the breeder, and these will be maintained indefinitely by random-mating within the population.

This approach establishes a plant variety by shifting allele frequencies in a population through selection. Alternatively, the variety may be produced as a hybrid between two parents, or as a mixture of hybrids between several different parents. By careful selection of the parents it is possible to ensure that the population of hybrid plants contains the mix of alleles and the qualities required by the breeder.

## Clonal propagation

It is not necessary for plants that are propagated vegetatively (see Chapter 10) to be homozygous since there is no gamete formation and hence no re-assortment of alleles that might generate new genetic variation within the variety. Thus plants selected for particular characteristics, and propagated vegetatively, will produce progeny plants that are true to type, even if they are highly heterozygous.

Breeding clonally propagated plants consists largely of making new hybrids, selecting for improved types in subsequent generations and then fixing the best-performing types by propagating them vegetatively.

## F1 hybrid breeding

It is possible to produce a uniform population of plants from seed even though the plants themselves are heterozygous. This is so-called F1 hybrid breeding and is based on Mendel's original observation that hybridisation of two different inbred lines results in a uniform F1 generation. The disadvantage of F1 breeding is that the F1 plants do not breed true, so that each *F1 seed* generation has to be recreated by hybridising the two parent lines. This means that the grower cannot retain seed year to year and that the seed is more expensive to produce than seed from inbred or open-pollinated breeding systems. Despite this double expense to the grower, F1 breeding is employed because, in many species, heterozygous plants are more vigorous and robust than homozygous types. For example, F1 hybrid ornamentals may produce significantly larger flowers than normal, while F1 hybrid vegetables may crop more heavily. This phenomenon is called *hybrid vigour* or heterosis. Its genetic basis is complex and still debated. In practice, heterosis means that if a comparison is made of the performance of the offspring from pairwise hybridisations among a group of parental lines that are typically, but not always, inbred, then certain parent combinations will produce offspring with higher levels of vigour (heterosis) than others. The specific level of heterosis between particular pairs of parents

is called specific combining ability (SCA). By creating hybrids between parents with high SCA, breeders are able to produce F1 seed that results in plants with better performance than either of the parent lines or the best inbred lines of the crop.

A major problem in F1 hybrid breeding is the need to ensure that the seed harvested from the parent plants is indeed hybrid seed and not the product of self-pollinations. Several techniques are therefore employed to prevent self-pollination and force hybrid-seed production. In tomatoes the flowers of the maternal parent are hand-emasculated. In onions a *cytoplasmic male sterility (CMS)* system (discussed below) is used to ensure that the female parent does not produce pollen, while a *self-incompatibility (SI)* system is used in some brassica vegetable breeding. At various times chemical emasculation using selective gametocides has been tested, but the effective use of a chemical spray to disrupt pollen development in maternal parents has proved difficult in practice.

As its name suggests, CMS prevents plants from making viable pollen. Consequently, CMS plants will only set seed if the stigma receives pollen from another, male-fertile, plant. The CMS character is determined by DNA in the mitochondria organelles present only in the cytoplasm. This type of male sterility is relatively common, and has been identified in sweet pepper (*Capsicum* spp.), sunflower (*Helianthus annuus*), carrot (*Daucus carota*), petunia (*Petunia* spp.) and tomato (*Lycopersicon esculentum*). Genetic male sterility is determined by nuclear genes, and is found in fennel (*Foeniculum vulgare*), lettuce (*Lactuca sativa*) and broad beans (*Vicia faba*).

Self-incompatibility (SI) is a genetically determined system to prevent or deter self-fertilisation. There are some subtle differences in how different SI systems work but essential to all of them is the principle that only certain combinations of SI alleles in the male and female gametes, or parent plants, permit successful fertilisation. A crude analogy is of a lock-and-key system, where only some lock-and-key combinations permit fertilisation. Pollen is unable to develop properly on an 'incompatible' stigma and so fails to fertilise an egg cell. The genetics of the system usually means that an individual's own pollen and stigmas are incompatible, precluding self-pollination.

In either case, the cultivation of a mixture of the two parents, usually as separate rows or blocks, will ensure that seed set on a female (CMS) parent or, in the case of an SI system, both parents, will be hybrid and not a result of self pollination.

## Wide hybridisation

Hybridisation is central to designing plants as it is the means both to generate new allele combinations and to introduce new alleles into the existing pool. Among plants there are several natural barriers to hybridisation. Pollen exchange between otherwise compatible plants may be prevented by geographic separation, non-coincidence of flowering times or adaptation to different pollinators, so that the size and shape of the flowers precludes natural cross-pollination. Plants also have specific biological mechanisms for recognising the compatibility/incompatibility of gametes and thus regulating cross-fertilisation. These incompatibility reactions may occur on the stigma at pollination, as described above, in the style immediately prior to fertilisation, or in the developing seed after fertilisation.

In seeking to produce new allele and gene combinations that might lead to improvements in plants, several procedures have been developed to overcome such barriers and so generate novel hybrids. For example, treating the stigma with dilute salt solution and then carrying out a controlled pollination with genetically incompatible pollen can defeat the brassica SI system. This system can also be overcome by treating flowers with elevated levels of carbon dioxide after pollination, or by pollinating the stigmas of unopened buds. In some plants, for example *Alstroemeria* and *Lilium* spp., it is possible to graft a 'compatible' style onto an 'incompatible' ovary and thus overcome incompatibility barriers that operate at the pollen/style interface.

Although pollination and fertilisation of distantly related plants may be successful, in some species the embryo or the endosperm aborts during seed development, resulting in a non-viable seed. Excising the immature hybrid embryos and growing them on a culture medium (see Chapter 10 for techniques) may allow continued embryo development and permit the production of a novel hybrid plant. Numerous hybrids have been produced by such embryo rescue, for potential use in breeding programmes. For example, the lily hybrids *Lilium speciosum* × *Lilium auratum* and *Lilium lankongense* × *Lilium davidii* do

not produce viable seed but can be recovered by culturing immature embryos. Hybridisation of *Brassica campestris* × *Brassica oleracea*, to produce synthetic versions of the natural hybrid *Brassica napus* (oilseed rape and swedes), is only successful if the embryos are rescued in culture. *Lycopersicon peruvianum* can be hybridised with *L. esculentum* (tomato) but again the hybrid embryos must be rescued to obtain plants.

## Somatic variation

The production of plants *in vitro*, using *tissue culture* techniques, is not restricted to embryo rescue, and new plants may be induced to grow from fragments (*explants*) of many different plant tissues (see Chapter 10). In some instances it is possible to stimulate the cells of the explant to produce embryos, rather than shoots and roots, and these too may be grown on to produce whole plants.

Populations of plants produced from tissue culture (with the exception of *meristem culture*; see Chapter 10) often show a range of new genetic variation. This somatic variation, sometimes called somaclonal variation, occurs even when the explants are from genetically homozygous plants. The causes of somatic variation are still debated, but it is generally agreed that mutations in the cells of the explant are important. It has also been suggested that the tissue culture process itself in some way induces mutations (see Chapter 10 for details of the effects of repeated subculture). Some somatic variation is unstable, and with time the variant plants revert to the parental type or to another variant type. Stable somatic variation is, however, being explored as a technique for use in plant breeding.

## Somatic hybridisation

An extreme means of overcoming barriers to sexual hybridisation is to use *protoplast fusion* techniques, also called *somatic hybridisation*. The walls of plant cells can be removed by treating tissue pieces with a mixture of enzymes that specifically break down cellulose and other major cell wall components. The naked protoplasts so released may be induced to re-grow a cell wall if they are cultivated in an appropriate mixture of nutrients and plant hormones.

These cells, again with the appropriate culture technique, may be induced to grow and divide, eventually producing new plantlets.

If protoplasts isolated from two different plant species or genera are mixed, it is possible, by applying a chemical or electrical shock, to cause them first to stick together, and then to fuse. The fusion process is typically applied to millions of protoplasts at one time and is largely random. Consequently, of the fusions that take place, a proportion will be between protoplasts of the same species; others will involve large numbers of protoplasts; and in some instances the fusions will be of single protoplasts from different parents. In the latter case the hybrid protoplasts will contain two copies of the genome from each of the parents. It is, in theory, possible to induce such hybrid protoplasts to regenerate a cell wall, divide and grow to produce plants that are hybrids between the parental species. In practice there are a number of obstacles to be overcome and some hybrid combinations are not viable.

Nevertheless, somatic hybridisation has been used to generate new hybrids of *Dianthus chinensis* × *Dianthus barbatus*, and between various sexually incompatible species of tomato (*Lycopersicon* spp.), potato (*Solanum* spp.) and tobacco (*Nicotiana* spp.), for use as sources of novel genetic variability in breeding programmes.

## RECOMBINANT DNA TECHNOLOGY

Finally, recombinant DNA (rDNA) technology, regarded by some as the ultimate in plant breeding, will be considered. The development of this technology over the past 20 years has provided the tools to identify, isolate and characterise single genes from plants, animals and microbes, and to transfer them from one individual to another. Three new opportunities of potential interest to plant breeders are now available, as detailed below.

## Marker-assisted breeding

It is now increasingly easy to detect the presence or absence of particular DNA sequences in the genomes of plants. This has enabled the development of *marker-assisted breeding* techniques. If a plant breeder can identify a particular DNA sequence that is always present in plants with a particular phenotype, it becomes possible to screen progeny of a cross for the presence/absence of the marker DNA sequence, rather than for the phenotype itself. This may initially seem counter-intuitive, as it should be easier to look for a phenotype than to isolate DNA and test for the presence or absence of a particular sequence. Generally this is true, but the identification of some important phenotypes is difficult to achieve reproducibly. For example, disease-resistance phenotypes may be difficult to identify because the incidence of disease depends on weather conditions and the level of pathogen inoculum carried over from the previous season. Also, some traits may be determined by several genes/alleles that individually have relatively small effects. Recognising which alleles have significant, beneficial effects on a particular trait and then identifying DNA markers for each allele means that a breeder need only select for plants that include all the markers to be able to produce a plant that performs well with regard to that particular characteristic.

The production of DNA markers for particular genes and alleles is an expensive process and the markets occupied by many horticultural plants are of relatively low value. It is therefore unlikely that this technology will receive a great deal of use for garden plants, unless its cost can be reduced dramatically.

## Genome sequencing

In December 2000 an international scientific team reported the completion of the first total *genome sequence* for a plant, thale cress (*Arabidopsis thaliana*). This common, insignificant and ephemeral weed, found around the world in many different habitats, often goes unnoticed in the garden. *A. thaliana* is a wonderful research tool, however, since it combines three important characteristics: it is small, so very large numbers of individuals may be grown in a relatively small space; it has a rapid life cycle and prolific seed set, so several generations may be produced in a year; and most importantly, it has one of the smallest plant genomes known. Nonetheless, *A. thaliana* proved to contain 114 million letters of DNA code and 26 000 genes.

*A. thaliana* is a member of the Brassicaceae and so a distant relative of the brassica crops, but its potential usefulness extends far beyond this single group of plants. Even before its genome sequence was complete, scientists were beginning to extract information from it that shed new light on plant genetics and biology in general. For example, about 50 years ago a genetic mutation was discovered in wheat that dramatically affected plant height by preventing the normal elongation of the cells in the stem, with the result that plants with the mutation were dwarf. The mutation causes insensitivity to gibberellic acid, a plant hormone that promotes cell elongation (see Chapters 2 and 11). The genetic complexity of wheat prevented isolation of the gene so that its action might be studied more closely. However, a mutation with the same phenotype and biological activity was found in *A. thaliana*, where the relative simplicity of the genome made it easier to isolate the gene. This dwarfing gene was then used to find the equivalent gene in wheat. When the mutant gene was introduced, by genetic modification (see below), into tall varieties of rice (*Oryza sativa*) and chrysanthemum it caused dwarfing, demonstrating that it controls a process fundamental to plant development. Similarly, genes that control the time taken for plants to flower, or a response to cold treatments (Chapter 13), have been identified in, and isolated from, *A. thaliana* and then used to explore both the basic biology of the genes and their role in important crop species.

## Genetic modification

*Genetic modification (GM)*, sometimes called *genetic engineering*, is an important tool that makes possible the isolation and modification of plant genes and their reintroduction into plants of the same or, in some cases, a different species. GM may be used to study gene function and is also a route to the production of new genetic variation in cultivated plants.

There are several methods for introducing genes into plants. Protoplasts may be mixed with DNA and then subjected to an electrical or chemical shock that causes them to absorb DNA fragments. Some of these become integrated into the DNA of the protoplast, and subsequent regeneration of a plant

(see above) will result in a genetically modified individual.

Alternatively, tiny beads of gold or tungsten may be coated with the gene of interest and then shot into pieces of plant tissue. Some of the cells in the tissue will be hit and survive, and in a few of these some of the DNA from the beads will be integrated into the genome. Unfortunately, both of these methods are imprecise and inefficient.

A more efficient, and therefore more popular, method for introducing DNA into plants is to use a natural soil-borne plant pathogen, *Agrobacterium tumefaciens*, the causal agent of crown gall disease on fruit and other trees. These bacteria enter the host plant through wounds, and are then stimulated to transfer into the cells at the wound surface a specific piece of their DNA, which is then integrated into the DNA of the plant.

The genetic material of bacterial cells usually consists of a single circular chromosome and multiple copies of plasmids, which are small circles of DNA carrying a few genes. *A. tumefaciens* contains a particular type of plasmid, the Ti (tumour-inducing) plasmid. It is a small piece of DNA from the Ti plasmid, called the transfer DNA (T-DNA), that is transferred into the plant cells exposed in the wound. This T-DNA carries a few genes that code for the production of plant hormones, which induce the cells in the wound to divide and grow to produce a mass of callus, the crown gall. The T-DNA also includes genes that code for the production, by the plant host, of specific molecules, called opines, which the plant itself cannot use, but which the bacteria use as sources of carbohydrates and nitrogen as they live and grow in the developing gall.

Using GM techniques, the bacterial genes in the T-DNA may be replaced with genes that a breeder wishes to transfer into plants. The Ti-plasmid, with its modified T-DNA, is then reintroduced into the *A. tumefaciens*, where it replicates, resulting in multiple copies in each bacterial cell. The modified *A. tumefaciens* is then grown in large numbers and used to infect plant cells by dipping pieces of excised tissue (*explants*) into the bacterial culture. The *A. tumefaciens* responds to the cut surfaces of the explant as if they were natural wounds and transfers T-DNA into the plant cells, where it becomes integrated into the plant DNA.

The cells in the explants treated with *A. tumefaciens*, or in tissues that have had genes 'shocked' or 'shot' into them, may be cultured on nutrient media, where they divide and eventually produce new shoots (see Chapter 10). Gene transfer is a random process so, whichever method is used, only a few of the plants regenerated from treated tissues contain the gene(s) introduced by the breeder. Selectable marker genes, attached to the same small piece of DNA as the gene(s) of interest, are therefore used to ensure that only cells that contain the introduced gene(s) will give rise to new plants. The most common kinds of markers are genes coding for resistance to an antibiotic or tolerance to a herbicide that would normally kill plant cells. If the induction of shoots from treated explants is carried out in the presence of the antibiotic or herbicide, the growth of cells not containing the marker gene for resistance is prevented, and only those containing the marker gene linked to the gene(s) of interest divide and eventually form plants.

GM is being explored as a means of changing specific characteristics in a range of ornamental and vegetable species. For example, studies are in progress on improving the storage characteristics of pea (*P. sativum*), insect resistance in *Chrysanthemum* hybrids, virus resistance in *Osteospermum* spp., fungal disease resistance in *Petunia* spp. and increased flower production and axillary-bud breakage in *Chrysanthemum* hybrids.

## CONCLUSION

As understanding of plant genetics and biology has increased, so has knowledge of how, through human intervention, natural phenomena may be adapted and exploited to select and breed plants for ornament, for the production of food and medicines, and for other uses. Over the years, with increased understanding of genetics and biology, such interventions have become increasingly focused. Yet however sophisticated the technology, the challenge for modern breeders and those who grow plants remain the same: to generate the allelic combinations that will result in plants which are better adapted to their intended end use in the managed environments of fields, greenhouses and gardens.

## FURTHER READING

Coen, E. (1999) *The Art of Genes – How Organisms Make Themselves*. Oxford University Press, New York.

Gonick, L. & Wheelis, M. (1991) *The Cartoon Guide to Genetics*. HarperCollins, New York.

Griffiths, A.J.F., Miller, J.H., Suzuki, D.T., Lewontin, R.C. & Gelbart, W.M. (1976) *An Introduction to Genetic Analysis*. W.H. Freeman and Company, New York.

Henig, R.M. (2000) *A Monk and Two Peas*. Weidenfeld & Nicholson, London.

Simmonds, N.W. (2000) *Principles of Crop Improvement*, 2nd edn. Blackwell Science, Oxford.

Tilney-Bassett, R.A.E. (1986) *Plant Chimeras*. Edward Arnold, London.

# 6

# Soils and Roots

## SUMMARY

This chapter considers the requirements of plants from soils. It goes on to discuss how soils are formed, how to recognise their key features, and the physical, chemical and biological properties of different soil types. Finally, the structure and growth of roots is described, with special emphasis on their role as absorbing organs.

## INTRODUCTION

### Why do plants need soil?

Most plants of interest to gardeners are grown in soil of some kind. However, plants can and do grow perfectly well in the absence of soil. Examples include floating aquatics and plants raised in nutrient solutions in commercial hydroponic systems. These examples are, however, quite specialised. Submerged aquatics rely on an internal anatomy that allows the rapid movement of oxygen through the plant (see Chapter 8), whereas commercial hydroponic systems rely on sophisticated techniques for the analysis and maintenance of the growing solutions. Soils, in contrast, provide the essential requirements of plants with relatively little attention, provided that they are properly managed. Because the responses of shoots and roots are closely linked, gardeners should know something of the root environment so that they can manage it effectively to produce the plants they want.

### What do plants want from the soil?

Soils supply several of the requirements needed for healthy plant growth, including anchorage for roots to stabilise the plant, water, oxygen to allow roots to respire, nutrients, and a buffer against changes of temperature and pH. The importance of soils for anchorage is readily appreciated after a storm, when trees growing on shallow chalk soils may be blown over.

As a general rule, it is rare for roots to occupy more than 5% of the total soil volume, even in the upper 10–15 cm where they are generally most abundant. The volume occupied decreases very rapidly with depth and is often no more than 0.01% at 50 cm. This means that only a small fraction of the soil within the rooting zone is in direct contact with roots. The implication of this for gardeners is that managing soil fertility is essentially about providing a medium in which roots can grow and proliferate to form a network capable of harvesting water and nutrients.

## HOW SOILS ARE FORMED

Soil is a dynamic medium and changes with time. Some changes may occur in a matter of minutes, such as the production of a hole by an earthworm; others, such as the formation of humus from plant remains, take a few years; while yet others, such as the weathering of rocks and the formation of clay minerals, may take several centuries. Temperature and rainfall affect the rates at which rocks and minerals weather to form the basic solid materials that constitute a soil. These are essentially long-term processes, but temperature and rainfall also have short-term effects, such as on the leaching of nutrients and the rate of biological processes, particularly the decomposition of organic materials.

Differences in the types of rocks and sediments which constitute the parent materials of a soil affect the end products, especially the types and sizes of mineral particles. In the UK, the variety of rocks exposed to weathering covers almost the whole geological time course, but the soils at a particular place are rarely dominated by the underlying geology. This is because glaciers in past ice ages have moved materials around, and wind erosion from continental Europe and North Africa has deposited materials over the original rock deposits. Thus in many places drift and soil materials eroded from elsewhere are the dominant precursors of the soil mineral matter. The topographic location also influences the type of soil that forms, mainly because of the drainage conditions. Generally, flat areas allow more rain to permeate the soil than sloping areas, where lateral movement is possible. At the base of slopes water may accumulate, either at depth or at the surface, to form waterlogged soils, known as gleys. If surface water persists for prolonged periods, peat bogs are formed.

Soils also contain a vast number of living organisms, most of which are not visible to the naked eye (see below). The interaction of these living organisms with the dead *organic* materials produces a vast array of chemicals, some of which promote weathering of soil mineral particles, while others join particles together, or form a reserve of nutrients which biological activity can make available to plants.

A feature of soils is that they usually show some horizontal banding, most obviously of colour, but also of other properties. Usually the upper 20 cm or so has the darkest colour because of the presence of organic matter. The lower parts of the soil frequently accumulate smaller particles and other materials washed downwards. These horizons of the soil indicate the dominant soil-forming processes. Most gardeners only look at the top horizon, the topsoil, but it is the second and subsequent horizons, the subsoils, that often make a difference to the growth of long-lived plants because of their contribution to root growth and the supply of nutrients and water.

The end result of all these interacting factors is to produce soils of almost infinite variety. Even in gardens of fairly modest size there are likely to be areas noticeably different from one another that need to be managed individually if the best results are to be obtained. In addition to the soil-forming factors must be added the special influence of one particular organism, humankind. Gardeners contribute in many ways to the formation of new soils by altering the drainage, moving soil materials to create deeper soils, adding waste products such as manure to promote biological activity, and adding fertilisers and manures to improve the nutrient status. Less well appreciated are the off-site effects of human activities that also have an effect on the development of soils. For example, industry and the burning of fossil fuels result in the soils of most of Europe and North America receiving inputs of nitrogen (N) and heavy metals from the air. In southern England, the average input of N is about 25 kg per hectare per year; this is equivalent to applying 36 g (a large spoonful) of Growmore fertiliser to a square metre. Similarly, the cadmium content of the topsoil at the research institute at Rothamsted, Harpenden, Hertfordshire, has increased by 40% since 1850. These and other industrial inputs to soils will have significant consequences if they continue unabated.

In simple terms, soils are made up of mineral matter and organic matter, with the spaces between these solid materials filled with either water or air. The organisms live on the surfaces of the solid material and in the spaces. Soils, then, are complex media that have physical, chemical and biological properties. The aim of gardeners must be to manage these to the best advantage of plants. To achieve this effectively demands some understanding of the properties of the individual soil components and how they interact with each other.

## PHYSICAL PROPERTIES OF SOILS

Soils are frequently defined in terms of their texture. This is a summary term describing the relative quantities of mineral particles of less than 2 mm diameter. Particles larger than 2 mm in diameter are usually referred to as stones, gravel or pebbles, and have no part in the description. The distribution of particles less than 2 mm in diameter cannot easily be altered, and influences many other important soil properties such as aeration, water-holding characteristics and ease of cultivation. This explains why this method of definition is so widely used; it is a stable, shorthand description of a range of properties.

For convenience, soils are allocated to a textural class depending on their content of sand-, silt- and clay-sized particles. Sand-sized particles are typically 0.05–2 mm diameter, silt 0.002–0.05 mm and clay less than 0.002 mm diameter. With a little practice it is possible to determine the textural class by rubbing a small ball of moist soil between the fingers and thumb (see Fig. 6.1). It is important to have removed roots, stones, insects, earthworms and slugs; substantial amounts of organic matter are especially misleading, since silt particles and finely divided organic matter both feel silky to the touch.

Sand particles usually consist of quartz, are rounded or irregular in shape and are not sticky when wet. Sands possess good drainage and aeration, but may be prone to drought. Silt particles are intermediate in size between those of sand and clay and they too are dominated by quartz. They give the soil some cohesion and plasticity but may cause the soil surface to become compacted and form a crust when dry. The clay-sized particles are dominated by clay minerals rather than quartz. They have a very large surface area in relation to their unit mass, about 10 000 times more than that of sand. This allows the adsorption (binding) of water and nutrients, so wet clays are sticky and cohesive.

Most soils are a kind of loam, with an ideal composition of sand, silt and clay particles in about equal proportions. The term loam implies nothing about the quantity of organic matter present, although in everyday language gardeners often erroneously refer to soils with moderate amounts of organic matter as loams. Another factor that can lead to misunderstanding is that the word clay is used here to describe the size of particles; however, it is also used (see below) to describe a wide range of different minerals, with different chemical properties. The terms light, medium and heavy are also commonly used to describe texture, and relate to sandy, loamy and clayey soils respectively. These descriptions reflect the water content of the soils when drainage has ceased.

The texture of a soil is not easily modified and, apart from the incorporation of large amounts of sand

to improve the physical properties for horticultural purposes, is not changed by cultural management. However, other physical properties of a soil can be substantially altered by management practices such as cultivation, drainage and manuring, through the effects on soil structure, especially in the topsoil. The term structure describes the way in which the sand, silt and clay particles are grouped or arranged together into units known as aggregates. Essentially, it is only these aggregates that are altered by most gardeners.

Within a soil, different structures are often present in different horizons. Spheroidal structures (granules and crumbs) are typical of many surface soils, particularly those under grass, and soils high in organic matter. In the subsoil, plate-like, columnar, prismatic and blocky structures usually dominate. Between the structures are spaces called pores, which are normally filled with air and water. The exact process of how the structures form is unknown, but living organisms play a key role. Plant roots tend to compress soil particles into small aggregates as they elongate through soils, and may even compress larger aggregates. Similar compression and contraction occurs as plants take up water, and soils are wetted and dried.

Plant roots and microorganisms also exude sticky organic compounds, mainly long-chain sugars known as polysaccharides, that can bind particles together. The decomposition of organic materials by microbes also produces other organic compounds that interact with clay-sized particles to cement them together. Thus, clay particles and fine organic materials are important components of aggregates (see Fig. 6.2), especially in topsoil, where living organisms and organic matter are most frequently found. In the subsoil, downwardly moving organic materials, clay minerals, oxides of iron and aluminium, and salts such as calcium carbonate can all act as cements.

The pores are an integral part of a soil and are filled with water and air. The proportions of water and air change as rain falls, as plants take up water, and as evaporation occurs from the soil surface. The size, shape and continuity of the pores determine how much of the rain falling on a soil drains away, is held for plant use, or is held so firmly that plants cannot use it. Table 6.1 summarises a useful classification of pores in terms of size and function. In crude terms, if a pore can be seen with the naked eye it will not hold water in a freely draining soil.

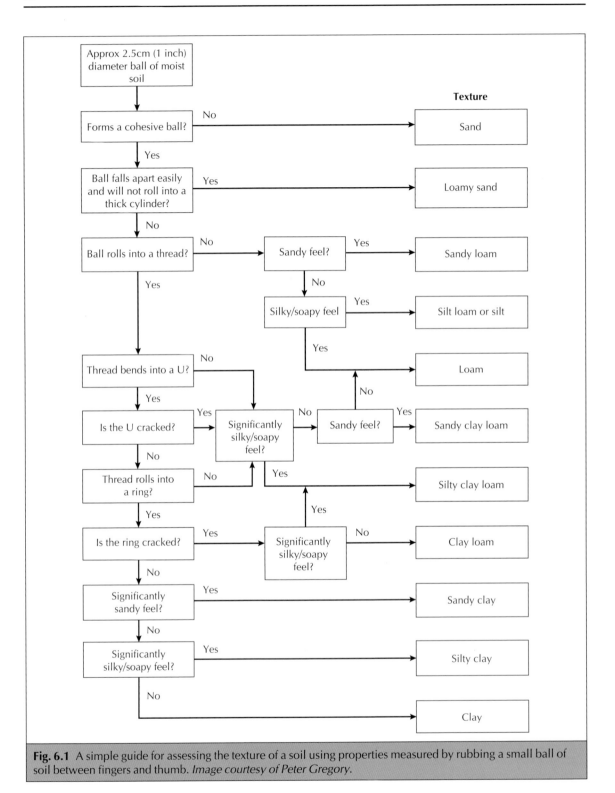

**Fig. 6.1** A simple guide for assessing the texture of a soil using properties measured by rubbing a small ball of soil between fingers and thumb. *Image courtesy of Peter Gregory.*

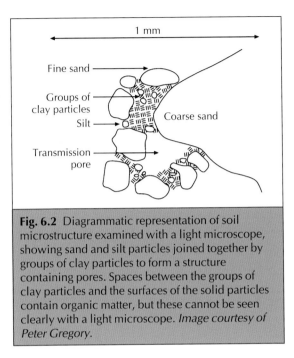

**Fig. 6.2** Diagrammatic representation of soil microstructure examined with a light microscope, showing sand and silt particles joined together by groups of clay particles to form a structure containing pores. Spaces between the groups of clay particles and the surfaces of the solid particles contain organic matter, but these cannot be seen clearly with a light microscope. *Image courtesy of Peter Gregory.*

plants. In contrast, clays do not drain readily, hold more water available for use by plants, and contain water even after prolonged dry periods. The critical limits for the three categories of pores are as follows. If transmission pores form less than 10% of the soil volume, drainage may be a problem; if storage pores form less than 15% of the volume, water availability is likely to be restricted; and if residual pores form more than 20% of the volume mechanical difficulties will be experienced, because the soil will be sticky when wet.

Water is held in soils against the force of gravity because of forces acting between water and the surfaces of soil particles. These forces result in water being held in small pores, at the necks of larger air-filled pores, and as thin films on particles enclosing air-filled pores (see Fig. 6.3).

## CHEMICAL PROPERTIES OF SOILS

In general, sandy soils have about 35–40% pore space, loams 45–55%, clays 50–70% and peats 80%. However, it is the sizes of the pores making up the pore space that differ significantly among soils of different textural classes. In sandy soils, most of the pores are typically the larger transmission pores, with only a few of the very small residual pores and some storage pores. This contrasts with clayey soils, which have fewer transmission pores, but three to four times the volume of residual pores. Sandy soils drain easily, hold only modest amounts of water available to plants, and retain little water after drying out by

The sand- and silt-sized components of the soil consist mainly of quartz and are therefore chemically inert, apart from the finer materials that are smeared across their surfaces. It is the clay-sized fraction, less than 2 $\mu$m ($\mu$m means micrometre; 1000 $\mu$m = 1 mm) in diameter, together with some components of the organic matter, that is chemically active. The clay-sized particles are electrically charged, enabling them to adsorb ions (charged atoms and molecules), including some that are plant nutrients; to adsorb water, with consequent swelling and plasticity; and to disperse, as a consequence of repulsion between

**Table 6.1** Classification of pores by size and function.

| Pore class | Pore diameter ($\mu$m) | Pore function |
|---|---|---|
| Transmission (macropore) | >50 | Drainage after saturation; aeration when soil is not saturated; both aeration and drainage require interconnection of pores; root penetration; root diameter is typically <200 $\mu$m |
| Storage (micropore) | 50–0.2 | Store water available for plant use |
| Residual (micropore) | <0.2 | Hold water so strongly that it is not available to plants; to a large extent this water controls the mechanical strength of the soil |

From Rowell (1994).

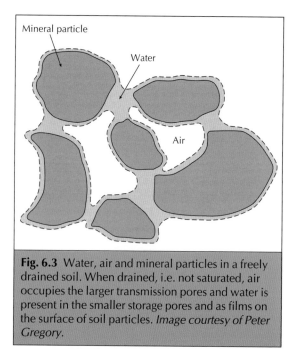

**Fig. 6.3** Water, air and mineral particles in a freely drained soil. When drained, i.e. not saturated, air occupies the larger transmission pores and water is present in the smaller storage pores and as films on the surface of soil particles. *Image courtesy of Peter Gregory.*

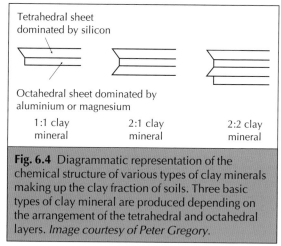

**Fig. 6.4** Diagrammatic representation of the chemical structure of various types of clay minerals making up the clay fraction of soils. Three basic types of clay mineral are produced depending on the arrangement of the tetrahedral and octahedral layers. *Image courtesy of Peter Gregory.*

charged surfaces. Together, the various processes producing clay minerals result in a suite of materials with different properties, and for this reason not all clayey soils behave in the same way.

The clay-sized fraction consists of colloidal material. When mixed with water, the fine particles of colloidal materials may form a so-called dispersed solution, in which the particles remain in suspension. This is intermediate between a true solution, in which the particles are completely dissolved, and a suspension, in which coarser particles eventually form a sediment under the influence of gravity. There are three types of colloidal material in the clay-sized fraction: namely clay minerals, other alumino-silicates and sesquioxides of iron and aluminium. Each clay particle is made up of a series of plate-like layers stacked one on top of the other, like the pages in a book. Every layer comprises horizontal sheets of silicon, aluminium, magnesium and other positively charged ions, surrounded and held together by oxygen ($O_2$) and hydroxyl (OH) groups. The sheets are of two types: tetrahedral sheets composed of silicon bound to oxygen atoms; and octahedral sheets composed of either aluminium or magnesium, each bound to six oxygen atoms or hydroxyl ions. These sheets are combined together into layers of clay minerals of three main types, as illustrated in Fig 6.4.

All clay minerals are small in size and this results in a large surface area that can adsorb ions, water and gases. However, there is another important property that distinguishes clay minerals, and which is of vital importance to plants. Due to the weathering processes that led to their formation, impurities are often found in both the tetrahedral and octahedral layers and this produces clay minerals with a permanent negative charge. These charges are always balanced, however, by positively charged ions, called cations, that are adsorbed onto the surfaces of the minerals. In general, calcium, magnesium, potassium and hydrogen balance the negative charge, but they are interchangeable, depending on their availability in particular soils. Such ions are known as exchangeable cations, the total charge as the *cation-exchange capacity* (*CEC*), and the process as cation exchange. The CEC and the balance of ions are important to gardeners because they determine such important features as the ability of a soil to replenish some nutrients to the soil solution and the *pH* (acidity or alkalinity; see Chapter 7) of the soil.

Unlike the clay minerals, the other alumino-silicates and sesquioxides of iron and aluminium do not have a permanent negative charge, although it is possible for a charge to develop associated with the hydroxyl groups at the edge of the particles. This charge is variable, depending on the pH of the surrounding soil solution. In alkaline soils the charge is negative, but in acid soils the charge is positive and the soil has an *anion-exchange capacity* (*AEC*), the opposite of a CEC, satisfied by negatively charged ions such as chloride, sulphate and nitrate. In temperate soils the AEC is rarely evident, but soils in tropical gardens may have this property.

Humus is a mixture of complex compounds that have been synthesised by microorganisms in the soil from the breakdown products of plant and animal remains or by alteration of such materials. It is typically brown or dark brown in colour, and gives many temperate soils their characteristic brown to olive appearance. Humus colloids are not crystalline and are composed essentially of carbon, hydrogen and oxygen. They have a pH-dependent negative charge, never a positive charge, so that humus, like clay minerals, adsorbs cations.

## BIOLOGICAL PROPERTIES OF SOILS

A teaspoon of moist garden topsoil will contain several million bacteria, a million or so fungi and ten thousand or so protozoa. These microorganisms live mainly on each other and on plant and animal residues which they convert into many chemical forms. The decayed and decaying organic matter in soils results from the combined activity of the soil fauna and flora, and this, together with the burrowing and mixing activities of live animals, means that the soil is a continually changing, living medium.

Living organisms are an essential part of the soil-forming process, and the carbon compounds produced by the photosynthesis of plants are broken down by them. In soils that have been formed for some while, a vast array of organisms is present, living on each other and on dead organic matter returned to the soil from plants. Microorganisms convert organic residues into a resistant residue, humus. This combines with living organisms, dead cells and mineral particles to form the solid component of a soil. The term *soil organic matter* is used to refer to all the organic material in soil, including humus. Soil organic matter comprises living material (soil animals, plant roots and microorganisms) and dead material, including shoot and root residues and humus.

The amount of organic matter present in a soil depends on its physical properties, the climate and the type of vegetation. Thus clayey soils generally contain greater amounts than sandy soils; temperate regions generally have greater amounts than Mediterranean regions; and woodlands and grasslands generally have more than cultivated soils.

In the UK and much of northern Europe only about 2–5% of a cultivated topsoil is organic matter, and of this only a small proportion is living. For example, a cultivated topsoil may contain 4.5 kg of organic matter per cubic metre, of which only about 2.5% (112 g) will be living microorganisms, called the microbial biomass. Although small, this component is vital, because almost all the plant and animal residues must pass through it if nutrients are to be released for plant use.

Some of the more important groups of organisms present in soils are shown in Fig. 6.5, and an indication of numbers and biomass is given in Table 6.2. The microorganisms are the most numerous and have the highest biomass. Since activity is generally related to biomass, the microorganisms, together with earthworms, dominate biological activity in the soil, with 60–80% of soil metabolism attributable to the microflora. In some areas ants and termites are important, and in some tropical soils the activities of termites and their associated fungi play a major part in soil regeneration.

Bacteria are single-celled organisms. Their length rarely exceeds 4–5 μm and the smaller ones are about the same size as a clay particle (2 μm). They exist in clumps or colonies, and different types use different food sources. Most soil bacteria are heterotrophic, meaning that their energy and carbon sources come directly from organic matter; together with fungi and actinomycetes (filamentous relatives of bacteria), they undertake the general breakdown of

**Table 6.2** Numbers and live-weight biomass of the organisms commonly found in surface soils.

| Organism | Number per $m^2$ | Biomass (g $m^{-2}$) |
|---|---|---|
| Bacteria | $10^{13}$–$10^{14}$ | 45–450 |
| Actinomycetes | $10^{12}$–$10^{13}$ | 45–450 |
| Fungi | $10^{10}$–$10^{11}$ | 110–1100 |
| Algae | $10^9$–$10^{10}$ | 5.6–56 |
| Protozoa | $10^9$–$10^{10}$ | 1.7–17 |
| Nematoda | $10^6$–$10^7$ | 1.1–11 |
| Other fauna | $10^3$–$10^5$ | 1.7–17 |
| Earthworms | 30–300 | 11–110 |

After Brady (1990).

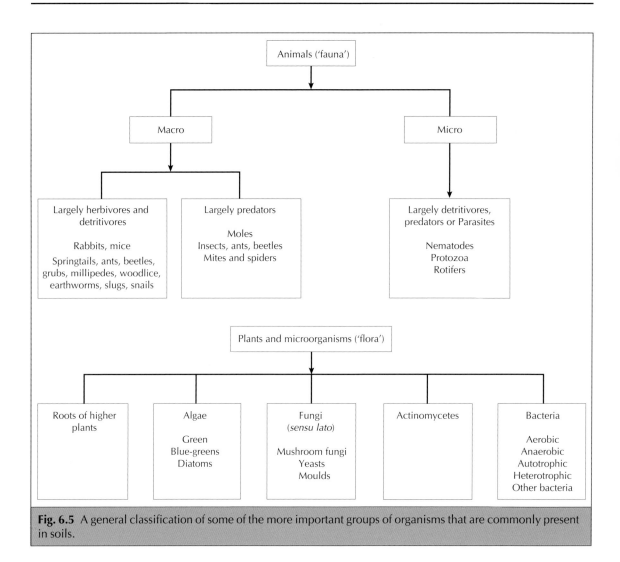

**Fig. 6.5** A general classification of some of the more important groups of organisms that are commonly present in soils.

organic matter in soils. A smaller group of bacteria obtain their energy from inorganic substances such as ammonium, sulphur and iron, and most of their carbon from carbon dioxide. This group plays a vital role in controlling the availability of nutrients, especially nitrogen, to garden plants (see below).

Actinomycetes, like bacteria, are unicellular, but are filamentous and frequently branched. They are similar to bacteria in size and, like bacteria, their main food source is soil organic matter which they break down. They appear to be more capable than many bacteria of breaking down some of the more resistant compounds produced by plants. However, whereas bacteria can survive across a wide range of

pH values, the optimum for actinomycetes is between 6.0 and 7.5.

Most soil fungi are multicellular organisms forming long hyphae normally 5–20 μm in diameter, although a few species of non-cellular slime moulds and unicellular yeasts also occur. The soil fungi can use a wide variety of carbon substrates as a source of energy. Some can utilise only simple carbohydrates, alcohols and organic acids, while others can decompose polymers (see Chapter 2) such as cellulose and lignin. Some are parasites of plants (see Chapter 15) while others, such as mycorrhizas, have a symbiotic relationship with their host (see Chapter 2). Overall, fungi are the most versatile group of microorganisms and, because of their role in humus formation and aggregate stabilisation, play a significant role in all soils.

Much has been written on the importance of earthworms. They ingest leaf and root material along with mineral particles, which are subjected to digestive enzymes and grinding processes within the animal. The mass of material passing through the body of a worm daily may equal its own body mass, and the casts produced have significantly greater nutrient content and structural stability than the bulk soil. Earthworms must have organic matter as a food source, and prefer a well-aerated but moist habitat. Most thrive when the pH of the soil is between 6 and 7, although a few species tolerate lower pH values. Many native species of earthworm are killed by the disruption of their food sources and burrows that results from cultivation. Where earthworms are not present, for example in acidic, dry or very wet conditions, the accumulated litter is broken down only very slowly.

From the description above it is clear that inputs of new organic materials are essential for maintaining an active population of living soil organisms. It should also be clear that adding organic materials will not necessarily increase the organic matter content of a soil. Much may simply decompose to carbon dioxide, which is lost to the atmosphere, and inorganic ions, which may be lost by drainage.

## ROOTS AND SOILS

Roots, which were introduced in Chapter 2, have several functions. They absorb water and nutrients from the soil and transport these from the site of absorption to the shoot; they form the site of synthesis for several plant hormones which may act in either the root or the shoot or both; they may act as storage organs as, for example, in root vegetables such as carrot; and they serve to anchor the plant.

A typical young root is shown in Fig. 6.6. It comprises a *root cap*, behind which are the dividing cells of the *meristem*, which later differentiate into cells forming the *epidermis*, the *cortex* and the *stele*. During this differentiation the cells increase in volume and elongate to push the root through the soil. Behind this zone of elongation, the cells of the stele themselves differentiate to form the *endodermis*, *phloem* and *xylem*. The xylem is essentially a mass of non-living pipes that transports water and nutrients upwards to the shoot, while the phloem consists of living cells that allow bi-directional movement of assimilated materials. The endodermis gradually becomes thickened with deposited corky material, *suberin*, as the plant ages. The suberisation of the roots strengthens them, but also prevents the movement of some nutrients to the xylem and thence to the shoot.

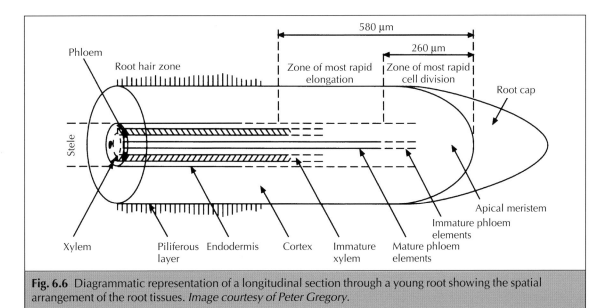

**Fig. 6.6** Diagrammatic representation of a longitudinal section through a young root showing the spatial arrangement of the root tissues. *Image courtesy of Peter Gregory.*

**Fig. 6.7** (*left*) Mucilage production at the root tip of a maize (*Zea mays*) root; and (*right*) the binding of soil particles to a root by mucilage and root hairs to produce a *rhizosheath* that is very difficult to remove. The production of rhizosheaths is common in many species, especially as the soil dries. *Photographs courtesy of Peter Gregory.*

The outer cells of the root cap may exude materials which coat the root tip with mucilage (see Fig. 6.7, left). This mucilage is very noticeable in some plant species, for example grasses; it ensures good contact between the root and the soil, and aids the transfer of water and nutrients to the plant. So effective is this material that it is impossible to detach soil particles from some roots, and they exist within a *rhizosheath* (Fig. 6.7, right). The root hairs also act to increase contact between the root and the soil particles. In addition, because they effectively extend the surface of the root outwards, they are able quickly to mine a zone of soil that would otherwise take some time to deplete. This process is particularly important in the acquisition of nutrients, such as phosphate, that otherwise move slowly through the soil. Mycorrhizal fungi may serve a similar purpose (see Chapter 2).

However, most roots do not look like or function like young roots. Root hairs are ephemeral and generally last for only a few days. As roots age, the epidermis and cortex are attacked by other organisms, become desiccated during dry periods and eventually disappear. Older roots of annual species may have the endodermis as the outer layer for a significant part of the growing period. Roots of most dicotyledonous plants undergo secondary thickening in which a corky *periderm* is formed and the endodermis, cortex and epidermis are sloughed off to leave a woody root; this is most noticeable in trees and shrubs.

Water can be taken up along the whole length of the root system if it is available in the soil and if contact between the root and soil is good. The water directly surrounding a root system is only a small fraction of the total needed to supply the amount which transpires during the course of a sunny day. Consequently, the liquid water must flow from some distance away to reach the root surface, often as much as 10 mm in a day. This cannot happen if there is an air gap between the soil and the root.

Two different situations have to be managed by gardeners. The first is on sandy/silty soils with high organic matter content where puffiness, the presence of large amounts of air, may lead to poor soil–root contact. This can be overcome by firming the soil, manually or with a roller. The second situation is more difficult to manage because it occurs where there are only a few cracks in the subsoil and the roots are limited to following them. As the soil dries, the roots may lose contact with the soil, and hang in the air between the cracks.

The activity of roots in taking up nutrients is rather more complex than that of taking up water, because of the role played by the endodermis. Older books on this subject often suggest that nutrient uptake is restricted to an absorbing zone close to the root tip. This is now known to be incorrect. Ions such as nitrate, ammonium, phosphate and potassium can be taken up along the whole length of a root and transported to the xylem for onward transportation to the shoot. For calcium and iron, however, while limited uptake can occur along the length of the root, uptake is greatest nearest the root tip and translocation to

**Table 6.3** The maximum depth of rooting and length of root systems of some vegetable crops grown on a silt loom.

| Crop | Maximum rooting depth (m) | Total length (km m$^{-2}$) |
|------|---------------------------|----------------------------|
| Broad bean | 0.8 | 1.76 |
| Cauliflower | 0.8 | 11.9 |
| Lettuce | 0.6 | 2.37 |
| Onion | 0.6 | 1.83 |
| Parsnip | 0.8 | 5.65 |
| Pea | 0.7 | 5.46 |
| Turnip | >0.8 | 15.42 |
| Grass | 1.5 | >40 |

From publications at Horticulture Research International, Wellesbourne, UK.

shoots is confined to this zone. The reason for this differing behaviour by different nutrients lies in the pathways they use to move from cell to cell within the root. Nitrate, ammonium, phosphate and potassium move mainly inside the living part of the cell, the protoplast, while calcium and iron move initially in the non-living cell walls, a route which may be blocked by the suberised endodermis.

Little research has been done on the roots of garden plants and there are only few data available on the roots of vegetables, mostly in soils with a sandy texture. Pulling up a plant to examine the roots often gives a false impression of the depth and size of the root system. The roots of most garden vegetables normally grow to at least 50 cm (see Table 6.3), and the roots beneath a lawn often penetrate to 1.5 metres if the soil is deep enough. Beneath each square metre of ground there will be several kilometres of roots, with about 40 km m$^{-2}$ in the case of grass. In fact, the values shown in Table 6.3 are something of an underestimate, because the methods used to obtain them ignore the very fine roots and root hairs.

The weight of a root system is also a substantial component of the total plant weight, although its relative size generally decreases with time. For example, in many crops of peas (*Pisum sativum*) and beans (*Vicia faba*) up to 40% of the total plant weight may consist of roots, prior to flowering. After flowering, as the pods start to grow, root growth occurs at a much slower rate, and the roots typically form only 15–20% of the total plant weight. In the final stages

of pod filling, the plant allocates most of its resources to the pod and only a small amount to the roots, so the root system begins to degenerate.

In general, most of the root length is located in the top 15 cm of soil, and the quantity decreases very rapidly with depth. The zone of maximum root length coincides with the most chemically and biologically fertile zone of the soil. Water, however, is more evenly distributed in the soil profile, so that the depth of rooting and the roots growing in the subsoil are important for the acquisition of water. A deep root system is one mechanism that plants have to tolerate drought (see Chapter 8).

Among the many soil conditions that can impede root growth, the most important that can be managed by gardeners include compaction, a shortage of oxygen due to waterlogging or poor drainage, dryness, low nutrient supply and a pH that is too low.

## CONCLUSION

Soils may be very diverse, even within the space of a very small garden, having been derived from a wide range of rock types, drift materials and organic matter. Different mixes of sand, silt and clay particles give rise to soils of different textures. Soils are also populated by large numbers of micro- and macroorganisms, which play a central role in the breakdown of organic matter and the release of nutrients, and in maintaining and enhancing soil structure and fertility. Understanding the formation and properties of different soil types, the nutrient requirements of plants and the structure and growth of roots, is an essential pre-requisite for the development of appropriate soil-management strategies.

## FURTHER READING

Brady, N.C. (1990) *The Nature and Properties of Soils*, 10th edn. Macmillan Publishing Co., New York.
Gregory, P.J. (2006) *Plant Roots: Growth, Activity and Interaction with Soils*. Blackwell Publishing, Oxford.
Rowell, D.L. (1994) *Soil Science: Methods and Applications*. Longman Scientific & Technical, Harlow.
See also Chapter 7, Further reading.

# 7

# Soil Cultivation and Fertility

## SUMMARY

This chapter builds on the descriptions of the formation and features of soils and the structure of roots given in Chapter 6, and considers the management of soils in terms of their cultivation, water, pH and nutrients, including the use of fertilisers, manures and composts.

## CULTIVATING THE SOIL

Soils are cultivated for a number of reasons: to obtain a seed-bed; to kill weeds; to remedy damage done by previous traffic or cultivation; to increase the permeability of the surface soil or subsoil to water; to incorporate crop residues and manures; and to provide a hospitable medium for root growth.

In the particular case of double digging, this is done to increase the depth of cultivated soil, leading to an increased store of readily accessible nutrients and water through the encouragement of a more extensive root system. Given the constraints of time and labour, and the availability of fertilisers, this form of cultivation is practised much less than in the past. Many gardeners use single digging as their primary means of cultivation, but apart from its virtue as a form of exercise and a good reason for being out in the fresh air on a bright spring morning, it is vastly overrated, and although it is thought by some to be a good method of weed control and essential for many perennial weeds (but see Chapter 17), selective use of herbicides and many ground-cover plants can substantially obviate the need. Digging almost always destroys soil structure, and should rarely be done on clayey soils, where the frost in winter and drying in summer will regenerate the

structure naturally. On sandy and silty soils some digging may be beneficial but, except for vegetable production, a light cultivation with a fork, hoe or rake will often suffice.

The essential requirement of a good seed-bed is that the tilth produced by cultivation should be sufficiently fine and firm to allow the seeds to be placed at a uniform depth, and to have sufficient contact with the soil to take up water. The soil around the seeds must have sufficient pore space to maintain good aeration, while the soil above and below the seed must be sufficiently loose to allow both the shoots to emerge into the air and the roots to elongate. Weeds should be absent, so that there is no competition for soil resources. On clayey soils it is difficult to achieve such conditions but on sandy and silty soils it is practicable. However, many silty soils have poor structural stability when wet, so that the surface aggregates collapse after wetting and then form a hard crust on drying. This can substantially reduce the number of seedlings that emerge, unless the crust is disrupted by a light hoeing.

The roots of most plants have a diameter greater than 0.1 mm, so must either grow through transmission pores or deform the soil to make a pore of sufficient size for elongation to proceed. One function of cultivation is to create transmission pores, or to loosen the soil sufficiently to allow the roots to elongate and create such pores. However, if

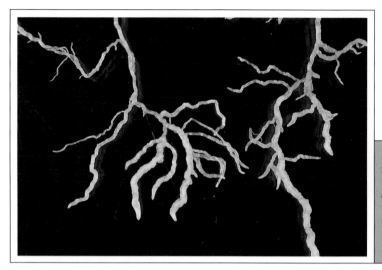

**Fig. 7.1** Effects of compaction on the growth of wheat (*Triticum* spp.) roots. Note the thicker and more distorted roots with sharp bends. These roots were grown in a sand with a pore space of only 30%. *Photograph courtesy of Peter Gregory.*

**Fig. 7.2** Effects of a cultivation pan, produced by cultivating at the same depth for many years at the topsoil/subsoil interface, on the growth of leeks. The crop was not marketable. *Photograph courtesy of Peter Gregory.*

cultivation is done in the wrong conditions or with the wrong implement, it can exacerbate the problem of poor root growth. An obvious sign of compaction is that the root diameter increases immediately behind the root tip. In extreme conditions, as in Fig. 7.1, the root becomes distorted in the attempt to find gaps between soil particles, and the root tip may eventually die.

In gardens, there are two frequent situations that may lead to soil conditions that impede root growth. The first is due to the impact of machinery and feet. Building work, laying pipes and walking on the top

of exposed subsoil may all lead to the formation of compacted layers that will reduce root growth. Essentially, these layers should be broken up with a spade or fork before planting. In addition, the use of powered cultivators in wet conditions may cause smearing (forming plate-like structures) of the soil at the depth of cultivation, which produces a solid layer and blocks large pores. In agriculture this is known as a 'plough pan'. Fig. 7.2 shows the results of such a pan on a crop of leeks; this crop was unmarketable.

The second set of problems arises when transplanting. Some soils, especially silty and clayey soils,

**Fig. 7.3** Roots of a strawberry plant restricted by 'dibbing in'. *Photograph courtesy of Peter Gregory.*

smear when wet, so that the use of a dibber or trowel may produce a lining to a hole that confines roots. Fig. 7.3 shows the roots of a strawberry plant confined to the dibber hole. A related problem may occur when transplanting a container-grown plant. Frequently such plants form a root mat against the inside of the container that needs to be teased apart if normal root growth is to be resumed. This is particularly important in drought-prone areas, because the volume of water available within the root mat is typically sufficient only for a day or two.

Why is it so important to get the physical aspects of soils right? As well as the obvious effects of constraints on the size of the root system, and therefore the root surface available to intercept water and nutrients, there are less obvious effects. For example, there is now ample experimental evidence that when a root experiences a constraint to growth such as drought, it rapidly produces a chemical signal (e.g. abscisic acid), which is transmitted to the shoot and reduces the rate of leaf expansion (see Chapter 8). In

this way, reduced expansion of the root system is matched by reduced canopy expansion, and the sizes of the root and shoot systems remain in equilibrium. Big shoots, therefore, cannot be produced if the root system is constrained.

## MANAGING SOIL NUTRIENTS

Plants require 14 essential mineral elements to produce healthy growth and to reproduce, as follows.

The macronutrients are nitrogen (N), phosphorus (P), potassium (K), calcium (Ca), magnesium (Mg) and sulphur (S).

The micronutrients are iron (Fe), manganese (Mn), copper (Cu), zinc (Zn), boron (B), chlorine (Cl), molybdenum (Mo) and nickel (Ni).

The *macronutrients* and *micronutrients*, sometimes called trace elements, are all essential nutrients, although the known requirement for chlorine and nickel is as yet restricted to a limited number of plant species. The distinction between macronutrients and micronutrients is somewhat arbitrary but is based on their typical concentrations in plant dry matter. The concentration of macronutrients is normally at least ten times that of the micronutrients and often the difference is much greater. For example, the concentration of nitrogen will be typically one million times greater than that of molybdenum and one thousand times greater than that of manganese.

Some elements are not essential to all plants but may be beneficial to some. For example, cobalt (Co) is essential for biological nitrogen fixation by bacteria, including those in the nodules of leguminous plants such as peas and beans (see below). Sodium (Na) is not essential for most plants, except the salt bush *Atriplex vesicaria*, but it is beneficial to the growth of sugar beet. Similarly, silicon (Si) is essential for some grasses and also for *Equisetum arvense*, but for most plants it is beneficial rather than essential. Other non-essential elements, for example vanadium (V), may also be present in plants.

Each mineral nutrient may perform several functions within the plant: as a part of an organic structure; as a catalyst of enzyme reactions (see Chapter 2); as a charged carrier to maintain electrochemical balance; or as a regulator of osmotic pressure. Generally the micronutrients, because of their low concentration, play little part in either osmotic regulation or in the maintenance of electrochemical balance. They mainly assist in the activation of enzyme reactions. Conversely potassium and chlorine, the only nutrients that are not part of organic structures, are very important in osmotic regulation and for electrochemical balance (see Chapter 2).

Nitrogen is an essential constituent of many organic components of plants, including proteins, nucleic acids, chlorophyll and hormones. Similarly, phosphorus in plants is present as part of the many organic compounds which play a very important part in the enzyme reactions that depend on phosphorylation. Magnesium is a constituent of chlorophyll and iron is required for synthesis of this pigment.

Probably the most important aspect of plant nutrition for the gardener is being able to recognise symptoms of deficiency, and in some cases toxicity, and to remedy them. Often the growth of plants will be considerably reduced before deficiency symptoms first appear, so any measures to correct a shortage of nutrients may be beneficial only to subsequent plantings. Whereas generalisations can be made about the deficiency symptoms of the macronutri-ents, the symptoms are often specific to the plant species. For micronutrients the symptoms are almost entirely plant-specific. Table 7.1 gives some general symptoms of deficiency. A gardener suspecting nutrient deficiency as a cause of poor growth would be best advised to consult a specialist textbook and to keep some plant leaves for laboratory analysis. In the UK, deficiencies of macronutrients are more common than those of micronutrients and if nitrogen, phosphorus and potassium are provided then there are usually sufficient amounts of other nutrients in soils to allow good growth. However, this is not the case elsewhere, especially on sandy soils and those where leaching of nutrients by excess rain occurs.

Toxicity may occur if there are excess amounts of manganese, boron, aluminium or chlorine in the soil solution. Chlorine toxicity may occur when water with a high salt concentration is used for irrigation, but the others are associated with soils of low pH (less than 5) and can be remedied by the application of lime.

All the essential mineral nutrients are obtained predominantly from the soil, although deposition on to leaves may lead to uptake of some nutrients in

**Table 7.1** Symptoms of nutrient deficiency.

| Nutrient | Visible symptoms |
|----------|------------------|
| Nitrogen | Uniform pale colour with yellowing especially of lower leaves. Feeble growth and lack of branching. |
| Phosphorus | Reduced growth especially soon after emergence. Often there are no other symptoms although some species show purple coloration of older leaves. |
| Potassium | Margins of older leaves scorched and curled up or down. |
| Sulphur | New leaves a uniform golden yellow or cupped and deformed. Foliage frequently stiff and erect. |
| Calcium | Cupping and burning of leaf tips with blackening of young leaves. |
| Magnesium | Yellowing between the veins of older leaves to give a mottled appearance. |
| Iron | Yellowing between the veins of older leaves and the production of nearly white young leaves. |
| Manganese | Yellowing between veins of new leaves but may quickly spread to others. |
| Zinc | Specific to individual species. |
| Copper | Very dependent on species. New leaves become greyish-green, yellow and sometimes white. |
| Boron | Brittle tissues which crack easily. Death of growing point and production of side shoots. |
| Molybdenum | Rare except in cauliflowers where new leaves become progressively twisted and reduced until only the midrib appears (known as whiptail). |
| Chlorine | Wilting. |

polluted industrial areas. The nutrients are obtained in the following ways.

1. Potassium, calcium and magnesium are taken up as the positively charged *ions* $K^+$, $Ca^{2+}$ and $Mg^{2+}$ from the soil solution.
2. Nitrogen, phosphorus and sulphur are taken up as the negatively charged ions $NO_3^-$, $H_2PO_4^-$, $HPO_4^{2-}$ and $SO_4^{2-}$ from the soil solution. Nitrogen is also taken up as ammonium ions ($NH_4^+$) from the soil solution. Atmospheric nitrogen gas is not available as a source of nitrogen to higher plants but is utilised by some bacteria.
3. Other elements, such as copper, $Cu^{2+}$, are obtained as ions from the soil solution.

Two essential features emerge from all this. First, plants take up most of their nutrients from the soil solution, and not from the solid soil materials. Second, the nutrients are taken up predominantly as inorganic ions, not as organic ions or attached to organic matter. A further feature of nutrient uptake by plants is that it is generally highly selective. Membranes in root cells are selective about what they allow into the plant. Some ions, for example phosphate, are present in very much greater amounts inside the plant than would be anticipated from their concentration in the soil solution while others, for example calcium, are present in much smaller amounts. In turn, this selectivity at the root surface means that some ions, for example calcium and magnesium, tend to accumulate there as a consequence of the mass flow of soil solution to meet the demand of water for transpiration while other ions, such as phosphate and nitrate, are depleted, resulting in a concentration gradient which permits diffusion towards the root surface.

## SOURCES OF THE MAJOR PLANT NUTRIENTS

Sections of Chapter 6 showed that the clay minerals and the organic matter in the soil have negatively charged surfaces. This property is very important in maintaining chemical fertility, especially for the retention of positively charged nutrient ions such as potassium. Potassium ions are held on the negatively charged surfaces of the clay minerals and replenish the soil solution as potassium is removed by plant roots. Most soils contain quite large quantities of potassium in minerals such as feldspar, but this is unavailable to plants. It only becomes available as it is weathered from the crystalline minerals and held on the negatively charged surface of clay minerals. For potassium, then, the replenishment of the soil solution is essentially an inorganic, chemical process.

Phosphate in soil solution is present as two negatively charged ions, $H_2PO_4^-$ and $HPO_4^{2-}$, over the usual range of soil pH values (see below), so it is not held by the negatively charged surfaces of clay minerals. The main feature of soil phosphate is its very low solubility, because these ions bind strongly to iron and aluminium sesquioxides and calcite to give very low concentrations in soil solutions. Typically, the concentration of phosphate in solution is about 10–100 less than that of potassium and nitrate. As a result, phosphate supply is a major limitation to plant growth worldwide, and plants have evolved many mechanisms, including the release of organic acids and solubilising enzymes, and the development of mycorrhizal associations (see Chapter 2), to help in its acquisition.

Phosphorus is also present in soil organic matter and this is released to the soil solution by microbial decomposition. In some tropical regions the organic forms of phosphorus are a major contributor to uptake of this element by plants, but in most places, although the total amounts of inorganic and organic phosphorus are similar in a soil, the inorganic minerals dominate in providing phosphorus that is available for uptake.

In contrast to both potassium and phosphorus, the processes controlling the availability of nitrogen are biologically mediated. One process that brings nitrogen into soils and plants is *nitrogen fixation* (see Chapter 2): nitrogen gas in the soil atmosphere is 'fixed' to form ammonia, which is then incorporated into organic compounds. The process occurs both in bacteria living in soil, which fix about 5–20 kg nitrogen (N) $ha^{-1}$ per year in the UK, and also, most effectively, in bacteria living inside nodules associated with the roots of legumes such as peas (*Pisum sativum*) and beans (*Phaseolus* spp.), where 40–200 kg N $ha^{-1}$ may be fixed per year. In the UK, fixation by bacteria in legumes is typically 50–100 kg N $ha^{-1}$ per year, although only a small proportion of this enters the soil if the pods are harvested.

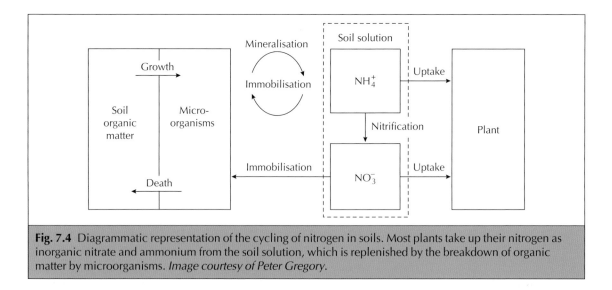

**Fig. 7.4** Diagrammatic representation of the cycling of nitrogen in soils. Most plants take up their nitrogen as inorganic nitrate and ammonium from the soil solution, which is replenished by the breakdown of organic matter by microorganisms. *Image courtesy of Peter Gregory.*

Although soil minerals contain some nitrogen, it is the soil organic matter that dominates the supply of nitrogen to plants. This nitrogen is part of a very dynamic system that moves through a variety of forms as a result of the activities of plants and microorganisms. Plants and microbes take up ammonium ($NH_4^+$) or nitrate ($NO_3^-$) ions from the soil and convert it into organic nitrogen. When the organisms die, the organic materials returned to the soil are, in turn, decomposed, and the organic nitrogen converted back to ammonium in a process known as *mineralisation*, which normally is then followed rapidly by nitrification to nitrate by specific nitrifying bacteria (Fig. 7.4). The nitrate produced is available for uptake by plants, but if the microorganisms cannot satisfy their needs for nitrogen from the organic compounds on which they are 'feeding', they will use this nitrate themselves, and re-convert it to organic nitrogen. This process is known as *immobilisation*. Microbes and plants thus compete for the inorganic nitrogen ($NH_4^+$ and $NO_3^-$), depending on the nitrogen-richness of the soil organic materials available to the microbial population.

This dynamic set of processes, mediated by the soil microorganisms, is at the heart of nitrogen management in the garden. The organic matter in soils, which includes the microorganisms, has a conservative *carbon/nitrogen ratio (C/N ratio)* of about 10:1 to 12:1. The carbon/nitrogen ratio in plant materials returned to soil is typically 20:1 to 30:1 in green plants and farmyard manure, about 100:1 in straw, and as high as 400:1 in sawdust. In microorganisms the ratio is more stable and much lower at between 4:1 and 9:1. It follows that there will be keen competition between microorganisms for soil nitrogen. This has two important practical consequences for gardeners.

First, most compost heaps contain plant materials both rich in nitrogen, such as early spring grass cuttings, and low in nitrogen, such as autumn prunings, so mixing will promote more even decomposition. Because the carbon/nitrogen ratio is likely to be higher than 10:1, a compost enhancer, essentially a source of nitrogen, will promote decomposition by favouring the microbial population. Second, the addition of fresh organic material such as farmyard manure to the soil will, in the short term, reduce the availability of nitrogen to plants, as nitrate is taken up and immobilised by the population of microorganisms.

A final practical point is that because the activities of soil microorganisms dominate the processes of mineralisation that produce plant-available nitrogen, it is impossible at present to predict the rates and amounts of inorganic nitrogen that will be produced. There is no soil test available to assist with this prediction. Maximum rates of mineralisation are achieved during the warming of moist soils in the spring, during re-wetting of warm, dry soils in the autumn and by the addition of fresh plant material.

Soil organic matter also contains phosphorus and sulphur in a combination of about 100C:10N:1P and

1.S, so as nitrogen is mineralised, smaller quantities of phosphorus and sulphur are made available for uptake by plants and microbes.

# FERTILISERS

The purpose of adding fertilisers, manures and compost to soils is primarily to provide a supply of essential nutrients for plants; a subsidiary purpose is to maintain the biological activity of the soil. Most garden soils in the UK and northern Europe contain ample resources of micronutrients, so attention can be focused on nitrogen, phosphorus and potassium (N, P and K). In other parts of the world micronutrients are often a more important issue, and specialised products exist to remedy the deficiencies.

A *fertiliser* is defined as any material that provides a concentrated source of one or more essential nutrients. Fertilisers are often divided into organic or inorganic products in an attempt to differentiate them by source. Organic indicates that the material is of plant or animal origin, for example bone meal or hoof and horn. Inorganic refers to materials which have been chemically synthesised, such as ammonium nitrate, or are the result of processing mineral deposits, for example superphosphates. Fertilisers are available as straight fertilisers, which supply only one or two nutrients (see Table 7.2), and as compound fertilisers, which provide a more complete range of the elements essential for plant growth, especially N, P and K. An example of an inorganic compound fertiliser is Growmore, which supplies mainly nitrogen, phosphorus and potassium. In contrast, fish, blood and bone may be considered to be an organic compound fertiliser, being made up of fish meal, dried blood and bone meal, and is designed to supply a broad range of major nutrients and micronutrients in an organic form.

Fertilisers may also be differentiated by the rate at which the nutrients they contain are made available for uptake by plants. Although there are exceptions, the straight inorganic fertilisers tend to be highly soluble and the nutrients they contain are in forms which are readily available to plants. In contrast, many organic fertilisers contain a large proportion of their nutrients bound up as complex organic compounds, which are made available for uptake only slowly. Common exceptions to these generalisations are fertilisers such as dried blood, which is rapidly mineralised, and controlled-release fertilisers, which release their inorganic nutrient load over an extended period, depending on temperature.

The objective of fertiliser application is to supplement the available nutrient reserves in a soil to ensure that optimum nutrient levels are available for plant

**Table 7.2** The composition of some common inorganic fertilisers.

| Name | N, P or K concentration (%) | Nutrients supplied |
|---|---|---|
| **Nitrogen fertilisers** | | |
| Ammonium sulphate | 21 | Ammonium plus some sulphur |
| Ammonium nitrate | 32–34.5 | Ammonium plus nitrate |
| Calcium ammonium nitrate (CAN) | 25 | Ammonium, nitrate plus some calcium |
| Urea | 46 | Converted to ammonium |
| **Phosphorus fertilisers** | | |
| Single superphosphate | 8–9 | Phosphate, calcium plus sulphur |
| Triple superphosphate | 20 | Phosphate plus calcium |
| Diammonium phosphate (DAP) | 23 | Phosphate plus ammonium |
| **Potassium fertilisers** | | |
| Potassium chloride (muriate of potash) | 50 | Potassium plus chlorine |
| Potassium sulphate | 42 | Potassium plus sulphur |

growth at the right time. Exactly which nutrients need to be supplied, and at what time of year, is dependent upon a variety of factors including the plants being grown, the type of soil they are growing in, the prevailing climate and the existing nutrient reserves.

Much research has been done on commercial crops with the aim of providing accurate recommendations about what fertiliser to supply, how much to use and when to use it. These recommendations are extremely valuable to professional growers, since under-application of nutrients can lead to reduced yields and products of poor quality, while over-application leads to pollution of water courses and excessive wastage of both the product and the energy used to produce it, with consequences for profits and the environment. For the individual gardener, fertiliser usage is relatively low, but it still makes environmental and economic sense to employ good judgement when using these chemical compounds.

Professional growers depend upon regular soil and plant analysis to ensure they are applying exactly the right amounts of fertilisers, but the cost of laboratory analyses means that most gardeners will have them done only occasionally. Perhaps the best time for having a soil analysis carried out is when buying a new area of land with an unknown cultivation history. Another relevant situation is where heavy vegetable cropping has taken place over a prolonged period, and nutrients and organic matter may be depleted. In practice, rather than spending money on soil testing, many gardeners make so-called insurance dressings of fertiliser just in case they are needed. This is understandable but wasteful, and can lead to toxic or unbalanced levels of nutrients developing in the soil.

When purchasing fertilisers it is important to read the small print on the label, both to ensure that the product contains the correct nutrients, and that it represents value for money. By law, the container must state the composition of the contents. Table 7.2 gives the N, P and K content of most common single fertilisers. Unfortunately, the law obliges manufacturers to specify the composition not as N, P and K but as N, $P_2O_5$ and $K_2O$. This is an unhelpful historical anomaly dating back to the times when chemists carried out analyses in terms of the mass of an oxide. The oxides can be converted to P by multiplying $P_2O_5$ by 0.44, and to K by multiplying $K_2O$ by 0.83. So, for example, Growmore is labelled

as 7:7:7 and contains 7% N:7% $P_2O_5$:7% $K_2O$, or 7% N, 3% P and 5.8% K.

There is generally much less need for P and K fertilisers now than in the 1950s and 1960s, which is because many gardens have received inputs of P and K over prolonged periods. Results of tests on garden soils sent in by members of the Royal Horticultural Society during 2000 showed that of the 289 samples supplied, 72% contained sufficient P to produce most vegetable crops without additions of P fertiliser. Only 30 samples contained low amounts of P. Relatively more emphasis should be placed on supplying adequate N at the appropriate times.

## MANURES

Whereas fertilisers are a source of mineral nutrients only, *manures* and *composts* have additional attributes that contribute to the fertility of a soil. Traditionally the term manure referred to any material applied to soil to improve fertility. Today we tend to use it when referring to animal dung and urine. Both these waste products are valuable sources of major nutrients and trace elements, so manures are useful fertiliser materials. However, because most manures are mixed with carbon-rich bedding materials such as straw and wood shavings, they can also supply significant amounts of organic matter to the soil. The relative value of a particular manure depends on several factors including the animal species from which it was obtained, the amount and type of bedding material mixed with the dung and the degree of decomposition that has already taken place.

In terms of their value as fertilisers, manures are a valuable source of N, P and K, as well as magnesium, sulphur and micronutrients. The quantities of these in manure vary (see Table 7.3). Most research has been done on cattle, pig and poultry manures, since these are the types most frequently used by farmers. However, the manure most frequently used by gardeners is horse manure, together with the waste products of domestic pets and poultry. Where nitrogen-rich manures are applied, such as those derived from poultry, or where there is little bedding material present to dilute the manure, there is a risk that ammonia gas may be released following application. Ammonia is toxic to plants and can cause leaf scorch

**Table 7.3** The composition of some natural fertilisers and animal manures.

| Material | Concentration (%) | | |
|---|---|---|---|
| | N | P | K |
| Dried blood | 12–14 | | |
| Bone meal | 6–10 | 8 | |
| Fish meal | 7–14 | 4–7 | |
| Wood ash | 0.1 | 0.13 | 0.8 |
| Farmyard manure (cattle) | 0.9 | 0.18 | 0.5 |
| Farmyard manure (pigs) | 0.6 | 0.26 | 0.33 |

Note that many natural fertilisers do not contain potassium.

and root death, especially in seedlings and young plants. Ammonia production is most likely to occur when the weather is warm and dry, and when the manure has been left on the surface rather than incorporated. It is more prevalent in sandy, alkaline soils and those low in humus.

Only a portion of the applied manure is of value as fertiliser in the season of application because of the complex microbial processes of mineralisation and immobilisation described above. Indeed, as described, the use of fresh straw-rich manure can initially reduce the amount of plant-available nitrogen, due to immobilisation. However, a large reserve of nutrients can gradually be established in the soil if manuring is introduced as an annual gardening activity.

## COMPOSTS

The term *compost* has two meanings for the gardener. It can refer either to a medium in which container plants are grown (a potting compost; see Chapter 14) or to the product of the composting process. The second meaning is employed here. Although any organic material can be composted, garden compost is usually a mixture of composted garden and kitchen waste, sometimes supplemented with animal manure. Compost supplies a low level of nutrients, and its main value is as a source of humus which may help to improve soil structure, support soil life and recycle nutrients.

Composting is the rapid aerobic decomposition of organic matter by microorganisms at elevated temperatures. It differs from natural decay processes both in terms of the speed of decay and the heat that is generated. Whereas leaves on a woodland floor may take months to decompose, decay in an active compost heap will occur in a matter of weeks.

For a compost heap to function effectively a large volume of undecomposed raw material has to be accumulated before the heap is built, and it needs to be a good mixture of soft, nitrogen-rich materials such as grass clippings and kitchen scraps (ideally 25–50%) and drier, carbon-rich materials such as woody prunings, wood chippings and straw (50–75%). When the compost heap is built, the large volume of organic material acts as an insulating layer, allowing heat generated by microorganisms to build up. Accumulating a large quantity of fresh material at the start provides an ample supply of material for the microorganisms to metabolise. This allows a massive flush of microbial growth over a short time, resulting in the characteristic rise in temperature.

The heat generated within a compost heap has two effects. Up to about 45°C, the increasing temperature speeds up the biochemical reactions taking place and kills off undesirable pathogens and weed seeds. The degree of pasteurisation increases as the temperatures continue to rise, but at temperatures greater than 55°C the microorganisms start to decline. Although some microorganisms will function at much higher temperatures, the rate of composting slows down. For this reason, commercial composting heaps are normally maintained at a maximum temperature of 55–65°C. The compost should also be turned, to prevent it becoming either too wet or too dry.

Making compost in gardens is often a small-scale process, using lower temperatures and involving the gradual addition of unselected materials over a long time. For a small-scale garden compost heap it is more realistic to expect a batch of compost to be ready in 8 to 12 months, rather than the shorter period anticipated in a commercial operation. Having said this, garden composting is still a very effective means of boosting soil fertility at no cost other than that of time. The resulting material is an excellent soil conditioner and mulch, although decomposition, particularly of tough vegetable matter, may remain incomplete and it will usually still contain viable weed seeds.

The following guidelines may be useful for improving the quality of compost produced in a typical compost heap.

1. Accumulate as much mixed raw material as possible for adding to the heap at one time.
2. Build the heap over bare soil and include some old compost with the raw materials to ensure rapid colonisation by the necessary microorganisms.
3. Mix all the materials together and chop up or shred large pieces to speed up decomposition.
4. Protect the heap from the rain with a lid or cover to avoid waterlogging, especially over the winter months.
5. Turn the heap once, or if possible twice, early in the composting process.

## MANAGING SOIL pH

Acidity, strictly defined, depends on the concentration of hydrogen ions in a solution. The *soil pH* is a measure of the hydrogen ion concentration in the soil solution. The scale used is logarithmic, so that a difference of one unit of *pH* reflects a ten-fold difference in concentration. By definition, the pH of pure water is 7 and is neutral, while acid solutions have a pH below 7 and alkaline solutions have a pH greater than 7. The cation-exchange capacity of clay minerals and organic materials allows hydrogen ions to be adsorbed and to exchange with other ions in the soil solution. Most soils will become acidic if they are exposed to rainwater for sufficient time and drainage occurs.

Acidification is a natural process; it occurs because water in contact with atmospheric carbon dioxide produces a solution of dilute carbonic acid with a pH of 5.6. In heavily polluted air the pH may be much lower, typically 4.4 over eastern Britain in the early

1990s, and even in unpolluted air rain picks up small quantities of naturally occurring acids, resulting in a pH of about 5. As the rain enters the soil it is affected by numerous other chemical processes, acting as both sources of and sinks for acidity. The presence of sinks for acidity, notably the presence of large quantities of basic minerals and exchangeable calcium in many soils, has the effect that the pH of the soil does not fall immediately to that of the rainwater, but is subject to *buffering*, i.e. kept at a near-constant value, for a period of time. Generally, clay soils and soils with appreciable amounts of soil organic matter are better buffered than sandy soils. Chalky soils will normally be alkaline.

The management of soil pH is important for gardeners, because as soils grow more acidic there are several associated changes in their properties (Fig. 7.5). The availability of many plant nutrients, such as calcium, magnesium and phosphate, is reduced, while that of manganese and aluminium (not a plant nutrient) may be increased to toxic levels. Also, the activity of many soil organisms is reduced, resulting in

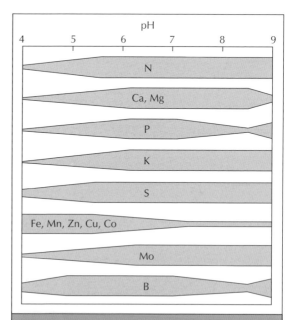

**Fig. 7.5** Relationships between the availability of plant nutrients and soil pH. The widest portion of the bands indicates the most ready availability of nutrients to plants. Overall, it is evident that a pH range of about 6 to 7 will ensure the best availability of nutrients. *Image courtesy of Peter Gregory.*

lower availability of nitrogen, phosphorus and sulphur and the accumulation of undecomposed organic matter. These considerations mean that the optimum pH for many common plants growing in soils is about 6.5.

Many groups of plants have evolved on soils with particular characteristics, and are consequently better suited to some soils than others. Important groups for gardeners are the *calcifuges*, plants which have adapted to growing on acid soils, and the *calcicoles*, which prefer chalky or calcareous soils (see Table 7.4). The low availability of iron and manganese in chalky soils can lead to deficiencies in

the plant, especially in calcifuge species such as *Rhododendron*. This results in lack of chlorophyll, especially in young leaves, a condition know as lime-induced chlorosis.

Although it is preferable not to try and grow acid-loving plants such as rhododendrons and azalea on alkaline soils, some benefit can be obtained by the application of elemental sulphur, which is converted to an acid, thereby reducing the pH and increasing the solubility of iron and manganese. As an alternative, foliar sprays or soil dressings of *chelated* iron, such as Sequestrene, may be applied. This renders iron and

**Table 7.4** Plants suitable for growing on acid or chalky soils.

| Plants that prefer acid soils (less than pH 7)* | Plants that tolerate chalky soils (more than pH 7)† |
|---|---|
| *Corydalis* spp. | Border carnations and pinks (*Dianthus* spp.) |
| Gentian (*Gentiana septemfida*) | Pasque flower (*Pulsatilla vulgaris*) |
| *Meconopsis* spp. | Poppy (*Papaver* spp.) |
| *Rhodohypoxis* spp. | Sage (*Salvia* spp.) |
| Bearberry (*Arctostaphylos* spp.) | Pincushion flower (*Scabiosa caucasica*) |
| Bottlebrush (*Callistemon* spp.) | Bearded rhizomatous iris (*Iris* spp.) |
| Ling (*Calluna vulgaris*) | *Clematis* spp. |
| *Camellia* spp. | Honeysuckle (*Lonicera* spp.) |
| *Clethra* spp. | Paper-bark maple (*Acer griseum*) |
| *Corylopsis* spp. | Horse chestnut (*Aesculus* spp.) |
| Lantern tree (*Crinodendron hookerianum*) | Judas tree (*Cercis siliquastrum*) |
| St Dabeoc's heath (*Daboecia cantabrica*) | Sun rose (*Cistus* spp.) |
| Chilean fire bush (*Embothrium coccineum*) | *Escallonia* spp. |
| *Enkianthus* spp. | Beech (*Fagus sylvatica*) |
| Heather (*Erica* spp., except *E.* x *darleyensis* | *Forsythia* spp. |
| and *E. carnea*, which are lime tolerant) | Holly (*Ilex* spp.) |
| *Fothergilla* spp. | Juniper (*Juniperus* spp.) |
| *Gaultheria* spp. | Batchelor's button (*Kerria japonica*) |
| *Halesia* spp. | Privet (*Ligustrum* spp.) |
| Witch hazel (*Hamamelis* spp.) | Daisy bush (*Olearia* spp.) |
| Calico bush (*Kalmia* spp.) | *Photinia* × *fraseri* 'Red Robin' (evergreen) |
| Tea tree (*Leptospermum* spp.) | Bird cherry (*Prunus avium*) |
| *Leucothoë* spp. | Sumac (*Rhus* spp.) |
| *Magnolia* spp. | Mountain ash (*Sorbus* spp.) |
| *Photinia beauverdiana, Photinia villosa* (deciduous) | Lilac (*Syringa vulgaris*) |
| *Pieris* spp. | *Viburnum* spp. |
| Rhododendrons and azaleas (*Rhododendron* spp.) | *Weigela* spp. |
| Blueberry (*Vaccinium corymbosum*) | |
| Cranberry (*Vaccinium vitis-idaea*) | |

* Although they cannot tolerate chalky conditions, many of the plants listed will grow successfully in neutral soil.
† These plants will grow well in chalky soil but most also grow well in neutral or acid conditions. However, acid soils can adversely affect growth in some plants (e.g. below pH 6.0 for beans, pH 5.9 for peas and pH 4.9 for potatoes).

**Table 7.5** The amounts of lime to add to bring the soil to the optimum pH (6.5 for most soils but 6 for peat) for the growth of most vegetables and flowers. The pH shown on the left is the measured pH of the soil before the addition of lime.

| Measured pH | Amount of lime to add (kg m$^{-2}$) | | | |
|---|---|---|---|---|
| | Sandy soils | Loamy soils | Clayey soils | Peats |
| 6.5 | 0 | 0 | 0 | 0 |
| 6.2 | 0.3 | 0.4 | 0.4 | 0 |
| 6.0 | 0.4 | 0.5 | 0.6 | 0 |
| 5.8 | 0.5 | 0.6 | 0.7 | 0 |
| 5.5 | 0.7 | 0.8 | 1.0 | 0.8 |
| 5.2 | 0.9 | 1.1 | 1.2 | 1.3 |
| 5.0 | 1.0 | 1.2 | 1.4 | 1.6 |
| 4.8 | 1.1 | 1.3 | 1.5 | 1.9 |
| 4.5 | 1.3 | 1.5 | 1.8 | 2.4 |
| 4.2 | 1.5 | 1.8 | 2.0 | 2.9 |
| 4.0 | 1.6 | 1.9 | 2.2 | 3.2 |

Source: from Ministry for Agriculture, Fisheries and Food (now the Department for the Environment, Food and Rural Affairs) publications.

manganese soluble despite the pH of the soil itself. The treatments need to be repeated at intervals.

Where soils are too acid, the chief means of managing soil pH is through the application of lime (calcium carbonate, $CaCO_3$). It is usually offered for sale as garden lime, but may also be referred to as carbonate of lime, ground limestone or ground chalk. Water-soluble calcium hydroxide (hydrated lime) may also be used for managing soil pH, but is not recommended since it is unpleasant to handle and easily overapplied. The amount of lime needed depends on the soil pH, the buffering capacity of the soil and the target pH. Strictly speaking, a laboratory test is required for accurate results, but Table 7.5 provides useful guidance based on extensive experience of field trials in England and Wales. It shows, for example, that if the measured pH of a sandy soil is 4.8, then lime should be applied at 1.1 kg m$^{-2}$ to bring the pH to the optimum value of 6.5. In the east of England, losses of lime by leaching will be smaller than in the higher-rainfall areas of the South-west. Consequently, liming needs to be more frequent in western England, or plants that are more tolerant of acid soils (see Table 7.4) must be grown. Autumn is the best season for adding lime to soils, as it can take effect over the winter months and will not damage young growth.

A final point is that ammonium fertilisers such as ammonium sulphate and ammonium nitrate produce hydrogen ions as they are nitrified by microorganisms to nitrate. For every 1 kg of fertiliser nitrogen applied as ammonium, about 4 kg of lime is required to neutralise the acidity produced. This is not to say that lime should be added at the same time as fertiliser (it should not, because there is a strong possibility that the ammonium will be converted to ammonia gas and be lost to the atmosphere), but rather that the natural processes of acidification will be enhanced when ammonium fertilisers are used.

## MANAGING WATER

In Chapter 6 it was stated that soils contain a network of interconnected pores in which the largest pores do not retain water, the smallest pores retain water too strongly for the water to be available to plants, and only the medium-sized pores (50–0.2 μm in diameter) retain water available to plants (see Table 6.1). A well-structured soil should contain a wide range of pore sizes, however, and these fulfil a range of functions in addition to supplying water to plants, as follows.

1. Animal burrows allow rapid movement of water during periods of flooding ('mole' drains are common in clay soils).
2. Earthworm channels allow for the rapid movement of water when water-filled and provide an easy supply of oxygen for aerobic respiration when open.
3. Root channels also drain freely, and therefore help with aeration.
4. Storage pores hold water in close association with particle surfaces, allowing the rapid replenishment of nutrients when uptake occurs.
5. Storage pores provide habitats for bacteria and other microorganisms, most of which require a water film and a supply of oxygen to function normally.

Water management of soils is essentially about ensuring that the largest pores can operate to remove excess water during wet periods and that the storage pores are re-filled with water during dry periods. The management of water in relation to plants and their function is discussed in Chapter 8.

Excess water can be removed from garden soils by the installation of a drainage system. The necessity for such a system is largely confined to clayey soils where swelling of clay minerals occurs on wetting, reducing the volume of transmission pores. The solution is to install a network of pipes or channels which provides a stable system of pores to direct the water away from the site. The outlet of the drainage system must, however, be above the level of any water table or water in an open watercourse.

Occasional waterlogging may also occur on other types of soil. This is frequently the result of past mismanagement of the site as, for example, during construction work where the use of machines or repeated treading on an area smears over the transmission pores, causing them to close. If topsoil is then placed on top of this, waterlogging will often ensue because the interconnectedness of the transmission pores is disrupted at the interface. Sometimes this problem resolves itself, as when roots dry the soil in summer, re-forming a network of cracks, but light cultivation of the interface with a fork to recreate some large pores is a more reliable and faster method of amelioration.

Controlling excess water is important not only because of the influence of the water on the aeration status of the soil and its consequence for roots (see Chapter 8), but also because it has an effect on the speed with which soils warm up. Clayey soils warm up more slowly than loams in spring in a temperate climate because their high water content and high thermal capacity mean that, for a given input of energy from the sun, the temperature increase is less. Clayey soils are sometimes colloquially described as 'cold' for this reason and their slowness to warm up in spring may limit early root growth.

The water available to plants in a freely drained soil is determined by the volume of the storage pores. Measurements of this volume have been taken in many soils, so it is possible to make generalisations about the likely storage capacity of soils with different textures (see Table 7.6). For many loam soils the amount of water available to plants, termed the *available water capacity* (*AWC*), is about 20–22 mm for every 100 mm depth of soil. In sands and loamy sands it is about half this value, and in clays, silty clays and sandy loams it is roughly three-quarters of this value (see Table 7.6).

The frequency of watering required in the garden depends on the amount of water stored in the soil and how quickly it is depleted. As shown above, the size of the store is governed principally by soil texture and also by the depth of rooting. So a plant with its roots confined to the topsoil, usually about 250 mm deep, has only half the amount of water available to a plant that can root to a depth of 500 mm (half a metre). This is one reason why gardeners should pay attention to the subsoil. The time for which the store will last depends on the rate of evaporation to the atmosphere, a rate that is increasing with climate change. At present, in the UK, this varies from about 1–2 mm per day on a bright sunny day in April to about 4–5 mm per day on a bright sunny day in June and July. Days are rarely wholly sunny, so that a reasonable long-term approximation for summer is 3.5 mm per day. This means that a plant with a rooting depth of 500 mm growing on loam in a UK summer has sufficient water for 31 days [$(22 \times 5)/3.5$] before it will wilt permanently. In reality, the plant will start to suffer reduced growth long before all the available water is exploited (see Chapter 8). Plants like potatoes and many vegetables and flowers are adversely affected once about a quarter of the available water has been used, while grass growth is reduced when about half of the available water is used.

In general, plants will need to be watered about every 3 to 7 days to prevent water shortage restricting

**Table 7.6** The amount of water available in different soils, the number of days taken to deplete soil water to a point where growth of vegetables and flowers might be adversely affected and the amounts of water required to replenish the soil store. The values in column 3 were calculated assuming that the rooting depth was 0.5 metres, the potential rate of evaporation was 3.5 mm per day and that only 0.25 of AWC could be depleted before growth is adversely affected.

| Soil texture | Available water (millimetres of water per 100 mm depth of soil) | Days to deplete stored soil water | Volume of water required to replenish soil water store (litres per square metre) |
|---|---|---|---|
| Sand | 10 | 3.5 | 12.5 |
| Loamy sand | 12 | 4.0 | 15.0 |
| Sandy loam | 14 | 5.0 | 17.5 |
| Other loams | 22 | 7.5 | 26.0 |
| Clay and silty clay | 16 | 5.5 | 20.0 |

growth, depending on the type of soil and on the plant (Table 7.6). Table 7.6 also shows the volume of water that needs to be added to replenish the store of water once it has become depleted. The quantities are substantial, and are often underestimated by gardeners. Essentially, watering should be conducted infrequently in large amounts, rather than frequently in small amounts. This is because small amounts of water stay in the soil close to the surface, and can easily be lost by evaporation without ever passing through the plant. Also, if water is only present in the surface layers of soil, deep rooting is discouraged.

Besides replenishing the store by watering, there are other ways of influencing soil water content. Where soils are bare, applying a mulch to the surface will reduce the rate of water loss from the soil. The mulch needs to be coarse enough and thick enough to ensure that the hydraulic continuity of the soil pore system is broken between the soil and the atmosphere. Mulches of coarse organic material some 5–10 mm thick can assist the endeavour but, unless they have a low carbon/nitrogen ratio (for example, bark), will be incorporated into the soil by organisms within 12 months, and thus lose their effectiveness.

The addition of organic materials to soil may increase the size of the soil water store. This is because organic materials contain quantities of storage pores themselves, and also increase the aggregation of soil materials to create storage pores. However, the application of fairly large quantities of material over sustained periods is needed for there to be any measurable impact at a particular site. For example,

at Warwick Horticulture Research International, Wellesbourne, UK, applications of 50 tonnes ha$^{-1}$ of farmyard manure applied twice a year for 5 years on a sandy loam soil increased the AWC of the upper 150 mm of the soil by only 5 mm, the equivalent of about 1 day's worth of evaporation in summer. Given the quantities of manure available to most gardeners, the effects on water storage are likely to be insignificant if the manure is applied broadly. Limited benefits may be gained by burying large volumes of organic material close to specific plants at the time of planting, although the benefits will disappear with time as the organic materials are decomposed.

An alternative means of managing soil water is to select plants that will survive and flower even if summer drought occurs, and a wide range of drought-tolerant flowering plants is now available, as will be shown in Chapter 8. With the current awareness of possible changes in climate and generally drier summers in southern Britain, horticulturists and the water-supply industry are both promoting this approach. So far, however, there have been few attempts to introduce drought-tolerant vegetables.

## CONCLUSION

Gardeners often underestimate the importance of good root growth for the overall growth and health of the whole plant. This chapter emphasises the importance of combining an understanding of the properties of different soil types and root structure and growth (Chapter 6) with a knowledge of the different soil-management procedures and their effect, to produce the most favourable conditions for the healthy growth of both roots and shoots.

## FURTHER READING

Marschner, H. (1995) *Mineral Nutrition of Higher Plants*, 2nd edn. Academic Press, London.

Scaife, A. & Turner, M. (1983) *Diagnosis of Mineral Disorders in Plants. Volume 2: Vegetables*. The Stationery Office, London.

Wild, A. (1993) *Soils and the Environment: an Introduction*. Cambridge University Press, Cambridge.

See also Chapter 6, Further reading.

# 8

# The Plant's Environment: Light and Water

## SUMMARY

The control of gas [carbon dioxide ($CO_2$), oxygen ($O_2$) and water vapour ($H_2O$)] exchange by the plant and the structure and functions of stomata are described. The mechanisms that enable plants to detect lack of light are outlined, and shade tolerators and shade avoiders compared. The problems for plants posed by water stress and the physical and biochemical adaptations, notably C-4 and crassulacean acid metabolism, that enable them to grow in dry conditions are described. A survey of how plants deal with excess water leads to a discussion about aquatic plants. Plants adapted to saline conditions are also discussed. Finally, advice is given on watering practice and choosing garden plants able to cope with drought.

## INTRODUCTION

Because they are unable to move and find new and better habitats, plants have evolved a number of ways to grow in a variety of different conditions (see Chapter 1). This makes it possible for gardeners to choose plants that can grow well in the environment they occupy. Selecting the best plant for a site is an important factor in growing healthy plants with the minimum of attention. To some extent the site may be ameliorated, for example by making peat beds for acid-loving plants or installing irrigation systems in dry habitats, but the better and more environmentally friendly option is to live with what you have and try to match the plant to the conditions in which it is to be grown (see also Chapter 18). Among the most important local factors are the amount of light and the availability of water, and these are discussed in the present chapter. Discussion of seasonal factors such as day length and temperature is deferred until Chapter 13, whereas soil factors such as alkalinity and acidity have already been considered in Chapters 6 and 7.

## LIGHT

Green plants are dependent on receiving sufficient energy from sunlight for the manufacture of organic compounds. Gardens often have areas of shade, either from trees or from surrounding buildings, and it is important to choose plants that will grow well under these conditions. However, whereas shade-tolerant plants for such areas may also grow happily in full sun, provided there is plenty of water available, sun-loving plants will only grow well in areas with plenty of light, and will often tolerate some degree of water shortage.

### Light and photosynthesis

The main reactions and products of photosynthesis, including the structure and pigment content of the chloroplasts, have been discussed in Chapter 2 and here it is only necessary to state the overall photosynthetic reaction:

$$CO_2 + 2H_2O \rightarrow CH_2O \text{ (sugars)} + O_2 + H_2O$$

In this chapter, the ways in which plants interact with their aerial environment to obtain the essential ingredients of photosynthesis, namely light and carbon dioxide, will be considered. Water is the hydrogen donor in photosynthesis, but only about 2% of the water taken up by the plant is used in this way. The other 98% is required to keep the cells turgid, to provide an appropriate environment for metabolic reactions, to carry dissolved substances about the plant and, by evaporation from the surface, to cool the plant (see below). The main garden problems concerning water are drought and waterlogging of the soil, and these are discussed below.

## Controlling gas exchange

The surface of leaves is covered with a waterproof layer, the *cuticle*, largely composed of *cutin*, a hydrophobic, fatty molecule (see Chapter 2). Although there may be some microscopic cracks and fissures in the cuticle, by and large it provides a very effective means of restricting the loss of water from the leaf. However, these waterproofing properties also create problems, for the leaf must be able to take up carbon dioxide for photosynthesis and lose the excess oxygen not required for respiration. It must also be able to take up oxygen for respiration when photosynthesis is not taking place, such as during the night or when the level of sunlight is low. Moreover, evaporation of water from the leaf surface, a process which consumes heat energy, also helps to cool the leaf, an essential requirement since sunlight contains large amounts of infrared radiation and the exposed leaf is therefore subject to a considerable heat load. This process of evaporation is why a lawn feels cool if walked on in bare feet. The solution to these problems has been the evolution of specialised pores in the cuticle and *epidermis* – the *stomata* – with variable apertures so that gas and water movement can be controlled.

The apertures of these pores are controlled by the guard cells (see Chapter 2), the walls of which are thickened in such a way that the cells change shape as they take up and lose water, thus opening or closing the pores. These changes are achieved by the movement of potassium ions into and out of the guard cells under the influence of environmental and hormonal stimuli (see below).

## Distribution of the stomata

The spacing of stomata on the leaf surface is very precise, being ten times the maximum pore diameter. There are good reasons for this. Diffusion of any gas through a perforated surface is assisted by sideways diffusion towards the rims of the pores. Provided that the pores are not so close together that the areas of collection or loss overlap, there is, for each gas, an optimum spacing of pores such that the presence of the barrier between them has a negligible effect on diffusion of the gas. A spacing of ten times the maximum diameter is optimum for carbon dioxide ($CO_2$) diffusion. So, although the area for diffusion of $CO_2$ into a leaf is reduced a hundred-fold by the presence of the cuticle, the spacing of the stomata ensures that when they are fully open the diffusion of this gas occurs almost as readily as if no cuticle were present.

Water vapour diffuses more rapidly than $CO_2$, so the spacing of stomata in leaves is less optimal for water loss, which occurs only at a rate roughly proportional to the area of the pores themselves. The result is that as stomata begin to close, loss of water vapour is restricted far more than $CO_2$ gain. This difference is very important for plants with a limited water supply, for water can be conserved by partial closure of the stomata without reducing the supply of $CO_2$ for photosynthesis by too great an amount.

Also, the atmospheric concentration of carbon dioxide is low, being about 380 ppm (parts per million; although it is currently increasing due to the accumulation of greenhouse gases; see Chapter 18), so the concentration gradient between carbon dioxide outside and inside the leaf is shallow and the diffusion rate is rather slow. In contrast, the concentration of water vapour inside the leaf in the sub-stomatal cavities approaches 100%, while the atmospheric concentration may be very low, especially during dry periods. The resulting steep diffusion gradient means that the loss of water vapour to the atmosphere may be rapid.

In most higher land plants the stomata are located on the lower surface of the leaf, where there is relative protection from the drying effects of wind and sun. Exceptions to this are found in the floating leaves of aquatics and in certain plants that are adapted to arid conditions.

## Environmental effect on photosynthesis

Other things being equal, the rate of photosynthesis, and consequently the yield, increases with an increase

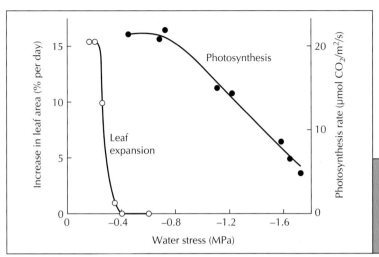

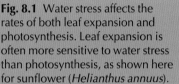

**Fig. 8.1** Water stress affects the rates of both leaf expansion and photosynthesis. Leaf expansion is often more sensitive to water stress than photosynthesis, as shown here for sunflower (*Helianthus annuus*).

in the amount of light being received by the leaves. There are, however, only a few ways in which the amount of light can be influenced by the gardener. These include the choice of shady or sunny locations, shading the plant to reduce water loss as discussed below and giving extra light in the greenhouse during winter when the natural light is low (see Chapter 14). Increasing the ambient $CO_2$ concentration also increases the rate of photosynthesis, but again this can only be achieved under glass (see Chapter 14).

High temperatures increase the rate of respiration (see Chapter 2), so may reduce the net uptake of $CO_2$, although the rate of photosynthesis is itself increased by increasing the temperature when light is not limiting. However temperature control, other than by shading, which itself reduces the rate of photosynthesis, is only possible in the greenhouse environment. Finally, shortage of water may also drastically reduce the rate of photosynthesis (Fig. 8.1).

## RESPONSES OF PLANTS TO SHADE

In response to their need for light in photosynthesis, green plants have evolved a range of mechanisms to enable them to adjust to the prevailing light conditions. For example, leaves are often larger and thinner in the shade, a response designed to maximise

the collection of light. Internal changes, such as an increase in the amount of light-harvesting chlorophyll (see Chapter 2), also take place under low light conditions and increase the ability of the plant to absorb light. Plants in the shade can also divert their energy into stem growth, which enables them to compete for available light and outgrow their neighbours to reach better light conditions. Such plants, for example many arable weeds, which grow taller in response to shade, are said to be *shade avoiders*. Competitive elongation would be of little value to plants growing on the woodland floor, however, and these are termed *shade tolerators*. They do not grow taller in shade like shade avoiders, but have other responses which improve their competitiveness under shady conditions, such as larger and thinner leaves. Photosynthesis in the shade avoiders is usually relatively inefficient under poor light conditions, but they have the ability to redirect their growth into stems at the first sign of shading. Some plants, such as the runner bean (*Phaseolus vulgaris*) maximise their exposure to light by turning their leaves towards the sun's direction during the day, a process known as heliotropism.

How are plants able to detect whether or not they are in a shady habitat? Under natural conditions, the most important shade is from leaf canopies, and the shade-avoidance response depends on the fact that the light quality is profoundly altered when it passes through leaf tissues. In order to understand this, it is first necessary to look at the spectrum of

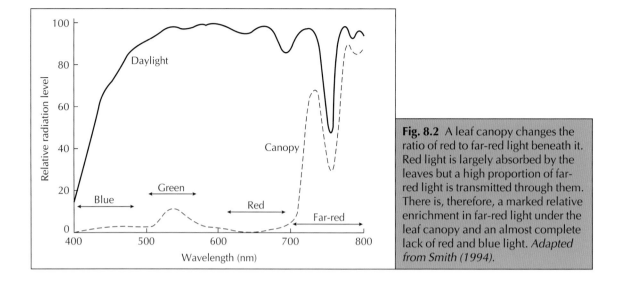

**Fig. 8.2** A leaf canopy changes the ratio of red to far-red light beneath it. Red light is largely absorbed by the leaves but a high proportion of far-red light is transmitted through them. There is, therefore, a marked relative enrichment in far-red light under the leaf canopy and an almost complete lack of red and blue light. *Adapted from Smith (1994).*

sunlight and how this changes as light passes through the leaves.

Sunlight contains both red (R) and the longer far-red (FR) wavelengths in approximately equal amounts and the effect of leaves, mainly due to the absorption of light by chlorophyll, is to remove the red wavelengths (Fig. 8.2). This means that under the leaf canopy the relative amounts of red and far-red light are changed. Shade avoiders respond to canopy shade because they can detect changes in the ratio of R to FR light through the pigment *phytochrome*. The major characteristics of this pigment and its multiple forms are discussed in more detail in Chapter 9 in relation to seed germination, the most important from the point of canopy shade being the red/far-red reversible reaction. Red light produces the biologically active form of phytochrome, Pfr, whereas far-red light converts Pfr back to the biologically inactive form Pr. Thus both forms are present in light that contains a mixture of red and far-red wavelengths, as does sunlight. Furthermore, the amount of Pfr that is present depends on the relative amounts of red and far-red wavelengths in the light (the R/FR ratio). For example, in sunlight about 55% of the phytochrome is present as Pfr. With increasing amounts of canopy shade, the proportion of red light decreases and so the amount of Pfr also decreases. Since Pfr (the form

which actually controls the response) inhibits stem elongation, the result is that stem length increases as the amount of shade increases and the proportion of red light decreases (Fig. 8.3).

Early experiments demonstrated that the red/far-red reversible reaction of phytochrome controls many responses of plants to light, including germination (see Chapter 9) and the changes that occur when dark-grown plants are initially exposed to light, collectively known as *de-etiolation* (Fig. 8.4). It was observed that the responses brought about by a short exposure to red light could be prevented when this was immediately followed by a short exposure to far-red. However, the relevance to natural conditions was not immediately apparent, since plants are not normally exposed to brief sequences of R and FR light in this way, and some scientists even doubted whether such a pigment actually existed. Indeed, in some early papers it was suggested that phytochrome might be a 'pigment of the imagination', and it was also dubbed a 'red/far-red herring'. It was several years before phytochrome was finally purified and shown to be a protein attached to a light-absorbing, non-protein part. Today, largely due to work at Leicester University, it is understood that the red/far-red reversible reaction of the pigment is a sophisticated way of measuring the degree of

**Fig. 8.3** The effect of the ratio of red (R) to far-red (FR) light on elongation growth in the potato. The plant (right) growing in a *low* R/FR ratio (equivalent to deep shade) is markedly elongated. The plant (left) growing in a *high* R/FR ratio (light from fluorescent lamps containing little FR) is considerably shorter than the one (centre) growing in a *medium* R/FR ratio (equivalent to sunlight, in which the R/FR ratio is approximately 1.0). *Photograph courtesy of Brian Thomas, University of Warwick.*

**Fig. 8.4** The effect of darkness (*etiolation*) and light on seedling growth. Seedlings of mung beans (*Vigna radiata*) grown in darkness (left) are tall, with apical hooks, small leaves and no green or red pigmentation. Exposure to light (right) reduces elongation, unfolds the apical hook and increases leaf expansion; it also results in the formation of green (chlorophylls) and red (anthocyanins) pigments. Many of these responses depend on exposure to red light (see text). *Photograph courtesy of Daphne Vince-Prue.*

shading by a plant canopy. As the canopy becomes thicker and the amount of light decreases, the R/FR ratio of the light also decreases. Thus, the thicker the leaf canopy and the denser the shade, the longer the stems.

In addition to the canopy effect, the way in which leaves absorb light results in an increased sideways reflection of FR wavelengths towards neighbouring plants (Fig. 8.5). By sensing the FR light reflected from their neighbours, some species are able to detect the presence of a neighbour at a distance of 20–30 cm and thus obtain an early warning of the competition for light that they are likely to encounter in the future. This enables them to adjust their growth patterns before the shading actually occurs.

The shade-avoidance response is controlled by the form of phytochrome (see Chapter 9) known as Phy B, but another form, Phy A, appears to limit the amount of elongation that takes place. Thus genet-ically modified plants (see Chapter 5) without Phy A still respond to a low R/FR ratio by growing taller, but they actually elongate more than wild-type plants with Phy A.

Not all plants respond to the ratio of R/FR light in this way. The ones that do so are termed shade avoiders, since this response enables them to compete with their neighbours by out-growing them. Such a

**Fig. 8.5** The transmission, reflection and absorption of red and far-red light by leaves. Light from the sun contains both red (R) and far-red (FR) wavelengths. Red light is absorbed by the leaf but far-red light is transmitted through the leaf and is also reflected to neighbouring plants. *Image courtesy of Brian Thomas, University of Warwick.*

The basic features of water movement in the plant and its loss through transpiration are discussed in Chapter 2. Here, it is only necessary to state the central problem. Plants require a supply of carbon dioxide for photosynthesis. This means the provision of entry pores, the stomata, for the gas and it follows that water vapour is lost to the atmosphere through these open channels. The plant must therefore be able to maintain an adequate supply of carbon dioxide to the photosynthesising cells in the leaf, while preventing a net loss of water. In this context, it is important to note that the amount of water contained in the majority of land plants is small compared with the potential loss of water by transpiration on a dry, sunny day. As a rough rule of thumb, an area of continuous vegetation like a lawn or a fully planted border can loses about 25 litres of water per square metre every 7–10 days. If water is lost from the leaves at a greater rate than it can be supplied from the roots, *turgor pressure* (see Chapter 2) is lost, wilting occurs and photosynthesis is depressed.

response is undesirable in the garden, since rapid stem growth results in 'leggy' plants with weak stems. For shady conditions, therefore, shade tolerators, which originated in woodland habitats, are more desirable. As described above, shade-tolerant plants often have mechanisms that improve their ability to collect light under shaded conditions: energy is directed into leaf expansion to produce a larger light-collecting area, and more light-harvesting chlorophyll develops. This too is a response to light, but appears to be largely controlled by a pigment system that responds primarily to the amount of blue light received by the plant, which is reduced by canopy shading (see Fig. 8.2), rather than to the R/FR ratio.

Shade from neighbouring buildings also reduces the amount of light available for photosynthesis but, unlike shade from trees, does not alter the R/FR ratio. Nevertheless, many plants show typical shade responses, probably detecting the shade conditions through their blue-light-sensitive pigment system.

## How water is lost from the leaf

When the guard cells around the stomata lose water and become flaccid, their physical structure means that the outer surfaces move together and close the pores. Obviously this reduces the loss of water vapour from the leaf, especially where the cuticle is thick and impervious to water vapour, and restores the water balance as long as there is available water present in the soil (see Chapter 6). The guard cells then absorb water again and re-open. The way in which the opening and closing of stomata is controlled has been found to be quite complex and involves both physical and biochemical factors. This is not surprising, since the regulation of water loss by stomatal movements is a major factor in determining the survival of the plant. If the stomata remain open, the plant can wilt and may die from dehydration. If the stomata remain closed, there is no photosynthesis and the plant will ultimately die from starvation.

# Opening and closing of the stomata

Stomata open because the guard cells absorb water. Many studies have shown that they absorb water from the surrounding cells by *osmosis*, because there is an increase in the amount of dissolved substances present inside the guard cells. It is now clear that the main increase is from potassium ions that move into the guard cells from the surrounding cells. Fundamentally, therefore, the opening and closure of stomata depends on the movement of potassium ions into and out of the guard cells.

The stomata of most plants open in the light, allowing the entry of carbon dioxide for photosynthesis, and close in darkness to reduce water loss. They are mainly sensitive to blue light, which results in a larger and faster response than when they are exposed to red light. Exposure to light causes potassium ions to build up inside the guard cells. A reduction in carbon dioxide in the air also increases the content of potassium in the guard cells. Thus stomata open in the light in part as a direct response to light and in part because the concentration of carbon dioxide within the leaf is reduced by photosynthesis. Darkness and an increase in the amount of carbon dioxide have the opposite effect.

# Water stress

An insufficient supply of water to the roots, or a too rapid loss from the leaves, leads to water stress, a common problem in horticulture. When leaves are subjected to water stress, the hormone *abscisic acid* (*ABA*) begins to build up in their tissues. The artificial application of ABA results in closing of the stomata. On this basis, it was originally thought that ABA is produced in the leaves when they experience a water deficit, resulting in closing of the stomata and a reduction in transpiration. More recently, however, it has been found that ABA can control water loss before any measurable water stress occurs in the leaves. It appears that, when they begin to encounter dryness in the soil, the root tips generate a signal which is transported to the leaves and results in partial closing of the stomata and a reduction in leaf expansion, both of which improve the water economy of the plant. The general consensus is that this signal from the roots is ABA and it is probably true to say that the regulatory role of ABA is essential for the survival of land plants in all but the least stressful environments. This is supported by the discovery of so-called wilty mutants, which are deficient in ABA and are always partly wilted, unless this hormone is applied artificially.

There is now good evidence that ABA from the roots plays a central role in the control of the closing and opening of the stomata under natural conditions. The ABA signal is produced as the soil begins to dry out, thus allowing the plant to take corrective action to improve the water economy of the leaves before wilting occurs. This is a good example of the ability of plants to 'detect what is going to happen next', so that they can react before the major stress arrives. (In a similar way, the perception of a day-length signal allows plants to prepare for the subsequent stress of low temperature; see Chapter 13.) Once water stress occurs and plants wilt, recovery may take some time even when the water balance is restored. This is an important fact in relation to watering and irrigation practice. Many cell functions are depressed when plants are subject to water stress, leading to an often severe reduction in growth and yield (Fig. 8.1).

Does the ABA signal from drying soils have any practical application? It has been found that, if only half of a root system is allowed to dry out while the remaining half is plentifully supplied with water, the stomata still close, presumably in response to the ABA signal from the roots located in the drying soil. This observation has already found some practical use on irrigated vines, where water is supplied to alternate sides in turn. Irrigation is rotated on a two to three week cycle so that first one side and then the other is drying out. This reduces the amount of water used and also results in more and better-quality fruit by reducing the excessive leaf growth found in a normally irrigated crop.

# Drought

Natural habitats often suffer from a shortage of water. This may occur in the form of a seasonal drought as in Mediterranean regions, where the annual rainfall (between 350 and 900 mm) occurs almost exclusively in winter so that the native plants of these regions experience a summer drought. Water is more permanently scarce in arid and semi-arid regions of low rainfall. Many strategies, both biochemical and physical, have evolved to allow plants to grow under

**Table 8.1** Some strategies for the adaptation of plants to water stress.

| Strategy | Examples/features |
| --- | --- |
| Drought escapers | Seeds or dormant structures |
| Water spenders | Deep roots |
| Water collectors | Succulents with CAM photosynthesis |
| Water savers | Sunken stomata, small leaves, hairs, shedding leaves, reflective leaves, grey/blue leaves |
| Tolerators of dehydration | Seeds, lichens, creosote bush |

CAM, crassulacean acid metabolism. Modified from Table 26.5 in Salisbury and Ross (1992).

such conditions, and an understanding of these enables the gardener to choose plants that are more tolerant of, or are actively resistant to, periods of water shortage. Some of these strategies are summarised in Table 8.1.

## Avoidance

Some plants survive by avoiding or escaping dry conditions. A good example of drought avoidance in nature is that of desert ephemerals and annuals, for example many South African members of the Asteraceae, such as Namaqualand daisies (*Dimorphotheca sinuata*), which escape the dry conditions as seeds. These only germinate when there has been sufficient rainfall for them to grow, flower and set seed, and the resultant synchronous flowering of deserts after rain is one of the most beautiful sights in the natural world (Fig. 8.6). In such plants, germination often depends on the leaching of inhibitors from the seed (see Chapter 9).

Other plants may become dormant during periods of drought, frequently surviving by producing underground storage organs (see Chapter 1) which lose only small amounts of water into the surrounding soil, an example being crown anemone (*Anemone coronaria*). Many grasses (e.g. *Poa bulbosa*) also survive drought by dormancy and leaf die-back.

Succulent plants that store water in their leaves or stems may also be considered as drought avoiders, since during periods of drought they survive on their stored water, which is replenished when conditions allow. Succulents have a mainly shallow root system which dies during periods of water stress, leaving only the supporting roots surviving. When the rains come, new absorbing roots form and refill the cells with water.

Some plants of arid regions avoid drought by having deep roots that reach the permanent water table. Examples of these include alfalfa (*Medicago sativa*) and mesquite (*Prosopis glandulosa*), where the roots can extend to 7–10 metres. Mesquite in particular is very deep-rooted and has become a pest in the semi-arid regions of the south-western USA, where its profligate water consumption has prevented the re-establishment of agronomically valuable grasses.

**Fig. 8.6** Drought avoidance. Annual composites of the South African Cape survive the summer drought conditions in the form of seeds, which germinate only after a sufficient winter rainfall. This results in synchronous flowering in the following spring, as shown here. *Photograph courtesy of Daphne Vince-Prue.*

The success of this strategy depends, of course, on there being enough water at some stage for roots to grow down to the deep water supply.

### Tolerance

The second major strategy that plants have evolved for dry conditions is to be able to tolerate them. True tolerators are those plants that can withstand a considerable degree of water stress in their cells. For example the creosote bush (*Larrea divaricata*), a desert shrub, can lose 70% of its fresh weight before the leaves die, whereas most leaves will withstand only 25–50% loss. Several mosses and some ferns also use this strategy, and the resurrection plant (*Selaginella lepidophylla*) can dry out almost completely and yet become metabolically active again on rehydration.

Other plants can tolerate dry conditions by reducing their water loss into the atmosphere. These plants can be considered as 'water savers' (Table 8.1) and they conserve water in a number of different ways, using both biochemical and physical strategies.

## PHYSICAL STRATEGIES THAT CONSERVE WATER

Water is lost from the leaves because of a gradient in water-vapour content between the internal spaces of the leaf and the surrounding air (see Chapter 2). Anything that affects this gradient will therefore alter the rate of loss of water vapour from the leaves. The air in the sub-stomatal cavity is normally saturated with water vapour (100% relative humidity) and so water vapour diffuses into the surrounding air when this has a lower content of water vapour. Water vapour is lost even into a saturated atmosphere when the leaf is at a higher temperature than the surrounding air. When water vapour diffuses into still air, a *boundary layer* with a higher content of water vapour forms between the leaf surface and the surrounding air. These facts explain many of the physical strategies that are designed to reduce water loss.

The presence of hairs helps to maintain the boundary layer and so increases the distance between the leaf cells and the external atmosphere. This increase in the length of the *diffusion pathway* means that it takes longer for water vapour to reach the external

atmosphere, and this reduces the rate at which water-vapour molecules are lost. Hairs also reduce the drying effect of wind, which tends to blow away the protective boundary layer and increase the rate of water loss. The holm oak (*Quercus ilex*), a characteristic species of the Mediterranean flora, has small leaves (see below) with dense hairs on the under-surface where most of the stomata are. Other drought-tolerant plants such as *Lavandula lanata* and *Phlomis fruticosa* have soft leaves, often with hairs on both surfaces.

The stomata of many drought-resistant plants are sunk below the main surface of the leaf. Examples include oleander (*Nerium oleander*) and many pines (e.g. *Pinus strobus*). This increases the length of the diffusion pathway from inside the leaf to the outside air and reduces the rate of water loss. The diffusion pathway is also lengthened in leaves that roll under in response to drought, for example in *Rhododendron* spp., because most of the stomata are on the lower surface of the leaf. In marram grass (*Ammophila arenaria*), which grows on sand dunes, not only are the stomata sunk in grooves, but the leaves are also tightly rolled.

Perhaps the most obvious way of decreasing water loss during periods of drought is to reduce the leaf area by shedding leaves. Drought-deciduous plants are found both in arid zones and in regions with seasonal dry periods such as those with hot dry summers. For example *Lotus scoparius*, a shrub native to California, sheds its leaves in summer and remains dormant until the arrival of the winter rains. Similarly *Fouqueria splendens*, a shrub from Mexico, remains leafless for most of the year but begins to expand new leaves immediately following rain. These are extreme examples and many plants shed some of their leaves, especially the older ones, in response to water stress. This also occurs in species that are not drought tolerant and often results in unsightly plants with bare basal stems and only a few leaves remaining at the tip of the shoot, a phenomenon well known to gardeners. It is thought that the shedding of older leaves in response to water stress is due to an increase in the production of the gaseous hormone ethylene by the plant, resulting in accelerated senescence.

Water loss is also decreased by any factor that reduces the heat load on the leaf. The obvious one is shade and shading plants in hot weather or when water uptake is reduced, as in cuttings or newly potted plants, is a common garden practice. Although shading may be disadvantageous because of the importance of light for photosynthesis, water stress and stomatal closure themselves severely depress the rate of photosynthesis (Fig. 8.1). Leaf surfaces that reflect at least some of the incoming radiation also act to reduce the heat load on the leaf. Hence many drought-tolerant plants have shiny leaves (*Quercus ilex, Ceanothus* spp.) or grey foliage (*Helictotrichon sempervirens*) that may also be covered with reflective hairs (*Lavandula lanata*). Leaf orientation may also reduce the heat load and some plants, such as soybean (*Glycine max*), are able to change the leaf angle away from the sun when the plants are water stressed.

Wind also affects the leaf temperature because the transfer of heat by convection from the leaf surface to the air is most rapid when the boundary layer is thin. In general the boundary layer is thinnest for small leaves and high wind velocities, so that smaller leaves have temperatures closer to the air temperature than do larger leaves, especially if there is a wind. That small leaves are a factor in drought tolerance is clear from the characteristic small leaves of many of the shrubs of the Mediterranean maquis (Fig. 8.7; e.g. Spanish broom *Spartium junceum*) and the fynbos of the South African Cape (e.g. *Erica* spp.). Where reduction of leaf area is carried to extremes by modification of leaves into spines, as in some desert cacti, the stems are often green and able to carry out photosynthesis.

Although wind helps to prevent a rise in leaf temperature and thus can decrease the rate of water loss, the overall effect of high wind is usually to increase the rate of loss, especially when the heat load is low. This is obviously a major factor in exposed sites. The drying effects of wind are particularly evident where the cuticle is thin, and does not itself offer a significant barrier to water loss. Plants with thick cuticles are therefore better able to withstand the effects of wind on exposed sites. These include many broad-leaved evergreens such as Portuguese laurel (*Prunus lusitanica*). Shelter from drying winds is obviously of the greatest importance for expanding foliage that has not yet developed a protective cuticle.

The usefulness of these different physical strategies for conserving water depends to a large extent on the conditions in which the plant is growing. When water is short, it is advantageous to the plant to conserve water and maintain a relatively low leaf temperature. Many drought-tolerant plants have small leaves and thin boundary layers, resulting in efficient heat transfer and leaf temperatures close to the air temperature. The rate of transpiration is further reduced by features that increase the length of the diffusion pathway for water vapour, such as sunken stomata and hairs, and by factors such as grey leaves that reflect much of the sun's radiation. When water is plentiful, however, high rates of transpiration cool the leaf by evaporation. Some drought-tolerant plants utilise this strategy by having long roots that can

**Fig. 8.7** Drought tolerance. Small leaves reduce water loss and are characteristic of the shrubs of the Mediterranean maquis shown here. Reflective or grey leaves reduce the heat load on the leaf and so reduce water loss. *Photograph courtesy of Daphne Vince-Prue.*

reach the water table and large leaves to maximise the rate of water loss. This strategy is only successful in air that has a relative humidity low enough for transpiration to continue, even when the leaf is several degrees below the ambient temperature.

## BIOCHEMICAL STRATEGIES THAT CONSERVE WATER

Two important biochemical strategies that conserve water are modifications to the normal process of photosynthesis. These are C-4 photosynthesis and crassulacean acid metabolism (CAM). Both are modifications that are presumed to have evolved in response to the stress imposed by low levels of soil water combined with high temperatures. Under such conditions it is necessary to close the stomata because the plant will die of dehydration if they remain open. Although partial closure of stomata is relatively effective in reducing water loss without significantly reducing carbon dioxide uptake, further closure begins to limit its availability and reduce its concentration in the leaf cells. This creates problems because ribulose bisphosphate carboxylase, the key enzyme in normal photosynthesis (see Chapter 2), does not function at low concentrations of carbon dioxide.

### C-4 plants

In contrast to normal *C-3 plants*, where the initial product of photosynthesis is a molecule with three carbon atoms (see Chapter 2), *C-4 plants* produce organic acids (malic and/or aspartic) containing four carbon atoms, which is why this type of photosynthesis is called C-4.

In C-4 plants, the organic acids formed in the palisade cells of the leaf are subsequently transferred to specially modified bundle sheath cells (see Chapter 2). These contain chloroplasts, unlike 'normal' bundle sheath cells. Here the four-carbon organic acids are broken down to release carbon dioxide in a concentrated form, and normal photosynthesis begins. Plants possessing such modified bundle sheaths are said to exhibit Krantz anatomy, after the person who first described them.

The C-4 pathway is an adaptation to environments with high light intensities and periods of water stress, and is commonly found in plants from desert and subtropical regions. The incorporation of carbon dioxide into organic acids occurs at high rates when the light intensity and temperature are high, the stomata are open and carbon dioxide is available. When water stress occurs and stomata close, the release of carbon dioxide from the four-carbon acids enables the plants to maintain a high concentration of carbon dioxide inside the leaf cells and continue to carry out normal C-3 photosynthesis.

Many, but not all, C-4 plants exceed most C-3 plants in their productivity, especially under high light and temperature conditions (see Fig. 8.8). The C-4 pathway is not, however, automatically better than C-3 photosynthesis. Indeed, C-4 photosynthesis is less efficient and uses more light energy to fix carbon dioxide. It is, however, particularly advantageous in high-light, high-temperature environments, where water is limiting.

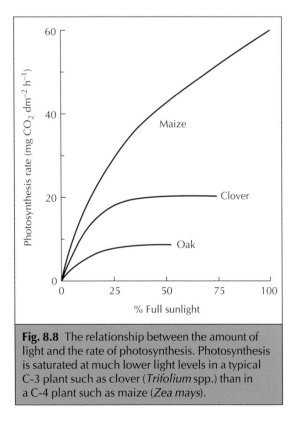

**Fig. 8.8** The relationship between the amount of light and the rate of photosynthesis. Photosynthesis is saturated at much lower light levels in a typical C-3 plant such as clover (*Trifolium* spp.) than in a C-4 plant such as maize (*Zea mays*).

C-4 metabolism is found in many tropical grasses, for example sugar cane (*Saccharum officinarum*) and maize (*Zea mays*), and in sedges (*Carex* spp.) as well as some dicotyledons. The North American prairie grasses that have recently become very popular in British landscapes are C-4 plants. These include, for example, little bluestem (*Schizachyrium scoparium*), Indian grass (*Sorghastrum nutans*) and *Sporobolus heterolepis*. They grow best in high summer and are drought tolerant. Several families, for example Chenopodiaceae and Amaranthaceae, contain both C-3 and C-4 species, suggesting that the C-4 photosynthesis pathway has arisen independently in different plant groups and is a recent evolutionary development.

## Crassulacean acid metabolism

Many drought-tolerant species have thick, succulent leaves or stems with a low surface-area-to-volume ratio. This, together with their thick cuticle, results in low rates of transpiration when the stomata are closed. Keeping the stomata closed during the daytime, when rates of water loss would be high if they were open, poses the problem of obtaining sufficient carbon dioxide for photosynthesis. Consequently many succulents have developed a biochemical strategy that allows them to open their stomata at night and still carry out photosynthesis. This mechanism was first investigated in members of the Crassulaceae and is, therefore, called *crassulacean acid metabolism (CAM)*. Its occurrence is, however, much more widespread than this and CAM has been found in at least 26 Angiosperm families and in some ferns. CAM plants normally inhabit areas where water is scarce or difficult to get at, such as arid and semi-arid regions, and salt marshes. Some epiphytic orchids that must obtain their water from the air also exhibit CAM.

CAM plants open their stomata at night in order to take up carbon dioxide, which is temporarily stored in the form of malic acid. This four-carbon organic acid accumulates in the vacuoles of the leaf cells and increases their osmotic concentration, allowing the plants to absorb and store water and thus leading to the observed succulence. During daylight, when the stomata are closed, the malic acid moves out of the vacuoles and is broken down to release carbon dioxide within the cells. Utilising light energy, the carbon

dioxide is then re-fixed into the usual products of photosynthesis, namely sugars and starch.

The capacity of succulents to store malic acid overnight is very limited, however, so their rate of growth is much slower than that of the C-4 plants, in which the four-carbon acid is not stored but regenerates its $CO_2$ immediately.

CAM is confined to certain plants and is determined by the plant's genes. However, many CAM plants, for example *Clusia rosea*, a tree that begins life as an epiphyte in the rain forest, can switch to normal photosynthesis under favourable conditions, when the stomata remain open longer during daylight hours. When CAM is occurring in these succulent plants there is little water loss during the day because of thick cuticles and closed stomata, but the absence of cooling by transpiration can lead to high cell temperatures which the plants must be able to tolerate.

## EXCESS WATER

Whereas lack of water may be the major environmental stress in many habitats, excess water can also pose problems. Hairy leaves become waterlogged in heavy rain and may rot, while most plants cannot tolerate waterlogged soils due to their lack of oxygen. Roots normally obtain their oxygen from gas-filled pores in the surrounding soil. When the soil is well drained and has a good structure (see Chapter 6) these pores permit the diffusion of gaseous oxygen to a depth of several metres, causing the oxygen concentration at that depth to be similar to that in humid air. In flooded soils, however, water fills the pores and oxygen must diffuse through liquid. This occurs much more slowly than in air. If the temperature is low and the plants are dormant, oxygen depletion in the soil is slow and may be relatively harmless. At temperatures above about 20°C, however, oxygen consumption by roots and soil microorganisms may totally deplete oxygen from the bulk of the soil water in as little as 24 hours. Growth and survival are then severely depressed in most species. In garden peas (*Pisum sativum*), for example, yield may be halved if the soil is flooded for 24 hours.

Specialised plants, such as marsh and bog plants and paddy rice, can survive for long periods in flooded soils. As with drought resistance, several different strategies have evolved to make this possible, although most involve some means of enabling oxygen to reach the roots rather than the ability to tolerate low oxygen levels within the cell. In most cases survival is aided by the development of *aerenchyma*, a tissue consisting of long files of interconnected gas-filled spaces. This allows for the diffusion of gaseous oxygen from the air, through the plant and into the roots. The roots of wetland plants are also often heavily suberised (i.e. covered with an impermeable corky layer), thus preventing the loss of oxygen into the soil.

Wetland species that grow at the margins of ponds and rivers, or in marshy conditions, are usually dependent on the presence of aerenchyma for their survival. The dormant rhizomes of salt-marsh bullrush (*Scirpus maritimus*) and cat-tail (*Typha* spp.) overwinter in anaerobic mud at the edges of lakes, but when the leaves begin to expand survival of the plant is dependent on the diffusion of oxygen through the gas-filled spaces of the aerenchyma tissue. Similarly, although the seeds of paddy rice can germinate at low oxygen concentrations, the roots are not tolerant of anaerobic conditions and their survival depends on the emergence of the *coleoptile* above the surface of the water. This acts as a kind of snorkel that allows oxygen to diffuse from the air, via the aerenchyma, to the roots. In other plants, such as the swamp cypress (*Taxodium distichum*) and mangroves, portions of the roots (known as knees in *T. distichum*) remain above the water or soil surface and allow oxygen to diffuse down to the submerged parts.

In flood-tolerant plants such as willow (*Salix* spp.) and sunflower (*Helianthus* spp.), little aerenchyma is present in plants growing in well-aerated soils. Aerynchyma develops in response to low oxygen concentrations in the soil. This is an interesting process that involves the gaseous hormone, ethylene (see Chapter 2). A low concentration of oxygen induces the production of ACC (a precursor of ethylene) in the roots, but the absence of oxygen blocks the final step in the pathway, the conversion of ACC to ethylene. However, the ACC is transported from the roots to the aerial parts of the plant, where the increased oxygen level results in the production of ethylene and the formation of aerenchyma.

The production of ethylene by the shoots of flooded plants has effects other than to induce the development of aerenchyma. In tomato (*Lycopersicon esculentum*), the formation of ethylene results in rolling under of the leaves, called *epinasty*, within 6–12 hours after flooding. In tomato, maize (*Zea mays*) and sunflower (*Helianthus annuus*) flooding leads to swelling at the base of the stem, probably also due to ethylene production. It is thought that such *hypertrophy* may increase the porosity in this region of the stem and so enhance aeration. However, although ethylene is a key hormone in the ability of some species to resist flooding, prolonged exposure to increased levels of ethylene leads to yellowing, senescence and ultimately death in sensitive species.

Waterlogging causes the formation of adventitious roots in some resistant species. These can replace old roots that have died off in response to flooding and so increase the possibility of survival. For example in *Rumex palustris*, a wetland species, waterlogging resulted in the development of 49 adventitious roots within 4 days, whereas under similar conditions *Rumex thyrsiflorus*, a dryland species, produced only eight.

Death of plants resulting from flooding of the soil is mainly caused by the lack of oxygen for normal aerobic respiration by the roots. However, other factors are also important. Aerobic soil microorganisms are replaced by anaerobic ones that produce toxic compounds such as hydrogen sulphide, a respiratory poison, and butyric acid. These are also partly the cause of the characteristic unpleasant odour of waterlogged soils. Moreover, increased production of ethylene gas in the aerial parts of the plants, which occurs as a result of low oxygen concentration around the roots, results in senescence and yellowing of the leaves, leading to their early death. Excess water also slows down the rate at which soils warm up; this is particularly important in the early spring when roots begin to grow again.

## SALINITY

In many arid zones of the world the accumulation of salt (sodium chloride) in the soil can be an important stress factor for the plants growing in them. Soil salinity may also be a problem in temperate regions, and large areas have gone out of production because salt from irrigation water has accumulated in the soil. Other areas where salinity is a problem include brackish marshes, usually close to the sea. Plants face two problems in such situations. The high salt content in the soil water increases its osmotic pressure, leading to decreased water uptake, while sodium ions are toxic to many plants, although they may be required by or beneficial to others.

Some plants, known as halophytes, are able to tolerate high levels of salt in the soil and different physiological mechanisms have evolved to increase their salt tolerance. Others are able to excrete salt from specialised glands in the leaf (e.g. *Acanthus spinosus*); this often leads to salty encrustations on the leaf surfaces. Other plants, known as salt accumulators, have evolved the capacity to store large quantities of sodium ions in their vacuoles without toxic effect. Many have succulent leaves and are able to store large volumes of water in their vacuoles and so keep the salt concentration low (e.g. *Carpobrotus edulis*). Salt regulators, in contrast, are able to limit the uptake of salt into the roots.

Some crop plants, such as beet (*Beta vulgaris*), tomatoes (*Lycopersicon esculentum*) and rye (*Secale cereale*), are much more tolerant than others, such as onions (*Allium cepa*) and peas (*Pisum sativum*). Also, new salt-tolerant cultivars are being developed by plant breeders.

## AQUATIC PLANTS

Well-managed ponds, which have a thriving population of oxygenating submerged aquatics, have more available oxygen than flooded soils, but because gases diffuse more slowly through water than through air, submerged leaves may experience difficulty in obtaining sufficient carbon dioxide.

There are several adaptations that increase the ability of plants to live in an aquatic environment. Some of these are immediately obvious. They include, for example, long petioles that allow leaves to be in an aerial environment while their roots remain in the soil at the bottom of the pond, and floating leaves such as those of water-lilies (*Nymphaea* spp.), that have most of their stomata on the upper surface, unlike land plants, which have most of their stomata on the lower surface in order to conserve water. Submerged leaves are often finely divided, with a high surface-area-

to-volume ratio, allowing more rapid diffusion of carbon dioxide into the leaf cells. Also, they may lack both a cuticle and stomata (e.g. *Elodea canadensis*). Several aquatic plants, such as water crowfoot (*Ranunculus aquatilis*), have undivided aerial leaves and finely divided submerged ones. The presence of aerenchyma (see above), which allows gases to diffuse rapidly through the plant, is also of major importance in aquatic species.

## WATERING PRACTICE

The golden rule for gardeners is to 'water well, not often' (see also Chapters 6 and 7). Enough water should be given to wet the soil down to 'root depth'. Roots are only able to grow in moist soil, and will therefore form mainly near the surface when only this layer is watered; since the surface layers of the soil dry out most rapidly, plants will suffer if roots are only present here.

Applying water beneath a canopy shade reduces drying out of the surface layers, while watering close to the stem bases of herbaceous plants allows water to reach the roots more rapidly. Weeds should be limited as far as possible, as they take up and transpire precious soil water.

The amount of water needed to re-wet the soil to its maximum storage capacity will vary with the soil texture (see Chapter 7), while the frequency of watering needed is highly dependent on the weather conditions.

## CHOOSING PLANTS FOR PARTICULAR CONDITIONS

Water, wind and light are major factors in determining survival of plants in a particular habitat and plants have evolved many strategies that enable them to tolerate, or avoid unfavourable conditions of water stress, excess water, wind and/or shade. Some of these are readily observable. For example, grey reflective leaves, hairy leaves, succulence, thick cuticles or small leaves are usually indicative of a greater resistance to water stress, while low-growing plants with small leaves are likely to be more tolerant of wind. Other strategies, such as an increase in the amount of light-harvesting chlorophyll during acclimation to shade, are not immediately apparent. Although the plant's natural habitat gives some indication of the conditions in which it will thrive, the choice of plants that are most suitable for any particular situation often has to depend on knowledge from experiments and trials. Many gardening books give lists of such plants and some examples are given in Table 8.2.

## CLIMATE CHANGE

As a result of climate change (see Chapter 18) the summer soil water in some areas of the UK may be reduced by up to 40% over the next few decades (see Chapter 18). Hotter drier summers and milder but wetter winters are predicted for many places, with an increased frequency of droughts and severe storms leading to temporary flooding. Consequently, the need to grow drought-resistant plants, such as those

**Table 8.2** Some plants for particular habitats.

**Dry and sunny**

| | | |
|---|---|---|
| Agave (*Agave* spp.) | Broom (*Genista* spp.) | Marjoram (*Origanum* spp.) |
| Wormwood (*Artemesia* spp.) | Halimium (*Halimium* spp.) | Shrubby potentilla (*Potentilla* spp.) |
| Sun rose (*Cistus* spp.) | Hyssop (*Hyssopus* spp.) | Rosemary (*Rosmarinus* spp.) |
| Broom (*Cytisus* spp.) | Shrubby veronica (*Hebe pinguifolia*) | Stonecrop (*Sedum* spp.) |
| Eryngium (*Eryngium* spp.) | Rhizomatous iris (*Iris* spp.) | Spanish broom (*Spartium junceum*) |
| Wallflower (*Erysimum* 'Bowles' Mauve') | Lavender (*Lavandula* spp.) | Thyme (*Thymus* spp.) |
| Treasure flower (*Gazania* spp.) | Catmint (*Nepeta* spp.) | Yucca (*Yucca* spp.) |

**Dry and shady**

| | | |
|---|---|---|
| Spotted laurel (*Aucuba japonica*) | Ivy (*Hedera helix*) | Many spring-flowering bulbs |
| Box (*Buxus* spp.) | Dead nettle (*Lamium* spp.) | |
| Cyclamen (*Cyclamen coum*) | Mahonia (*Mahonia aquifolium*) | |
| Mezereon (*Daphne mezereum*) | Portuguese laurel (*Prunus lusitanica*) | |
| Wood spurge (*Euphorbia amygdaloides* var. *robbiae*) | Periwinkle (*Vinca* spp.) | |
| Herb Robert (*Geranium robertianum*) | Sweet box (*Sarcococca* spp.) | |

**Wet and shady**

| | | |
|---|---|---|
| Chokeberry (*Aronia arbutifolia*) | Cardinal flower (*Lobelia cardinalis*) | Ferns, including Hart's tongue (*Asplenium scolopendrium*), lady fern (*Athyrium filix-femina*), buckler fern (*Dryopteris affinis*), shuttlecock fern (*Matteuccia struthiopteris*) |
| Astilbe (*Astilbe* spp.) | Bog myrtle (*Myrica gale*) | |
| Dogwood (*Cornus alba*) | Willow (*Salix daphnoides*) | |
| Aconite (*Eranthis hyemalis*) | Spiraea (*Spiraea* × *vanhouttei*) | |
| Dog's tooth violet (*Erythronium* spp.) | Wood lily (*Trillium* spp.) | |
| Hepatica (*Hepatica* spp.) | Blueberry (*Vaccinium corymbosum*) | |
| Summer snowflake (*Leucojum aestivum*) | Guelder rose (*Viburnum opulus*) | |

described here, is likely to become more important in the future. In areas prone to flooding the choice of plants may be limited to species such as *Myrica gale*, *Cornus* spp. or *Vaccinium* spp., that can tolerate waterlogging, at least for short periods.

Water conservation is likely to become more important in the future (see Chapter 18) because of the probability of more severe summer droughts and, in many places, greater evaporation due to higher temperatures. The choice of suitable plants is clearly also essential. Many come from regions with a Mediterranean-type climate and are already adapted to growing in a climate of hot, dry summers and warm wet winters. However, Mediterranean winters are relatively frost-free and many of the species are not sufficiently frost-tolerant to be grown in the UK, except in the milder parts of the country. This may change as the climate warms. Commercial nurseries are also increasing the import of new drought-tolerant plants (but see Chapter 18 for information about invasive non-native species).

Lawns pose particular problems because of the need for frequent watering (see Chapter 18) and alternatives to lawns, such as gravel gardens planted with drought-tolerant plants, are being more widely used. Other possibilities include the use of so-called prairie planting, which uses a range of drought-tolerant plants, and wild-flower meadows, with infrequent mowing (see Chapter 19). Many gardening books give suggestions for developing these. Potential problems with the supply of water, especially in the south of England, has led scientists to study the possibility of improving drought-resistance by target breeding. Since lawns require especially large amounts of water, much of the research has been directed towards increasing the drought tolerance of rye-grass (*Lolium* spp.) and fescue (*Festuca* spp.) by incorporation of a so-called green gene.

The possibilities for conserving water by changing watering practice are also being investigated experimentally. For example, it has been found that applying water only during critical growth periods did not affect the size and flowering performance of a crop of poinsettia (*Euphorbia pulcherrima*) growing under glass. Using this treatment, known as regulated water-deficit, up to 50% water saving could be achieved without compromising yield. Changes to the watering of vines, leading to significant savings of water, have already been discussed. Both approaches are being investigated for crops.

A major cause of climate change is the accumulation of carbon dioxide in the atmosphere (see Chapter 18). Since $CO_2$ is a major substrate for photosynthesis, an increase in its concentration would be expected to increase yield. Increases in the rate of photosynthesis in crops grown under glass with artificially increased $CO_2$ concentrations have been observed in many investigations. Overall yields, however, have been found to vary according to the crop studied. Increases in yield with rising concentrations of atmospheric $CO_2$ cannot, therefore, be predicted for all crops.

## CONCLUSION

Different plants have evolved a diversity of structures and metabolic processes to enable them to succeed with both normal and abnormal levels of light and water. Understanding the scientific basis of such structures and processes can help the gardener to plan with confidence planting schemes that are appropriate to prevailing environmental conditions and those predicted to result from climate change.

## FURTHER READING

Barnes, P. (1999) Ingenious strategies. *The Garden* **124**, 374–9.

Bidwell, R.G.S. (1979) *Plant Physiology*, 2nd edn, pp. 173–84. Macmillan Publishing Co., London.

Bisgrove, R. (2002) Weathering climate change. *The Horticulturist* **11**, 2–5.

Clevely, A. (2003) Making the most of moisture. *The Garden* **128**, 862–9.

Colborn, N. (2007) Perfectly adapted. *The Garden* **132**, 30–7.

Davies, W.J., Tardieu, F. & Trejo, C.L. (1994) How do chemical signals work in plants that grow in drying soils? *Plant Physiology* **104**, 309–14.

Grant-Downton, R. (1998) Dry Lazarus. *The Garden* **123**, 657–9.

Mansfield, T.A. (1994) Some aspects of stomatal physiology relevant to plants cultured *in vitro*. In *Physiology, Growth and Development of Plants in Culture*, P.J.

Lumsden, J.R. Nicholas and E.J. Davies (eds), pp. 120–31. Kluwer Academic Publishers, Dordrecht.

Prasad, M.N.V. (ed.) (1997) *Plant Ecophysiology*. John Wiley & Sons, New York, NY.

Salisbury, F.B. & Ross, C.W. (1992) *Plant Physiology*, 4th edn. Wadsworth Publishing Company, Belmont, CA.

Smith, H. (1994) Sensing the light environment: the functions of the phytochrome family. In *Photomor-phogenesis in Plants*, 2nd edn, R.E. Kendrick & G.H.M. Kronenberg (eds), pp. 377–416. Kluwer Academic Publishers, Dordrecht.

Vergine, G. & Jefferson-Brown, M. (1997) *Tough Plants for Tough Places*. David & Charles, Newton Abbot.

Vince-Prue, D. (1992) Phytochrome: a remarkable light sensor in plants. *The Plantsman* **13**, 203–14.

# 9

# Raising Plants from Seed

## SUMMARY

This chapter considers the scientific principles underlying the raising of plants from seed. It goes on to discuss the physiological mechanisms and ecological significance of seed dormancy and the methods that gardeners may use to break it. With few exceptions, however, the practical requirements for the successful germination of particular species have not been included as these are available in many gardening books and encyclopedias.

## INTRODUCTION

Flowering plants and their relatives usually reproduce sexually, developing seeds that are dispersed by wind, water or biological agents such as insects and birds. In the strict botanical sense a seed is a fertile and ripened ovule that contains an embryonic plant, usually supplied with stored food and surrounded by a protective coat called a *testa* (for details of seed structure see Chapter 3). Seeds are produced by two groups of plants, the *Gymnosperms* (conifers and their relatives) and the *Angiosperms* (flowering plants). A major difference between them is that the ovules, which become seeds, of Gymnosperms are not enclosed in a protective structure called an *ovary*; indeed the term Gymnosperm means 'naked seed'. In contrast, the *ovules* of flowering plants are enclosed in an ovary which develops into the fruit.

Although raising plants from seed is a relatively easy and cheap method of propagation, a number of problems may arise. These include infertile or non-viable seeds, failure to breed true to the parent type and in some cases the need for special conditions for germination, such as light or particular temperature regimes.

## SEED PRODUCTION AND GERMINATION

In Angiosperms, each unfertilised ovule contains an egg cell and a central cell with two nuclei, called the *polar nuclei* (Fig. 9.1). Both cells are *haploid*: thus

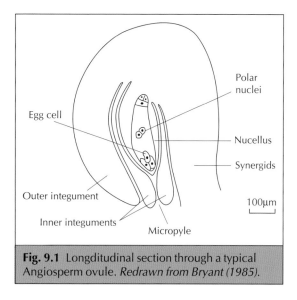

**Fig. 9.1** Longditudinal section through a typical Angiosperm ovule. *Redrawn from Bryant (1985).*

127

their nuclei contain only single sets of chromosomes, having been produced from normal *diploid* nuclei, which contain two sets of chromosomes, by a type of nuclear division called *meiosis* that occurs prior to sexual reproduction in plants (see Chapters 3 and 5). Pollination, followed by fertilisation, occurs when pollen grains from the anthers, the male parts of the flower, are deposited on the *stigma*, the female part. This process is usually carried out by intermediary agents such as insects, wind, water or even gravity.

Once on the surface of the stigma, each pollen grain germinates by producing a pollen tube containing three haploid nuclei: two sperm nuclei and a tube nucleus. The pollen tube grows down the *style*, the structure supporting the stigma, into the embryo sac. One of the haploid sperm nuclei then fuses with the haploid nucleus of the egg cell to form a diploid *zygote*, which divides to produce the young *embryo*, consisting of an embryonic root, the *radicle*, an embryonic shoot, the *plumule*, and a hypocotyl which connects the two. The second sperm nucleus fuses with the two haploid nuclei of the central cell, which then develops into a storage tissue, called the *endosperm*. Since this contains three sets of chromosomes, it is said to be *triploid*. In dicotyledons, the endosperm is short-lived and its storage function is taken over by two cotyledons. These are the first leaves of the embryo and are sometimes known as seed leaves as, for example, in beans (*Phaseolus vulgaris*). The period between pollination and the release of mature seeds varies from a few days to several months, depending on the species.

In Gymnosperms only one of the sperm nuclei in the pollen tube is functional, fusing with the egg cell nucleus to form the zygote. The other sperm nucleus degenerates. The tissue which surrounds the embryo remains haploid, but nevertheless develops to become the storage tissue (the *perisperm*) that nurtures the embryo during germination. In the pines (*Pinus* spp.), perhaps the most familiar of the Gymnosperms, the period between pollination and release of the mature seeds is usually 2 years, although in other genera this period is much shorter, as in Angiosperms.

Although the production of a seed is normally initiated by the sexual processes described above, a non-sexual process called *apomixis* occurs in some Angiosperms, the egg cell being formed from a diploid cell without meiosis. This means that the genomes of the offspring are identical with those of the parent plants, whereas with sexual reproduction each of the offspring has its own mix of the parental genomes. Apomixis is common in several grasses, such as annual meadow grass (*Poa annua*) and Kentucky blue grass (*Poa pratensis*), and also occurs in dandelion (*Taraxacum officinale*). In citrus fruits, several apomictic embryos may develop together with a normal, sexually produced embryo, leading to there being several embryos in the same seed, termed *poly-embryony*. However, poly-embryony may also arise by division of the sexually derived embryo, as in many orchids.

The embryos of all seed plants, whether produced sexually or apomictically, develop to a particular stage which is characteristic of the species. When the embryo reaches full size, the deposition of food reserves is completed, physiological maturity is attained and ripening or dehydration begins. By this stage, the embryo has become a seed, which is physiologically inert and will remain in a dormant state until the conditions are right for germination.

Seeds vary enormously in size and shape. Orchid seeds, for example, are microscopic, while the giant seed of the double coconut (*Lodoicea maldivica*) can weigh up to 10 kg. Some species, such as cocklebur (*Xanthium strumarium*), have seeds of more than one shape and are said to be polymorphic. Often the differently shaped seeds have different requirements for germination.

## Stored materials in seeds

Most seeds contain enough stored food to support germination and early growth until the seedling is able to photosynthesise. The main storage materials are carbohydrates (simple and complex sugars and starch), proteins and fats (Table 9.1). Seeds also contain mineral reserves, especially phosphates.

In the Gymnosperms and some Angiosperms, a tissue called the perisperm, derived from the remains of the *nucellus* (see Chapter 2 and Fig. 9.1), provides the main store of nutrients. It occurs in plants in which the endosperm does not completely replace the nucellus, such as members of the Caryophyllaceae and coffee (*Coffea* spp.). In Angiosperms it is more usual for the endosperm, as in onions and their relatives (*Allium* spp.), or the swollen cotyledons, as in legumes such as French bean (*Phaseolus vulgaris*), to be used as a food store. Occasionally other tissues

**Table 9.1** The major storage materials of certain seeds.

| Species | Percentage in air-dried seed | | | |
| --- | --- | --- | --- | --- |
| | Sugars | Proteins | Fats | Starch |
| *Zea mays* (maize) | 1–4 | 10 | 5 | 50–70 |
| *Pisum sativum* (pea) | 4–6 | 20 | 2 | 30–40 |
| *Helianthus annuus* (sunflower) | 2 | 25 | 45–50 | 0 |
| *Papaver somniferum* (opium poppy) | 25 | 20 | 34 | 0 |

may also act as food stores, for example the enlarged hypocotyl in Brazil nut (*Bertholletia excelsa*). Sometimes more than one tissue is involved, as in beet (*Beta vulgaris*), where there is both a well-developed endosperm and a perisperm.

The smallest seeds have few reserves, either because the seedlings are parasitic and live off their hosts, as in the case of broomrapes (*Orobanche* spp.) and witchweed (*Striga* spp.), or because, like epiphytic orchids, they form symbiotic associations with mycorrhizal fungi (see Chapter 2) that provide nutrients during germination. In nature most orchid seeds begin life in partnership with a symbiotic fungus. The fungal hyphae, which are present in the soil or in the bark of the host tree, invade the seed and enter the cells of the embryo. The orchid soon begins to digest this fungal tissue and obtain nutrients from it, thus using the fungus as an intermediary in obtaining nutrients from decaying material in the soil. Using micropropagation techniques (see Chapter 10) orchid seeds can be germinated and will develop successfully on a sterile, inorganic medium containing sugar.

## Sowing seeds

The most reliable way to sow seeds is under cover, where the conditions may be more carefully controlled. They should be sown thinly, and spread widely in the case of large seeds, in pots or small seed trays on seed compost that has been thoroughly saturated with water and then allowed to drain. It is important to ensure that the seed compost is never thereafter allowed to dry out. For most seeds, germination is better if light is allowed to reach them, so after sowing the seeds should be covered only very lightly, either with compost or a moisture-retentive

material such as vermiculite through which light can penetrate. After sowing, the containers should, ideally, be placed in a heated propagator and kept in a light place at an even, warm temperature (18–21°C). Alternatively the containers may be placed inside a loose plastic bag in order to retain moisture. Once germinated, the plastic cover needs to be removed to prevent fungal infection and the containers placed in as good light conditions as possible in order to prevent excessive elongation of the hypocotyl. Seedlings left on a window sill will grow towards the light (*phototropism*) and need to be turned frequently.

As soon as they are large enough to be handled easily, the seedlings should be pricked out into pots or modules. It is important to hold them by the cotyledons to avoid damage to the more fragile hypocotyl.

## Germination

Germination is the term used to describe the physiological and physical changes immediately prior to and including the first visible signs of growth of a seed. The dry seed is inert, but once in the soil, water becomes available and enters through openings in the seed coat, a physical process called *imbibition* which causes the tissues to swell. This ruptures the seed coat, allowing the entry of gases and the uptake of more water. As the cells become fully hydrated, enzymes are activated and begin to break down the storage tissue and transfer nutrients to the embryo. Most of the metabolic activity that ensues is directed towards embryo growth. Some, however, is utilised to repair those membranes of cells and their organelles damaged during the drying out of the maturing seed.

The large molecules of stored nutrients in the seed (Table 9.1) must be broken down by enzymes before they can be assimilated by the embryo to support the growing seedling until it is capable of photosynthesis. As germination proceeds, more water enters the seed and the storage nutrients are gradually assimilated by the embryo. This increases in size and undergoes a number of cellular changes, including an increase in several enzymes. Finally, the primary root begins to emerge, followed by emergence of the young shoot.

A key factor for the earliest stages of germination is thus the availability of water. Dry seeds imbibe water in three phases: an initial, extremely rapid uptake; a second, lag phase in which little or no water enters; and a third phase, less rapid than the first and associated with embryo growth and seedling emergence. Very dry seeds must be allowed to absorb water gradually, as rapid uptake into dry tissue can cause severe damage, and in some cases may even lead to death of the embryo. Because of this, the common practice of soaking seeds before sowing, allegedly to improve germination, is all too often detrimental and results in poorer germination than if the seeds were sown dry. The temperature at which water uptake occurs is also important; the lower the temperature the greater the damage.

Once the cells are sufficiently hydrated, a supply of oxygen for respiration is required. Most seeds will germinate fully in air, although in some cases, notably cocklebur (*Xanthium strumarium*) and certain cereals, a higher percentage of germination is achieved by increasing the oxygen level.

Temperature is also important, for the metabolic processes necessary for embryo growth only occur when the temperature is within a certain range. The optimum temperature is that at which the highest percentage of germination is attained in the shortest possible time. The range is defined by the minimum and maximum temperatures, which are the lowest and highest temperatures at which germination will occur. For example, too high a temperature is often the cause of poor emergence and establishment of summer-sown lettuces (*Lactuca sativa*) in the UK. Both the optimum temperature and the temperature range vary with species. For example, the maximum temperature for the germination of lettuce seeds is 25°C, whereas for leeks (*Allium porri*) it is 18°C. In the latter example, the maximum temperature may be increased slightly by changing the conditions during

seed maturation. Some seeds germinate best when exposed to alternating temperatures (see below).

## SEED VIGOUR

Vigour is a measure of seed quality and relates to the ability of the seed to germinate and establish under a wide range of environmental conditions. High-vigour seeds will germinate rapidly to produce a uniform stand, even in adverse conditions, while low-vigour seeds germinate erratically, producing a variable, patchy stand. Seed vigour affects various aspects of performance, including the rate and uniformity of germination, seedling emergence and growth, the ability of seedlings to emerge under unfavourable environmental conditions and the potential for storage.

Changes in growing techniques have made gardeners more concerned with the quality of seeds. Technologies such as the production of F1 seeds (see Chapter 5), precision drilling, seed pelleting and priming, the use of plant modules and once-over harvests all rely on high-vigour seeds capable of producing a uniform stand. Moreover, supermarkets now demand crops of specific sizes, achieved by establishing a particular spacing of the plants which, in turn, depends heavily on high-vigour seeds giving uniform germination over the entire crop.

As might be expected, seed vigour is influenced by the genetic constitution of the plant. For example, seeds of lettuce, onion (*Allium cepa*) and parsnip (*Pastinaca sativa*) typically have low vigour, whereas cereals and most pea seeds have high vigour. However, wrinkled and round-seeded peas differ in their vigour: round peas have starchy seeds and are usually much more vigorous than their wrinkle-seeded counterparts.

In addition to their genetic constitution, a number of factors influence the quality and vigour of seeds. To maintain high vigour, it is important that the mother plant receives an adequate supply of nutrients, especially nitrogen and phosphorus, and that the seeds are allowed to grow to maturity before being harvested. Although smaller seeds are not necessarily of poor quality, they are likely to be immature, and therefore of lower vigour. Other procedures that may be adopted to produce and maintain high vigour in seed lots include the use of protective seed dressings against pests and diseases and the adoption of new technologies such as pelleting and priming.

Priming involves the controlled hydration of seeds. Enough water is allowed to enter to enable pre-germination metabolic events to occur, but not so much as to allow radicle emergence. Thus the seed lot is made more uniform with all the seeds at the brink of germination when sown, resulting in rapid and uniform emergence. Primed seeds of leek, carrot (*Daucus carota*), tomato (*Lycopersicon esculentum*) and some flowers are easily obtainable. Pelleted seed is encapsulated in an inert coat, usually clay, to which pesticides may be added. The process rounds off seeds with irregular shapes like those of celery, carrots, parsnips and lettuce, and this allows them to flow easily through seed drills, resulting in more even stands.

Weather conditions during seed ripening may affect seed vigour adversely. Excessive rainfall may lead to an increase in fungal infection, while cycles of wet and dry periods may damage the seeds as they swell and contract under the changing moisture conditions. Damaged seeds are more susceptible to attack by pathogens and insects, resulting in reduced vigour.

## Loss of seed viability

Surprisingly, scientists do not know what causes loss of viability and ultimately the death of seeds. Aseptic seeds are known to last longer than those that are contaminated with microorganisms, but the former still lose viability eventually. Loss of viability also occurs at relative humidities below 65% when storage fungi are inactive and so it is concluded that although fungi may accelerate the loss of viability in stored seeds, they do not cause it. Other suggested reasons for loss of viability include the accumulation of inhibitory chemicals, the depletion of essential food reserves, or the break-down of proteins and nucleic acids, but so far there is little evidence to support any of them. However, it has been demonstrated that loss of viability is accompanied by genetic deterioration. Gardeners who grow peas, particularly from home-saved seeds, will be familiar with the occasional emergence of albino plants which lack chlorophyll. This is a direct result of a mutation which occurs during seed ageing. Background ionising radiation was once thought to be a cause of such genetic lesions, but natural levels are now known to

be too low to bring about this type of damage. It is evident that irreversible damage to the genome occurs as the seeds age, but it is not clear whether this is a cause or an effect of the loss of viability.

## Self-saved seed

Many amateur gardeners rely on saving seeds from their own plants, but this can lead to problems. For instance, plants may not set seeds at all, or only in some years. This may be due to pollination failure when the conditions are unfavourable for pollinators, such as bees during spring low temperatures, or simply due to the absence of suitable pollinators for exotic species. Dioecious species, in which male and female flowers are produced on different plants, as in *Salix* and *Ilex* spp., require that both types of plant are present in the vicinity. Complex hybrids, such as some fruit trees, may require the presence of more than one pollinator. Finally, poor weather conditions may prevent proper maturation of seeds.

Even when seeds are produced, they may be non-viable or may germinate poorly and lack vigour. This is often exacerbated by poor storage conditions (Chapter 17), although even in ideal conditions viability decreases with time (Table 17.1). It may, therefore, be advisable to test the viability of saved seeds by placing a sample on damp paper towelling, kept moist in a lidded plastic box at 18–25°C.

A further problem with self-saved seeds is that many plants do not breed true to type. Perhaps the most obvious example is that of F1 hybrid seeds which are produced by crossing two inbred lines. Self-saved seeds from these plants result in a range of types that differ from their hybrid parents. Apart from the special case of F1 hybrids, many common garden plants are complex hybrids that rarely breed true to type. Cross-pollination produces a range of individuals, each with its own mix of parental genes and results in a diverse population with the potential for the selection of better phenotypes. Selfing produces a more genetically homogeneous population, usually closer to the parental type, although many plants have mechanisms that prevent self-fertilisation. A few plants naturally produce seed from both cross- and self-fertilisation as, for example, violets (*Viola* spp.) which produce open flowers in early spring which may be cross-pollinated, and later produce closed flowers that are consequently self-fertilised.

For the gardener who relies on self-saved seed for the next crop, the best way to maximise the production of high vigour seed that stores well is to ensure that the mother plant has adequate nutrition and that the seeds mature fully before being harvested.

## DORMANCY

Not all viable seeds are able to germinate immediately, even when supplied with adequate water and maintained at temperatures that are favourable for growth. Such seeds are usually said to be dormant. There are several reasons why, in nature, a period of seed *dormancy* confers an advantage to a plant. For example, germination may be prevented until conditions are favourable for seedling growth, or germination may be spread out in time so that competition is reduced. The conditions needed to break dormancy usually relate to the natural environment and confer a selective advantage to the species. For gardeners, however, seed dormancy is often a disadvantage leading to delayed and erratic germination. Indeed, many tried and tested gardening techniques to improve germination are no more than dormancy-breaking mechanisms.

There are three types of dormancy, namely: innate, induced and enforced. In other words, 'some seeds are born dormant, some achieve dormancy and some have dormancy thrust upon them.'

### Enforced dormancy

*Enforced dormancy* occurs when seeds are capable of germination but are prevented by the immediate environmental conditions, such as lack of water or inappropriate temperatures. Germination occurs as soon as the restriction is removed and suitable conditions prevail. Such seeds should not be regarded as truly dormant, since their failure to germinate is not caused by factors internal to the seed itself.

### Innate dormancy

*Innate dormancy* is under genetic control, but is influenced by the prevailing environmental condi-

tions during seed maturation. Seeds of the vast majority of species have a period of innate dormancy, which begins immediately the developing embryo ceases to grow and while it is still attached to the parent plant. Such dormancy prevents the seed from germinating while still on the plant and usually persists for some time after the seed is shed or harvested.

The most usual cause of innate dormancy is that the embryo is insufficiently developed at the time of shedding, a common problem in members of the Apiaceae such as carrot and parsnip, in several palms, and in magnolia (*Magnolia grandiflora*). In other cases, chemical inhibitors, such as phenolic compounds and abscisic acid, may be present in the seed coat and germination cannot proceed until they have been removed by leaching. Such inhibitors are found in beet (*Beta vulgaris*) and in certain East Asian species of ash (*Fraxinus* spp.). Germination inhibitors are also often found in seeds of desert annuals. They are likely to be leached out only after enough rain has fallen to allow these ephemeral species to germinate and set seed without further rainfall occurring.

A further cause of innate dormancy is that certain biochemical changes must occur in the seed before germination is possible. This is known as *after-ripening* and is usually influenced by environmental factors such as temperature and light. Fresh seed of barley (*Hordeum vulgare*), for example, has to be stored for 3 months at 20°C in order to break its innate dormancy. The need for after-ripening may, in some cases, be associated with the presence of inhibitors of germination which may either be leached out as described above, or be broken down to inactive forms by chemical reactions within the seed.

Another type of innate dormancy is found in many species in a range of families, including the Papilionaceae (pea and bean family), Malvaceae (mallows), Convolvulaceae (bindweed family), Liliaceae (lilies) and Chenopodiaceae (goosefoot family). Here germination is prevented because the seeds have impermeable seed coats which present a barrier to the uptake of water. Under natural conditions the impermeable barrier in such *hard seeds* is removed by conditions which result in cracking, abrasion or breakdown of the seed coat, such as repeated wetting and drying, exposure to particular temperatures or digestion by animals.

# Induced dormancy

*Induced dormancy* occurs when seeds that are capable of germination are exposed to conditions that lead to the dormant state, such as particular temperature regimes or prolonged burial. As with innate dormancy, germination cannot take place until the seeds have been exposed to conditions that bring about a further metabolic change within the seed. Many of the environmental conditions that are effective in breaking innate dormancy are also effective in breaking induced dormancy, and in many cases it is not easy for the gardener to distinguish between the two types.

## SPECIAL REQUIREMENTS FOR GERMINATION

A number of special treatments are effective in breaking seed dormancy or are required for rapid germination. It is not really possible to distinguish between these two situations and any viable seed that fails to germinate when given an adequate supply of water and maintained at temperatures favourable for growth can be considered as being dormant. The factors that are effective in breaking dormancy and permitting germination confer a selective advantage on the species and are related to the conditions experienced in its natural habitat. These factors include light, temperature, fire, water and digestion by animals.

## Light

Many seeds will only germinate when exposed to light (Fig. 9.2), but they need to be fully or partly hydrated before they become responsive to this stimulus. A requirement for light has obvious advantages for the plant as it prevents the germination of deeply buried seeds until they are uncovered by cultivation or the activities of animals. Many of such species are *ruderals*. These are plants with a short life cycle which are poor competitors in a perennial sward but are able to grow, flower and set seed rapidly if the ground is disturbed and the competition thus reduced. In the garden environment, the

**Fig. 9.2** The effect of sowing depth on the germination of light-requiring cress seeds. The seeds were completely covered with soil (*left*), sown on the surface (*right*) or covered with the soil surface lightly scratched to expose some of the seeds (*centre*). *Photograph courtesy of Daphne Vince-Prue.*

majority of these are annual weeds (see Chapters 15 and 16). Garden plants with seeds that normally require exposure to light before they can germinate include cress (*Lepidium sativum*; Fig. 9.2), poppy (*Papaver* spp.), birch (*Betula* spp.), tobacco (*Nicotiana tabacum*), some cultivars of lettuce (*Lactuca sativa*) and several grasses. Such seeds must be sown at the soil surface. Although it is not common, germination is actually inhibited by light in a few plants; these include pansy (*Viola* spp.), love-lies-bleeding (*Amaranthus* spp.) and love-in-a-mist (*Nigella damascena*). When sown, such seeds must be well covered with soil or black plastic to exclude light. Some further examples of the responses of seeds to light are given in Table 9.2.

### Light quality

The kind of light to which the seed is exposed is also important and most light-requiring seeds are responsive to red light, although a small number will germinate only in blue light (e.g. *Nemophila* spp.). As sunlight contains both red and blue wavelengths, this difference is not something that normally impinges on horticulture. However, it was from studies of the responses of lettuce seeds to different wavelengths of

**Table 9.2** Some examples of the responses of seeds to light.

| Germination favoured by light | Germination favoured by darkness | Germination indifferent to light or darkness |
|---|---|---|
| Adonis vernalis | Ailanthus glandulosa | Anemone nemorosa |
| Alisma plantago | Aloe variegata | Bryonia alba |
| Bellis perennis | Cistus radiatus | Cytisus nigricans |
| Capparis spinosa | Delphinium elatum | Datura stramonium |
| Colchicum autumnale | Ephedra helvetica | Hyacinthus candicans |
| Erodium cicutarium | Euonymus japonicus | Juncus tenagea |
| Fagus sylvatica | Forsythia suspensa | Linaria cymbalaria |
| Genista tinctoria | Gladiolus communis | Origanum majorana |
| Helianthemum chamaecistus | Hedera helix | Sorghum halepense |
| Iris pseudacorus | Linnaea borealis | Theobroma cacao |
| Juncus tenuis | Mirabilis jalapa | Tragopogon pratensis |
| Lactuca serriola | Nigella damascena | Vesicaria viscosa |
| Magnolia grandiflora | Phacelia tanacetifolia | |
| Nasturtium officinale | Ranunculus crenatus | |
| Oenothera biennis | Tamus communis | |
| Panicum capillare | Tulipa gesneriana | |
| Reseda lutea | Yucca aloifolia | |
| Salvia pratense | | |
| Sueda maritima | | |
| Tamarix germanica | | |
| Taraxacum officinale | | |
| Veronica arvensis | | |

Based on data given in Rollin P. (1972) Phytochrome control of seed germination. In *Phytochrome*, K. Mitrakos and W. Shropshire (eds), pp. 229–54. Academic Press, London, with additions.

light that the light-sensitive pigment *phytochrome* was discovered.

The discovery of phytochrome stemmed from an observation in 1935 that the germination of lettuce seeds was suppressed by light of wavelengths in the part of the spectrum between 700 and 760 nm, now called far-red light (see Chapter 12 for the spectrum of light and its physical properties). Nearly 20 years later, in 1952, a classic scientific paper established that the germination of lettuce seeds was stimulated by exposure to a few minutes of red light, between 600 and 700 nm, and that this was prevented when the red light was immediately followed by a brief exposure to light in the far-red region. After a series of alternate exposures to red and far-red, the response depended on the last exposure in the sequence. When the sequence ended in red, most of the seeds subsequently germinated in darkness, but when the sequence ended in far-red many of the seeds failed to germinate (Table 9.3).

The authors concluded that the response must be mediated by a single light-sensitive pigment, which they called 'phytochrome', a word which means plant pigment. They also concluded that phytochrome could exist in two forms, Pr and Pfr, and that these could be converted from one to the other by exposure to the appropriate wavelength of light in the following reactions:

$$Pr + red\ light \rightarrow Pfr$$
$$Pr \leftarrow far\text{-}red\ light + Pfr$$

The Pr form of phytochrome absorbs red light and is converted to Pfr. In the reverse reaction, the Pfr form absorbs far-red light and is converted to Pr. Since the germination response required exposure to red light, it was further deduced that Pfr (formed by red light) is the biologically active form and that phytochrome must be synthesised in darkness in the inactive form as follows:

**Table 9.3** Exposing seeds of lettuce 'Grand Rapids' to red light promotes germination, while exposure to far-red light decreases germination.

| Light treatment | Percentage of seeds germinating |
| --- | --- |
| Darkness | 9 |
| 1 minute of **red light** | 98 |
| 1 minute of red, followed by 4 minutes of **far-red light** | 54 |
| 1 minute of red, 4 minutes of far-red, followed by 1 minute of **red light** | 100 |
| 1 minute of red, 4 minutes of far-red, 1 minute of red followed by 4 minutes of **far-red light** | 43 |

When the sequence of light treatments ended in **red**, germination was increased. When the sequence of light treatments ended in **far-red**, germination was decreased.

$$\text{Darkness} \to \text{Pr} \underset{\text{Far red}}{\overset{\text{Red}}{\rightleftharpoons}} \text{Pfr} \to \text{Response}$$

Phytochrome was purified in 1966 and shown to be a protein associated with a light-absorbing region, called a chromophore. Much more recently it has been found that plants contain up to five kinds of phytochrome, called PhyA–PhyE, and that these different molecules may have distinct but overlapping functions. For example, using *transgenic* plants, it has been found that PhyA operates to regulate germination, and both PhyA and PhyB operate to modify the response to shade.

Phytochrome-like substances occur in all green plants and in the Cyanobacteria (see Chapter 1). They have been found to operate quantiavely to limit the rates of response rather than to select between different developmental pathways. Phytochrome is now known to control a wide range of plant responses to light in addition to germination, such as the elongation of stems, branching and the expansion of leaves, as well as responses to day length (see Chapters 8, 11 and 13).

To return to germination, we now understand the reasons for the evolution of the photoreversible phytochrome system. Initially, scientists could not comprehend why such a pigment should have developed, as plants are not exposed to sequences of red and far-red light under natural conditions. However, red light is strongly absorbed by leaves, while far-red light passes through them (Chapter 8 and Fig. 8.2). Since far-red light strongly inhibits germination by converting phytochrome to the inactive form, germination is prevented under a leaf canopy close to the parent plant. Thus the presence of the inactive form in the dark prevents the germination of buried seeds, and the reversible reaction, which converts phytochrome to the inactive Pr form in far-red light, prevents germination in heavy shade where the seedlings would not flourish. As discussed in Chapter 8, the reversible property of phytochrome is also an important factor in the survival strategy of plants that are shaded by their neighbours.

### Light requirements for germination

Not all seeds require light to germinate. Nor indeed is the light requirement always satisfied by a brief exposure. In some cases, quite long durations of light, up to several hours each day, are necessary. This means that sun flecks passing between overhead leaves have no effect, and ensures that germination would only take place where there is a gap in the canopy with sufficient light to allow the new seedling to become established.

Measurements of the penetration of light into soils have shown that they are very opaque and only in the top few millimetres is there enough light for germination. For example, in field mustard (*Sinapis arvensis*) germination was reduced to zero when seeds were buried 10 mm deep, and to 20% at only 5 mm. For light-requiring seeds, therefore, it is important to sow them at, or very close to, the surface of the soil (Fig. 9.2). This applies to most turf grasses, leading to the well-known problem of birds eating the seeds that are at or near the soil surface.

Large seeds often germinate in the dark and the seedlings are able to reach the surface before their food reserves are exhausted. In contrast, a high proportion of small seeds need light and so only germinate when close to the soil surface, allowing them to emerge into the light and establish photo-synthesis before their small reserves are depleted. Most soils contain a large reservoir of seeds that are potentially capable of germinating but are prevented from doing so by lack of light until the soil is dis-turbed. Gardeners know only too well that cultivating the soil often leads to an explosion of germinating weeds (see Chapters 15 and 16). However, under natural conditions, the need for soil disturbance results in germination being spread over time and increases the chances of survival of the population as a whole.

The requirement for light may not be absolute. For example, seeds of *Ageratum* spp. germinate equally well in light or darkness when maintained at temper-atures below 15°C, but are light-requiring at temper-atures above 20°C. Unless germination is known to be favoured by darkness (Table 9.2) it is best to cover seeds thinly, usually with vermiculite, to allow light to penetrate.

## Temperature

### Low temperature

In many temperate- and cool-climate species, the hydrated seed must be exposed to a period of chilling before germination can occur. The effective tem-peratures are similar to those that break winter dorm-ancy in the buds of woody species and bring about vernalisation in biennial plants (see Chapter 13); the optimum temperature is usually about 5°C. Under natural conditions, a chilling requirement delays germination until the spring, thus allowing the long-est possible time for annual plants to complete their life cycle and for seedlings of perennial plants to become established before being exposed to low winter temperatures.

A chilling requirement is the underlying reason for the horticultural technique of stratification, which involves layering the seeds in moist soil or sand and maintaining them at low temperatures. The optimum temperatures (3–5°C) are in the range found in most domestic refrigerators. Stratification probably acts by changing the balance between growth inhibitors and growth promoters in favour of the latter. For example, in seeds of Norway maple (*Acer platanoides*), the amount of abscisic acid, a growth inhibitor, decreases

during stratification, while the quantity of gibberellins, which are growth promoters, increases. A decrease in abscisic acid has also been observed in seeds of ash (*Fraxinus americana*), walnut (*Juglans regia*) and hazel (*Corylus avellana*).

### High temperature

A requirement for high temperature to break dormancy is rather rare in temperate species. The best example is bluebell (*Hyacinthoides non-scriptus*). The normal habitat of this species is deciduous wood-land, so germination and growth must occur in autumn, winter and spring when the leaves are off the trees and light can reach the seedlings. The seeds are shed in early summer, but the low temper-atures prevent them from germinating immediately; they only do so in the autumn after being released from dormancy by exposure to the high temper-atures of the mid and late summer months. Contrary to popular belief, a requirement for exposure to high temperature does not underlie the use of boil-ing water to break dormancy. If this treatment is effective, it is because it softens or cracks the seed coat, allowing water to enter. Although popular, especially for seeds of parsley (*Petroselinum crispum*), it is not recommended as the seed may be damaged or even killed.

### Alternating temperatures

Many seeds will not germinate until they have been exposed to particular alternating temperature regimes. Alternations between low and high tem-peratures during each day promote germination in a number of species, including evening primrose (*Oenothera biennis*), tobacco (*Nicotiana* spp.) poppy (*Papaver* spp.) and many common weeds such as dock (*Rumex crispus*) and rough meadow grass (*Poa trivi-alis*). However, alternating temperatures are often only effective when combined with other factors such as light, as in celery (*Apium graveolens*). The stimu-lation of germination by alternating temperatures may result from mechanical changes in the seed or may possibly be due to specific temperature require-ments of sequential reactions that take place during germination.

Seasonal alternations of temperature are necessary for the germination of some seeds. An example is that of *epicotyl dormancy* in the guelder rose (*Viburnum opulus*). The epicotyl is the portion of the stem of an embryo or seedling above the cotyledons. Thus the root of guelder rose emerges when the seeds are maintained in warm temperatures of 20–30°C, but the shoot fails to grow until the well-developed root system has been exposed to cold. Epicotyl dormancy also occurs in *Actaea* spp., *Allium burdickii*, *Asarum canadense*, *Fritillaria ussuriensi* and *Hydrophyllum*, *Lilium*, *Paeonia* and *Viburnum* spp. In all these cases the radicle of the seed emerges in autumn, at warm temperatures between 10 and 30°C, depending upon species, while the shoot apex remains dormant. This is similar to bud dormancy in woody plants (see Chapter 13), and requires cold temperatures (0–5°C) to break dormancy and allow the shoot to grow. Cold stratification for 3–7 months over the winter is the usual way that gardeners break such dormancy, but soaking the seedlings in a solution of the growth hormone gibberellic acid (GA$_3$) may also be effective.

The significance of epicotyl dormancy is that it allows the plant to establish a root system following germination in the autumn, but prevents the emergence of the shoot until the following spring, so that damage from the cold winter temperatures is avoided. It also enables germination of seeds over several years, ensuring a reserve of ungerminated seeds in the soil.

Another example is that of *double dormancy*, which occurs in *Trillium* spp. and the so-called two-year lilies. Here the embryo has not completed its development at the time of shedding of the seeds, which then require exposure to warm temperatures to allow the embryo to develop. This must be followed by chilling to trigger the formation of the hypocotyl and roots. A second exposure to warm temperature allows these to grow until, once a certain stage of development is reached, a second exposure to cold triggers the emergence of the shoot.

Under natural conditions, seeds with epicotyl or double dormancy fail to develop into seedlings in one year and are, therefore, called biennial seeds. Some doubt exists concerning the possible ecological advantage of double dormancy. Species exhibiting it will apparently only produce fully developed seedlings in nature after two or more seasons. However, as soon as growth has started, they lose the protection which the 'uncommitted' state provides against the elements.

## Hard seed coats

Hard seed coats inhibit germination by preventing the uptake of water into the dry seed. They occur in many different species and are often found in members of the Papilionaceae, such as older sweet pea varieties (*Lathyrus odoratus*). Under natural conditions, hard seed coats are rendered permeable to water by the action of microorganisms in the soil or by being passed through the digestive tract of an animal. However, the gardener has a number of short-cuts available. These include scarification with coarse sandpaper or, in the case of larger seeds, such as those of sweet peas, removing a small section of the seed coat with a knife, a process called *chipping*. When chipping the seed, the cut should not be made close to the hilum (the scar where it was attached to the plant). Treatment with alcohol is particularly effective on members of the Caesalpinaceae.

Seed coats that are heavily thickened and cannot easily be made permeable by scarification or chipping, such as those of some cultivars of *Pharbitis nil*, may be treated by soaking in concentrated sulphuric acid for a duration that depends on the species and seed batch and can only be determined by trial and error. Because of the danger inherent in the use of concentrated acids, this treatment is not recommended for use by amateur gardeners. The dormancy-breaking action of abrasion, alcohol or sulphuric acid is directly related to an increase in the permeability of the seed coats to water, although such treatments probably also induce other changes such as permeability to gases, sensitivity to light or temperature and even a reduction in the amount of inhibitory substances present.

## Fire

Several trees and shrubs of sub-tropical and semi-arid regions have seeds that fail to germinate until they have been exposed to fire, which is a major natural environmental factor in habitats such as the Californian chaparral and the South African fynbos. Under natural conditions, such a response prevents seeds from germinating until a fire has burned down the overhead vegetation, and so reduces the competition for light, nutrients and water. Several species take advantage of this reduction in competition and are known to be rapid colonisers of burnt ground. The heat of the fire may also be necessary for the dehiscence of the seed from the follicles, as in *Banksia ericifolia*. Fire may also destroy accumulated inhibitors in the soil, thereby removing a possible cause of failure to germinate.

The charred remains of chaparral vegetation appear to stimulate germination in yellow bells (*Emmeranthe penduliflora*), a species found in arid habitats of south-western North America, suggesting that gases, including volatile oils, in the smoke or substances generated on burnt sites play a role in the post-fire germination of chaparral annuals. Indeed, smoke derived from the burning of plant material has been found to stimulate germination in over 170 species, representing 37 families and 88 genera and a range of plant types, all native to Mediterranean-type ecosystems, on three continents. In such ecosystems fire is an ever-present threat. Within the garden context it has been found that smoke-impregnated disks may stimulate germination, and such disks are now commercially available in the UK. These need to be saturated with water before sowing the seeds on them.

## After-ripening in dry storage

When seeds are after-ripened by storage following harvest in an air-dry state, they gradually lose their innate dormancy and become able to germinate, as with seeds of barley, sand dropseed (*Sporobolus cryptandrus*) and sedge (*Cyperus rotundus*). In commercial practice it is extremely common for dormancy to be broken in this way, although dry storage should be used with caution because viability may be adversely affected if the temperature or moisture content of the after-ripening environment increases (see Chapter 17). It is likely that at least some of the changes which take place during dry storage are a result of changes in the seed coats, reducing their tensile strength or possibly their impermeability to water or respiratory gases. In other cases it is likely that changes occur in the embryo itself or the immediate chemical environment of the embryo, which result in an increase in the potential of the embryo to exert growth forces against the mechanical restraints of the seed coats.

The need for after-ripening may also exert a protective role against precocious or accidental germination early in the season of maturation when conditions may not be favourable to seedling development, for example in cotton (*Gossypium hirsutum*), garden balsam (*Impatiens balsamina*), lettuce (*Lactuca sativa* 'Grand Rapids') and evening primrose (*Oenothera odorata*).

## Chemical treatments

A number of chemical are known to break dormancy and stimulate germination. Among these are naturally occurring plant hormones and synthetic growth promoters. Seeds may contain different amounts of hormones, accounting in part both for their varying depths of dormancy in such species as apple (*Malus domestica*), blackberry (*Rubus idaeus*), peach (*Prunus persica*), ash (*Fraxinus excelsior*) and rowan (*Sorbus aucuparia*) and for the differences in the way that they respond to hormonal treatments. Gibberellins have been found to replace light as the trigger for germination of several light-requiring seeds and may also replace a requirement for chilling. Some, such as those of maize (*Zea mays*) and barley (*Hordeum* spp.), respond to the application of gibberellic acid ($GA_3$) alone, although a mixture of gibberellins often gives a better response. Unfortunately, gibberellic acid is the only gibberellin available to amateur gardeners and treatment with this is seldom successful with garden plants. $GA_4 + GA_7$ are more effective, especially when used as a mixture, but are not generally available.

Other chemicals that stimulate germination in one or more species include hydrogen peroxide, sodium hypochlorite, thiourea, mercaptoethanol, nitrates, nitrites, acids, cyanide, azide, hydroxylamine, chloroform, ether and ethylene gas (Table 9.4). In general, their action is not understood, although it is interesting that some of them, such as thiourea, may also be effective in breaking bud dormancy (see Chapter 13). Many of these chemicals are hazardous and are not recommended for general use.

In conclusion, it is important to note that treatments that are said to promote germination often interact, and a combination of more than one treatment may be necessary to break dormancy. A classical demonstration of this can be seen in dock (*Rumex crispus*), where the effects of light or darkness, alternating or constant temperature and

**Table 9.4** Miscellaneous chemicals that promote germination.

| Chemical | Species in which germination is promoted |
|---|---|
| Hydrogen peroxide | *Daphne* spp.; carrot, *Daucus carota*; beetroot, *Beta vulgaris* |
| Sodium hypochlorite | Celery, *Apium graveolens* var. *dulce* |
| Thiourea | Lettuce, *Lactuca sativa* |
| Mercaptoethanol | Barley, *Hordeum vulgare* |
| Nitrates | Tobacco, *Nicotiana tabacum* |
| Nitrites | Rice, *Oryza sativa*; barley, *Hordeum vulgare*; shepherd's purse, *Capsella bursa-pastoris* |
| Acids | Sedge, *Cyperus rotundus*; pineapple, *Ananas comosus*; grape, *Vitis vinifera* |
| Cyanide | Lettuce, *Lactuca sativa* 'Grand Rapids' |
| Azide | Wild oat, *Avena fatua*; guava, *Psidium guajava* |
| Hydroxylamine | Tobacco, *Nicotiana tabacum* |
| Chloroform | Cashew, *Anacardium occidentale* |
| Ether | Upland cotton, *Gossypium hirsutum* |
| Ethylene gas | Lettuce, *Lactuca sativa*; peanut, *Arachis hypogaea*; clover, *Trifolium subterraneum* |

treatment with potassium nitrate were studied. Only the combination of alternating temperatures, light and potassium nitrate broke dormancy completely, while none of the other treatments, singly or in combination, produced 100% germination. Recent work on *Daphne longilobata* has demonstrated the same phenomenon. In this case alternating temperatures, hydrogen peroxide and gibberellic acid produced nearly 100% germination.

## NEW DEVELOPMENTS IN SEED TECHNOLOGY

### Terminator gene technology

The term terminator gene technology refers to the use of genetic manipulation (GM) to prevent the seeds of genetically modified crops from germinating. A promoter sequence, which is a DNA sequence that activates or switches on a specific gene, is artificially attached to a gene that prevents germination and this is then inserted into the seed of the target species. The genetically modified seeds themselves are capable of germination, but the plants that develop from them form only sterile seeds because the germination-inhibiting gene is activated during seed maturation. Terminator genes, it has been suggested,

would be an effective means of preventing the threat of genetic pollution by genetically modified crop and garden plants.

The technology has caused widespread concern, however, because it undermines a common premise of food production, namely the ability to self-save seeds, with farmers and gardeners being forced to purchase fresh seed from the producer each season. However, this is also true of the well-established practice of producing F1 hybrid seeds, which are the product of crosses between inbred parental lines, resulting in plants that do not breed true. Nevertheless, public outrage has significantly slowed the further development of terminator gene technology.

## CONCLUSION

This chapter demonstrates that a knowledge of the scientific processes that underlie seed formation, dormancy and germination may be of great value to the gardener wishing to produce and save seed and/ or wishing to achieve optimum germination of sown seeds and growth of the plants emerging from them. A knowledge of the ecological or evolutionary significance of phenomena such as seed dormancy adds a measure of interest for those frustrated by difficulties in germinating the seeds of particular species and may provide a basis for experimentation aimed at improving germination success. Much of the information contained in the chapter is relevant to an understanding of the ecology and control of weeds outlined in Chapters 15 and 16.

## FURTHER READING

Atwater, B.R. (1980) Germination, dormancy and morphology of seeds of herbaceous ornamental plants. *Seed Science and Technology* **8**, 523–73.

Baskin, J.M. & Baskin, C.C. (1998) *Seeds: Ecology, Biogeography and Evolution of Dormancy and Germination*. Academic Press, San Diego, CA.

Black, M. & Bewley, J.D. (2000) *Seed Technology and its Biological Basis*. Academic Press, Sheffield.

Bleasdale, J.K.A. (1973) *Plant Physiology in Relation to Horticulture*. Macmillan, London.

Bryant, J.A. (1985) *Seed Physiology*. The Institute of Biology's Studies in Biology vol. 165. Edward Arnold, London.

Hide, D. (2005) Off to a good start. *The Garden* **130**, 182–5.

Khan, A.A. (ed.) (1977) *The Physiology and Biochemistry of Seed Dormancy and Germination*. Elsevier, North-Holland.

Kozlowski, T.T. (1972) *Seed Biology*, vols 1 & 2. Academic Press, London.

Mayer, A.M. & Poljakoff-Mayber, A. (1982) *The Germination of Seeds*. Pergamon Press, Oxford.

Smith, H. (2000) Phytochromes and light signalling by plants – an emerging synthesis. *Nature* **407**, 585–91.

Thompson, P. (2005) *Creative Propagation*, 2nd edn. Timber Press, Cambridge.

Wray, N. (2003) Some like it hot. *The Garden* **128**, 51–3.

# 10

# Propagating Plants Vegetatively

## SUMMARY

First, general matters relating to vegetative propagation are considered, including the properties of clones, the mechanisms of cellular differentiation, with special emphasis on the role of plant hormones, the concept of juvenility and etiolation. The different types of cuttings (leafy shoot, hardwood, leaf/bud and root) and their management are next described. The techniques of layering and division are outlined, followed by propagation from specialised structures (bulbs, offsets, cormels and rhizomes/tubers and stem pieces). The principles and practice of grafting and budding are dealt with. Finally, the techniques of micropropagation are reviewed, and their significance assessed.

## INTRODUCTION

Under natural conditions in the wild, the majority of flowering plants and conifers increase their numbers by producing seeds (Chapter 9). In cultivation, however, this is often not possible and gardeners have to resort to some form of non-sexual or *vegetative propagation* to increase their stock. Many garden plants do not breed true from seed, while others do breed true but produce little or no seed, or seed that is difficult to germinate. In all such cases, the plants must be propagated vegetatively. Other reasons for carrying out vegetative propagation include avoiding a juvenile period by propagating from adult plants and the use of specialised *rootstocks* to obtain a desirable habit of growth.

A variety of vegetative propagation procedures is available to the gardener. The most widely used method is that of taking cuttings, which involves detaching portions of roots, stems, leaves or buds, as appropriate, inserting them in a suitable growing medium and maintaining them in an appropriate

environment to induce the regeneration of new roots (called *adventitious* roots if they arise from a stem), shoots and leaves. A related technique, layering, involves the induction of adventitious roots from a stem that is still attached to the plant. The rooted shoot is then detached and grown on as a new plant. An even simpler procedure, division, involves the breaking up or division of a mass of closely knit shoots in a clump or crown, usually of a herbaceous perennial, and using the resulting pieces to form new plants. Perennial plants that produce underground storage organs such as *bulbs, corms, rhizomes* or stem or root *tubers* may be propagated in a related manner by detaching such newly formed structures or, in some cases, parts of them, and replanting. The remaining procedures are more specialised, but highly effective. Grafting and budding require the removal of a shoot or single bud (the *scion*) from the plant to be propagated and attaching it in such a way to a cut surface of a rooted plant (the *stock*) that the tissues of the two grow together to form a permanent union. In *micropropagation*, sometimes called *tissue culture*, a very small bud, shoot or portion of tissue (an *explant*) is removed from a plant aseptically (i.e. after surface sterilisation, to remove surface bacteria and fungi) and grown *in vitro* (literally 'in glass'; in reality in a transparent glass or plastic container) on a specially formulated liquid or gelatinous culture medium where new plantlets develop. These can subsequently be detached, and after a suitable period of acclimatisation grown on to form new plants. These various procedures will be dealt with in detail later in the chapter. Before this, however, it is necessary to deal with a number of basic terms and concepts common to all methods of vegetative propagation.

## CLONES

Many popular garden plants have arisen by selecting a specimen with a desirable trait, such as good autumn colour or dwarf habit, and then propagating it vegetatively. The cultivars of most fruit plants and many ornamentals have originated in this way. The resulting plants constitute a *clone* and are genetically identical. However, they are not necessarily identical in all characteristics, the actual appearance of the plant, its *phenotype*, being the result of the interaction between its genes (its *genotype*; see Chapter 5) and the environment in which it grows. The maintenance of the clone requires, of course, that any increase in the number of individuals has to be by vegetative propagation.

The life of a clone is theoretically unlimited, since ageing individuals are usually re-invigorated by propagation. For example, the 'Williams Bon Chrétien' pear is known to have been grown for some 200 years and the 'Thompson Seedless' grape may have been cultivated for more than 2000 years. Deterioration does occur frequently in particular clones, however, the most common reason being infection with one or more viruses (see Chapters 15 and 16). Genetic mutations in the vegetative cells can also produce off-types that may reduce the value of the clone. However, such mutations sometimes lead to improved forms; for example, new colour sports in ornamental plants such as carnations and chrysanthemums often arise in this way. An important consideration during vegetative propagation is to maintain healthy stocks that are free from virus and true to type.

## CELL DIFFERENTIATION

Vegetative propagation requires the differentiation of new types of cells. Stem cuttings must produce root cells, leaf cuttings must produce both root and shoot cells, while budded or grafted plants must generate new vascular tissues in order to form a union between the scion and the stock to which it is attached. Development into different types of cell is under genetic control and most plants cells contain all the genetic information required to produce all the tissues of a new individual. They are therefore said to be *totipotent*.

Successful vegetative propagation also requires that specific cells in the plant can be induced to express the genes controlling the production of new roots and/or shoots. Pre-formed root initials are present in the stems of a number of easy-to-root species such as poplar. These remain dormant until the cutting is made and then emerge as adventitious roots. In plants with no pre-formed root initials, new roots are often induced to form by wounding. It is for this reason that the wounded area at the base of a cutting is sometimes increased, by removing an outer sliver of tissue or by cutting up into the cutting from its base.

### Plant hormones

Plant hormones have been shown to influence the division, enlargement and differentiation of cells and they clearly play an important part in the processes involved in the rooting of cuttings and the formation of graft unions. They are organic substances that in many cases are synthesised in one region of the plant and move to another, where at very low concentrations they elicit a physiological response. In other cases they act at the site where they are produced. Several classes of hormone have been identified in plants and they are known to influence a very wide range of responses (see Chapter 2).

Based on studies of the formation of new plants from callus tissue (the undifferentiated mass of cells formed at the site of a wound) or from pieces of tissue consisting of a few of the outermost layers of a stem, two classes of plant hormone, cytokinins and auxins, have been found to influence the way in which plants

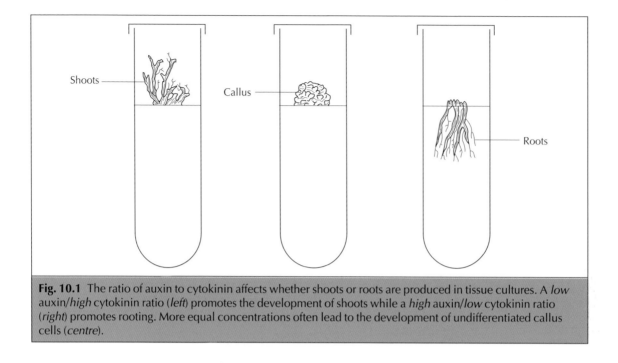

**Fig. 10.1** The ratio of auxin to cytokinin affects whether shoots or roots are produced in tissue cultures. A *low* auxin/*high* cytokinin ratio (*left*) promotes the development of shoots while a *high* auxin/*low* cytokinin ratio (*right*) promotes rooting. More equal concentrations often lead to the development of undifferentiated callus cells (*centre*).

develop. In general, it has been found that a high ratio of cytokinin to auxin favours the formation of shoot cells, whereas a high auxin-to-cytokinin ratio favours the formation of root cells (Fig. 10.1). The precise mechanisms through which these hormones control the direction of development are still unknown, although it is probable that they modify gene activity in some way. Auxins and cytokinins have considerable influence on the production of root and shoot cells, so information about their movement and effects within the plant helps the gardener to understand more about the management of various types of *propagule* (the name given to any part of a plant that is used to start a new plant).

## Auxins

This was the first category of plant hormones to be recognised. The major naturally occurring auxin is indole-3-acetic acid (IAA), but since its discovery in 1936 many synthetic auxins have been developed. Of these, the best known and most widely used are α-napthalene acetic acid (NAA), 2,4-dichlorophenoxy acetic acid (2,4-D) and 2-methyl-4-chlorophenoxy acetic acid (MCPA). These compunds do not occur naturally in plants, so they are not true hormones and are, therefore, called *plant growth regulators.*

Their effects are similar to those of IAA, but their behaviour in the plant is somewhat different. For example, whereas there are innate biochemical mechanisms for the destruction or inactivation of IAA when it has served its purpose in a plant, many synthetic auxins are not readily destroyed and so remain active in the tissues for a long time. This has important consequences for vegetative propagation since the effect of auxin is strongly dependent on its concentration. At high concentrations, auxins cause excessive growth (called *hypertrophy*) and ultimately death (for this reason, stable synthetic auxins such as 2,4-D are often used as weedkillers; see Chapters 15 and 16). At lower concentrations, auxins may modify tissue differentiation.

An important feature of plants is their *polarity*, which means shoots and roots can detect which way is up, even if they are removed from the plant. This is important for directing the movement of IAA within the plant. In stems it moves towards the root end regardless of the orientation of the shoot; therefore in stem cuttings IAA moves downwards from the young leaves and apex where it is produced and accumulates at the base of the cutting. Because a high auxin-to-cytokinin ratio favours root cell differentiation, new roots develop here. When stem sections rather than apices are used as cuttings, these are sometimes inserted into the soil upside down by mistake. Because of the polar movement of IAA, however, such cuttings are doomed to grow abnormally, attempting to produce roots into the air.

Root formation can often be increased by the application of auxins to the base of the cutting. IAA itself is not normally employed because it is easily converted to an inactive form, especially on exposure to air. The chemicals most commonly used are, therefore, the synthetic auxin NAA and the natural auxin indole butyric acid (IBA). The latter is closely related to IAA and has been found to occur naturally in several plants. Since synthetic auxins are not normally transported in a polar direction like IAA, they must always be applied to the base of the cutting where root initials are required.

We have already seen that many synthetic auxins are not readily degraded within the plant. The maintenance of high levels of auxins at the base of a cutting can inhibit root extension. For this reason, IBA is the most commonly used rooting hormone because it is gradually broken down, leading to a more optimal concentration for root elongation once the root initials have been formed. These different responses to auxins, depending on their concentration within the plant, illustrate the complexity of hormone action.

Although it is generally accepted that IAA plays a central role in root initiation, many plants are difficult to root and respond poorly to the external application of auxins, suggesting that other factors may also be involved. Seasonal variations in rooting performance are frequently observed and this may, in part, be due to changes in the amount of natural auxin present within the tissues. Both auxin and cytokinin levels can be influenced by day length, so the auxin/cytokinin ratio, which we have seen is important in cell differentiation patterns, may vary with season. A few experiments have shown that day length can influence the rooting behaviour of leafy cuttings (Table 10.1).

## Cytokinins

In the 1940s, coconut milk was found to be rich in a compound that increased cell division rates in plant tissues cultured on an artificial growth medium. Later the active ingredients were identified as *zeatin* and a related compound, *zeatin riboside*. Because of their effect on cell division (cytokinesis), these hormones

**Table 10.1** Day length affects rooting capacity in some plants.

| |
|---|
| **Rooting was increased when the parent plant was grown in long days**<br>Lorraine begonia, leaf cuttings (*Begonia* × *cheimantha*)<br>Flowering dogwood (*Cornus florida*)<br>Canadian poplar (*Populus* × *canadensis*) |
| **Rooting was increased when the parent plant was grown in short days**<br>Carnation (*Dianthus caryophyllus*)<br>Japanese holly (*Ilex crenata*)<br>Weeping willow (*Salix babylonica*) |
| **Rooting was increased when the cuttings were grown in long days**<br>Abelia (*Abelia* × *grandiflora*)<br>Canadian poplar (*Populus* × *canadensis* 'Robusta')<br>Weigela (*Weigela florida*) |
| **Rooting was increased when the cuttings were grown in short days**<br>Japanese holly (*Ilex crenata*)<br>Chinese juniper (*Juniperus chinensis*) |

Examples are taken from Vince-Prue (1975).

were called *cytokinins*. Zeatin and several other compounds that have since been shown to stimulate cell division are related to the chemical adenine. Most are not commercially available, with the exception of benzyladenine (BA), which occurs naturally but is not commonly found in plants, and kinetin, a synthetic cytokinin. Cytokinins occur in plants in young organs and root tips, and are probably produced there. Roots are known to be an important source of cytokinins and supply leaves, fruits and seeds via the xylem.

Experimental studies have shown that naturally occurring cytokinins are essential for cell division in plants, and therefore play an important role in tissue regeneration. For example, pith from tobacco produces a mass of unspecialised cells (*callus tissue*) when cultured in the laboratory on a growth medium with equal amounts of cytokinin and auxin. With a high cytokinin-to-auxin ratio, shoots develop, whereas a low cytokinin-to-auxin ratio favours the development of roots (Fig. 10.1). Cytokinins are not used in the propagation of cuttings because a high cytokinin-to-auxin ratio inhibits the formation of root initials. They are used, however, in micropropagation to stimulate the production of shoots.

Cytokinins from the roots also appear to be involved in the delay of leaf *senescence*. When leaves are removed from the plant, they rapidly become yellow, especially if kept in darkness. This ageing or senescence is greatly delayed, however, if adventitious roots form at the base of the petiole. Moreover, the effect of the roots can be partially replaced by the application of a cytokinin. Thus root formation on cuttings is important not only for water uptake, but also for the supply of essential growth factors to the shoot. Gardeners will be familiar with this phenomenon for, in the absence of roots, leafy cuttings rapidly yellow and die whereas, when roots develop, the leaves remain green and healthy.

## Juvenility

Rooting ability is markedly influenced by whether the plant is in its juvenile or adult phase of growth. *Juvenility* is the name usually given to an early phase of growth during which flowering cannot be induced by any treatment. It occurs in all plants but is most readily observed in trees because they have a long juvenile period often lasting many years, in contrast to many herbaceous plants where the juvenile period may last only a few days. Consequently juvenility can pose a special problem for tree breeding because of the extremely long period between successive generations.

Size appears to be important in the transition from the juvenile to the mature phase of growth, and in general conditions that promote growth reduce the duration of the juvenile period. Two possible explanations for the effect of size have been suggested. One is that a large-enough plant sends some kind of signal (perhaps a hormone and possibly from the leaves) to the growing point where the transition from juvenile to adult takes place. The second is that the shoot tip behaves independently of influences from the rest of the plant, perhaps because of some intrinsic ageing process in the dividing cells of this region. Currently, experiments are being carried out to investigate the role of specific genes in controlling the transition from juvenile to adult.

The important point for gardeners is that the change from juvenile to adult takes place at the shoot apex after the plant has reached a certain size. This fact has many implications for propagation, since the base of the plant always remains juvenile and there are many differences between adult and juvenile tissue, other than the fact that only adult plants can flower. *Axillary buds* retain the maturity level of their point of origin and those that develop low in the plant are initially juvenile. As they grow, however, individual branches undergo the phase change to maturity at their own shoot tips (Fig. 10.2). The change to the adult phase of growth is more or less permanent until new seedlings of the next generation revert again to the beginning of the juvenile phase.

**Fig. 10.2** Mature trees may retain their juvenile habit of leaf retention at the base. Buds at the base of the tree remain juvenile and, in some species, retain their juvenile habit of leaf retention. Here the outer branches are already adult and have shed their leaves. *Photograph courtesy of R. Heins, Michigan State University.*

It is important to distinguish between reversion to the juvenile phase (which occurs rarely and with difficulty in adult plants) and the re-invigoration of adult plants by improving their management or by taking cuttings.

## Morphology

Some plants have a characteristic morphology during their juvenile phase of growth, one of the best-known examples being the common ivy (Fig. 10.3). The juvenile growth habit is a creeping vine which clings by adventitious roots arising from the stem and has palmately lobed leaves. The adult plant is shrub-like with unlobed, oval leaves. Many experiments have

**Fig. 10.3** Adult and juvenile plants of ivy (*Hedera helix*). Flowering plants (*above*) show adult characteristics of entire leaves and shrubby growth. Juvenile plants (*below*) have lobed leaves, a climbing habit and adventitious roots. *Photograph courtesy of Daphne Vince-Prue.*

been carried out with ivy because it is relatively easy to handle in the adult phase of growth. Adult plants can be caused to revert to a juvenile habit by treating them with gibberellins (see Chapter 11). This effect can be prevented by the application of another hormone, abscisic acid (ABA; see Chapter 8), suggesting that a balance of gibberellins and ABA might normally be involved in the transition from one state to the other.

In some plants the juvenile foliage is considerably more attractive than that of adult plants. A good example is cider gum (*Eucalyptus gunnii*) which has bluish, circular leaves in its juvenile form and far less attractive dark green, elliptical leaves as an adult. It is grown commercially for its foliage and is also an ornamental garden plant. Since the transition to maturity takes place at the shoot tip while the base of the plant remains juvenile, repeated cutting back is practised both commercially and in the garden to maintain the juvenile foliage.

It is intriguing that cuttings taken from seedlings that are still in the juvenile phase may never undergo the normal phase change to maturity, and consequently retain throughout the life of the plant their juvenile foliage and other juvenile characteristics such as leaf retention (e.g. beech, *Fagus sylvatica*), thorniness (e.g. wattle, *Acacia* spp.) or lack of flowering. This is also seen in some conifers where the juvenile leaf form is fixed when they are propagated from seedling plants (e.g. in the Elegans form of Japanese cedar, *Cryptomeria japonica* Elegans Group). From the work with ivy, we know that the application of gibberellins may, in some instances, cause a reversion to the juvenile habit of growth but this still does not explain why the plants do not develop as normal and undergo a phase change to maturity. It is interesting and probably relevant that cuttings taken from horizontally growing branches often retain a prostrate habit of growth, at least for some time. This has been called *topophysis* and is the origin of some prostrate forms of garden plants.

## Rooting capacity

Another difference between juvenile and adult plants is their rooting capacity. Juvenile plants form adventitious roots quite readily, while in the adult phase rooting ability is usually diminished considerably and sometimes lost. Consequently, and especially in plants that are difficult to root, the age of the parent plant can be a very important factor. Several studies have shown that in some coniferous and deciduous trees known to root only with extreme difficulty, the most important single factor affecting root formation is the age of the tree from which the cuttings are taken. It has been suggested that the reason why some old gardening books list exotic species that could then be rooted from cuttings but are now considered to be difficult, if not impossible, to propagate in this way, is that such species were imported as seeds, and cuttings were subsequently made from young seedlings which, in their juvenile phase, were easy to root.

In one species of gum tree (*Eucalyptus*) there is a direct relationship between the decrease in rooting ability and the increase with age in the production of a natural root inhibitor at the base of the cutting. This may be so with other species too, but there are many instances where there is no obvious correlation between rooting ability and the amounts of promoters and/or inhibitors of rooting present in the tissues. This indicates that the decrease in rooting ability with age may, at least in part, be associated with a change in the sensitivity of cells to substances that influence rooting, rather than to the absolute quantities of the substances themselves.

Any treatment that maintains plants in the juvenile phase is of value in preventing the loss of their rooting capacity. One of these treatments is the repeated cutting back of the parent plants, a practice, known as *stooling* (see Chapter 11), that is employed commercially for the production of hardwood cuttings. In some tree species, leaves towards the base of the tree are retained late into the autumn and often through the winter, indicating the part of the tree that is still juvenile (Fig. 10.2). Cuttings taken from this region are still in their juvenile phase and so would be expected to root more readily. Suitable material for hardwood cuttings of beech and some other species such as oak and some fruit trees is also frequently produced by maintaining a 'juvenile' hedge system.

### Flowering

Juvenile plants do not flower and so the question of whether or not the parent plant is flowering does not arise. Cuttings from adult plants may, however, be taken from individuals that are either flowering or are vegetative. With easy-to-root species the presence or absence of flowers makes little difference, but for some difficult species experiments have shown that cuttings taken from flowering shoots root less well than those from non-flowering shoots. In some cases removing the flower buds increases the rooting potential, but this is not always so. It is prudent, however, to remove any flower buds that are present on a cutting just in case. Although it is evident that the presence of flower buds is often antagonistic to the formation of adventitious roots, the reason is by no means understood. The amounts of several hormones have been shown to change during the transition to flowering but none has been shown to correlate in a clear way with the reduction in rooting potential. Another possibility is that nutrients are diverted into developing flower buds, perhaps reducing their availability at the site of root production.

In woody plants, the juvenile phase with respect to flowering varies from only about a year in some shrubs to as much as 40 years in beech. Juvenile periods of 5 to 20 years are common in trees and pose serious obstacles to tree-breeding programmes, as well as leading to unacceptable delays in establishing fruiting trees. Various procedures have been shown to accelerate the transition to maturity. Since we already know that the time at which this occurs is a function of plant size, it is not surprising that most of these procedures involve treatments that increase the rate of seedling growth, such as maintaining them at higher temperatures and/or in long days to prevent dormancy (see Chapter 13).

In horticulture the problem can sometimes be overcome by taking material for propagation from the adult region of the tree, providing that the species is not one in which rooting is inhibited in adult material. In this way the juvenile phase (during which flowering and fruiting do not occur) can be bypassed. The problem does not arise with many fruit-tree clones, where generations of vegetative propagation has resulted in 'fixing' the adult phase and the plants are never truly juvenile. For this reason, they are often difficult to root and must be grafted. Recent work has shown that micropropagation can result in some reversion to juvenile characteristics, especially rooting capacity (see micropropagation, below).

## Etiolation

Several trials have shown that keeping plants in darkness (*etiolation*; see Fig. 8.4) is remarkably effective in increasing the production of adventitious roots in stem tissue. This was shown experimentally as long ago as 1864, but the reason is still not fully understood. The inhibitory effect of light on rooting is greatest in red light, suggesting that it is mediated through the pigment phytochrome (see Chapter 8), which is known to cause many changes in hormones and other biochemical constituents of plants. Complete etiolation would deplete the shoot of carbohydrates and depress rooting, so only the base of the shoot is maintained in the dark, either by insertion into the cutting medium or by mounding up the base of plants to be used as a source of hardwood cuttings. Experimentally, rooting has been increased by covering the entire parent plant for a period.

## TYPES OF CUTTINGS AND THEIR MANAGEMENT

### Leafy shoot cuttings

Leafy shoot cuttings consist of a portion of leafy shoot, usually including the apical bud. They are obtained from the current year's growth and may be taken during active growth (softwood), when they have become partly woody (semi-ripe) or when they have become fully mature and have set their terminal buds (hardwood). The latter are only from evergreens, since *deciduous* plants will already have shed their leaves by this stage and are treated differently (see below). As far as management is concerned, the difference is between leafy cuttings, where photosynthesis and water relations are important, and leafless hardwood cuttings, where the amount of stored carbohydrate is a major factor.

A large number of dicotyledons (see Chapter 1) can be propagated as leafy cuttings, including both herbaceous and woody plants. Following removal from the parent plant, auxin moving downwards accumulates at the base of the cutting and root cells are formed. As discussed above, rooting can often be improved by dipping the base of the cutting in a solution or powder containing an auxin, usually IBA. Many species have preformed root primordia at the *node* (the point at which the leaf joins the stem), which are arrested until stimulated by auxin. For this reason, cuttings are usually taken just below a node. In many cases, however, the position of the cut is not important, although roots will often still emerge mainly from the node. Where no pre-formed initials

are present, cell divisions in the outer layer of the phloem give rise to new roots.

Wounding the basal surface of the cutting by cutting into it or removing a sliver of bark can also increase its rooting potential, especially where there are no pre-formed initials. Removal of a few of the lower leaves is normal to reduce the incidence of rotting, which is more likely to occur in leaf tissue buried in the rooting medium. Reducing the leaf surface, by removing some of the lower leaves and/or halving large leaves, also helps to reduce water loss from the cutting.

## Management of leafy cuttings

The prime requirement in the management of leafy cuttings is to reduce water loss from them. The absence of roots means that water uptake is strictly limited, while water loss by transpiration from the leaves still occurs. Cell division requires metabolism, which depends on oxygen to support respiration and a source of energy supplied by sugars from photosynthesis. Consequently, for successful rooting, the base of the cutting where cell division is occurring must be in a well-aerated environment, water loss must be reduced to a minimum and photosynthesis should be optimised. It is important that cuttings are not allowed to wilt before inserting since this leads to prolonged closure of the stomata and a consequent loss of photosynthetic potential by cutting off the supply of carbon dioxide (see Chapter 8 for a discussion of the effects of wilting and the role of abscisic acid). Succulent plants, such as cacti and houseleeks, are an exception to this general rule because they have other mechanisms to avoid or minimise water loss (see Chapter 8) and cuttings are usually left for a short period, up to a few days, in a warm airy place to seal the base before they are inserted into the cutting compost.

### Temperature
Maintaining the base of cuttings at a higher temperature than the air (bottom heat) aids rooting and establishment. The higher temperature at the base stimulates metabolism and cell division activity at the site of root formation, while the lower air temperature retards shoot growth until roots develop, water uptake occurs and the water balance is restored. The optimum base temperature is between 18 and 25°C;

too high a temperature (more than 25°C) may inhibit rooting.

### Water

Leaves lose water to the atmosphere in the form of water vapour diffusing through the stomata (see Chapter 8). The absolute amount of water vapour present in the air is known as the *vapour pressure*, and like any other gas water vapour moves along a concentration gradient from a region of higher vapour pressure to a region of lower vapour pressure. This means that, to reduce water loss from the leaves, cuttings must be maintained in an atmosphere of high vapour pressure, i.e. in a high humidity.

The *relative humidity* is a measure of the amount of water vapour in the air compared with the amount that would be present if the air were saturated. When it is saturated, hot air holds more water vapour than cooler air. This means that even when the air both inside and outside the leaf is saturated (i.e. at 100% relative humidity), the amount of water vapour will be greater inside if the temperature inside is also higher. Leaves in sunlight are often hotter than the surrounding air and, consequently, a leaf can still lose water even when the air around it is saturated with water vapour.

The easiest way to cool the leaf and so reduce water loss is to shade it from direct sunlight. Although this reduces the amount of light for photosynthesis and consequently the supply of sugars for metabolism, photosynthesis would be more drastically reduced by wilting and stomatal closure. The most satisfactory way of cooling the leaf, however, is by spraying it with water. The subsequent evaporation of water, which consumes energy in the form of heat, results in cooling of the leaf surface. Water loss is thereby substantially reduced without the necessity of shading. A number of systems are available employing an artificial leaf that switches on spray jets each time its surface dries out. Such *intermittent mist systems* are widely used in commercial horticulture and are also available for the amateur gardener.

Injecting water vapour into the air around cuttings (known as *fogging*) is one way of maintaining a high humidity in the atmosphere surrounding cuttings. A more commonly used alternative is to enclose them in a glass or clear polythene chamber (this need be no more than a plastic bag placed over the pot). However, this practice itself can lead to a number of problems. High humidity and stagnant air favour fungal growth (see Chapter 14), and unless the chamber is permeable to carbon dioxide photosynthesis will be reduced as carbon dioxide levels fall. Some ventilation is therefore desirable. Most plastic bags have a low permeability to water vapour but allow the exchange of carbon dioxide and so need not be removed until rooting has occurred.

### Rooting media

Adequate aeration at the base of the cutting is ensured by the use of a freely draining rooting medium that retains water but does not become waterlogged. Cuttings rooted in clay pots are usually inserted around the edge since some oxygen can diffuse through the porous clay. Unfortunately this does not occur in today's plastic pots. Mineral nutrients are not necessary during the rooting stage itself, since they are not taken up in the absence of roots, but are required later when root growth begins. The most commonly used rooting media, therefore, consist of an inert material such a coarse sand, grit, vermiculite or perlite (to improve aeration ) mixed with soil or, more usually, with a commercially available compost.

### Seasonal effects

We have to recognise that even under ideal conditions some species have proved difficult or impossible to root from leafy cuttings for reasons that are not fully understood. Sometimes rooting potential is strongly dependent on season, as in some lilac cultivars (e.g. *Syringa vulgaris* 'Madame Lemoine'). Many factors such as stage of growth, light intensity, nutrition and temperature undoubtedly contribute to this window of opportunity. Some species are also sensitive to day length, perhaps operating through changes in their hormone content. Root initiation in leafy cuttings may be influenced by the day length experienced by the parent plant or by the day-length treatment given to the cuttings themselves (Table 10.1). However, many of the examples given in Table 10.1 have not been confirmed and amateur gardeners could provide valuable scientific information by carrying out their own experiments.

## Hardwood cuttings

Hardwood cuttings provide one of the easiest and most reliable ways of propagating many trees and

**Table 10.2** Examples of common garden plants suitable for propagation by hardwood cuttings.

**For rooting out of doors**
Butterfly bush (*Buddleja davidii*), Forsythia (*Forsythia* × *intermedia*), privet (*Ligustrum vulgare*), cherry plum (*Prunus cerasifera*), elderberry (*Sambucus nigra*), false spiraea (*Sorbaria tomentosa*), snowberry (*Symphoricarpos* × *doorenbosii*), Weigela (*Weigela florida*).

**For rooting in cold frames**
Abutilon (*Abutilon* spp.), barberry (*Berberis* spp.), Caryopteris (*Caryopteris* × *clandonensis*), flowering quince (*Chaenomeles japonica, C. speciosa*), dogwood (*Cornus* spp.), Deutzia (*Deutzia* spp.), St. John's wort (*Hypericum* spp.), Kerria (*Kerria japonica*), plus those listed above for rooting out of doors.

**For rooting in containers**
Actinidia (*Actinidia* spp.), fig (*Ficus carica*), Virginia creeper (*Parthenocissus quinquefolia*), Tamarisk (*Tamarix* spp.), vine (*Vitis* spp.), wisteria (*Wisteria* spp.), plus all those listed above.

**Evergreen shrubs suitable for winter propagation**
Box (*Buxus* spp.), Escallonia (*Escallonia* spp.), spindle tree (*Euonymus* spp.), broom (*Genista* spp.), ivy (*Hedera helix*), holly (*Ilex* spp.), sweet bay (*Laurus nobilis*), lavender (*Lavandula* spp.), privet (*Ligustrum* spp.), cherry laurel (*Prunus laurocerasus*), Portuguese laurel (*Prunus lusitanica*), rosemary (*Rosmarinus officinalis*), cotton lavender (*Santolina chamaecyparissus*), gorse (*Ulex europaeus*), Laurustinus (*Viburnum tinus*).

From Hide (1997a, 1998).

shrubs (Table 10.2). They usually consist of pieces of woody material of the current season's growth, about 20 cm long and slightly thicker than a pencil. Hardwood cuttings taken from deciduous plants are usually leafless but contain sufficient reserves of nutrients to maintain root growth until new leaves emerge in spring. Although timing is not crucial, and cuttings can be taken at any time between November and March provided that the shoot is not growing, the best times are just after leaf fall or just before growth begins in the spring. The cuttings

are normally trimmed below a node as for leafy cuttings, but the tip is often removed. When long stems are available, several cuttings can be made from a single shoot.

Hardwood cuttings from deciduous trees are leafless, so there is no problem of water conservation as there is with leafy cuttings and they are not normally enclosed. It is important, however, to avoid desiccation and cuttings in the open should be protected from drying winds. They are usually inserted into the soil with only the tip above ground, but for trees where a single stem is required the entire cutting is put below the soil in order to achieve a straight stem. Drying out is more of a problem with hardwood cuttings from evergreen plants, and these are often placed in a propagating frame either out of doors or in a cool greenhouse.

Stool beds maintained as a source of hardwood cuttings (for example, in the production of clonal rootstocks) are frequently earthed up to increase the rooting potential of cuttings by etiolation. Juvenility is maintained by repeatedly cutting back the parent plants.

Hardwood cuttings of some easy-to-root subjects may be inserted out of doors, usually with sharp sand at the base to increase aeration and drainage. In this case, rooting will not begin until the soil temperature increases in the spring and the cuttings should not be lifted until the following autumn. The range of plants that can be propagated successfully by hardwood cuttings is increased by giving them the protection of a cold frame or cool greenhouse (Table 10.2). Rooting will occur more rapidly if bottom heat is also provided. When rooting under glass it is important to ventilate on sunny days, to lower the air temperature and prevent premature bud growth before the new root system is established. When rooted over gentle bottom heat and given the protection of a greenhouse, cuttings taken in November will usually be well rooted by the end of February.

## Leaf and leaf-bud cuttings

Some plants can be regenerated from cuttings consisting of whole leaves, or sections of leaf, or even leaf discs. The latter have been used experimentally to examine the effect of the auxin/cytokinin ratio on the pattern of regeneration in leaves of Lorraine begonia (*Begonia* × *cheimantha*). As found with tissue cultures, the application of cytokinins increased the number of buds that developed, while auxins promoted the formation of roots. Plants frequently propagated by leaf cuttings include members of the Gesneriaceae, for example Cape primrose (*Streptocarpus* spp.) and the related African violet (*Saintpaulia ionantha*), which grow as rosettes and do not produce suitable stem material.

For this type of propagation to be successful, the specialised leaf cells must be capable of de-differentiating into actively dividing cells, which are then able to undergo *organogenesis* into new meristems that can produce shoots and roots. Leaves also have polarity and must be positioned so that the part of the leaf that was nearest to the parent plant is at the lower end of the cutting. New plantlets then arise at the base of the leaf. Although leaf cuttings are usually inserted vertically in the rooting medium, whole leaves are often simply laid flat on the surface. In this case, large veins on the under surface are partially cut through and new plants arise at these points.

It is important to note that if chimeras (see Chapter 5) such as the variegated mother-in law's tongue (*Sanseveria trifasciata* 'Laurentii') are propagated from leaf cuttings, the variegation will not be present in the new plants because they originate only from the green cells in the core of the leaf. A similar problem can occur with root cuttings (see below).

Many species that are unable to regenerate new plants from leaves do produce roots at the base of a leaf inserted into a growing medium. The failure seems to be in the ability of the leaf to produce a new shoot apex, since many plants unable to regenerate from leaves alone can do so from leaf-bud cuttings, which consist of a piece of stem bearing a single leaf with a bud in its axil. In this case, the lateral bud becomes the new apical shoot meristem. Honeysuckle (*Lonicera* spp.), ivy (*Hedera helix*), *Camellia* spp. and *Clematis* spp. are often propagated by leaf-bud cuttings. In some members of the Gesneriaceae (especially *Streptocarpus* spp.) an organised meristem is already present at the base of the *phyllomorph* (the leaf blade and petiole) when it is detached, and gives rise to new daughter plants in this region.

The management of leaf and leaf-bud cuttings is the same as for shoot cuttings.

## Root cuttings

Root cuttings consist of sections of root and are often used when suitable stem material is not available. Like stems, roots also have polarity so it is important that they are inserted into the rooting medium with the proximal end (nearest the parent shoot) pointing upwards, although in many cases it is satisfactory to place the root sections horizontally and so avoid confusion. Since they have no photosynthetic tissue, successful root cuttings must have a supply of stored carbohydrates to support metabolism until new shoots emerge above ground. Consequently, plants with thick fleshy roots are usually propagated more successfully by root cuttings than are plants with thin, fibrous roots. An exception is the tree poppy (*Romneya coulteri*), where the cuttings are made from the thinner, more active part of the root. Not all species can be propagated by root cuttings, but examples of those that propagate well by this method are given in Table 10.3.

Even where root cuttings are successful, they are not always appropriate, as in the case of *periclinal chimeras* (see Chapter 5), where an outer layer of mutated cells overlays an inner core of the original genotype. In the thornless blackberry, for example, the cells of the outer mutated layer contain the genes for thornlessness and plants grown from root cuttings

**Table 10.3** Some common garden plants that can be propagated by root cuttings.

Bear's breeches (*Acanthus mollis*)
Tree of heaven (*Ailanthus altissima*)
Alkanet (*Anchusa azurea*)
Horseradish (*Armoracia rusticana*)
Sea kale (*Crambe maritima*)
Oriental poppy (*Papaver orientale*)
Drumstick primula (*Primula denticulata*)
Stag's horn sumach (*Rhus typhina*)
Comfrey (*Symphytum ibericum*)
Mullein (*Verbascum* spp.)

revert to the thorny genotype because the new shoots arise only from the inner tissues.

## Layering

Layering often occurs naturally when shoots, especially of woody plants such as *Viburnum* spp., touch the surface of the ground. Rooting may then occur at one or more nodes, where pre-formed root initials are frequently found. The horticultural technique of layering was adapted from the natural layering of wild plants and was already used by the Romans in the propagation of grape vines in the first century BC. It represents an easy and convenient way of propagating, which may be especially useful for plants that are hard to propagate by other methods (Table 10.4).

Layering is carried out when the plants are dormant in the early spring or autumn, using wood of the current or previous year's growth that is still pliable enough to bend. In simple layering, a portion of stem near the tip is bent into a u-shape and pegged into a hole about 10 cm deep (Fig. 10.4). This is then filled with a suitable rooting medium and the tip of the shoot is tied to a cane to keep it upright. The likelihood of rooting is increased if the shoot is partially cut through where it is pegged into the ground, resulting in the accumulation of sugars and

**Table 10.4** Some common garden shrubs that can be propagated by simple layering.

> Camellia (*Camellia* spp.), winter sweet (*Chimonanthus praecox*), Corylopsis (*Corylopsis pauciflora, C. sinensis* and *C. spicata*), hazel (*Corylus avellana*), filbert (*Corylus maxima*), smoke bush (*Cotinus coggygria*), Daphne (*Daphne* spp.), spindle tree (*Euonymus*, deciduous species), witchhazel (*Hamamelis × intermedia, H. mollis* and *H. japonica*), hydrangea (*Hydrangea paniculata* and *H. quercifolia*), calico bush (*Kalmia latifolia*), magnolia (*Magnolia liliiflora, M. sieboldii, M. × soulangeana* and *M. stellata*), Pieris (*Pieris* spp.), cherry (*Prunus*, deciduous shrubby species), Viburnum (*Viburnum × bodnantense, V. carlesii, V. farreri* and *V. plicatum*)

From Goodwin (1999).

probably auxin above the cut. Wounding itself also increases the rooting potential, as discussed above. Covering the rooting zone with soil to exclude light also helps to promote establishment and prevents drying out. The new plant is cut from the parent when well rooted, which usually takes about a year, although rooting may take two years in some cases (e.g. *Rhododendron* spp. and cultivars).

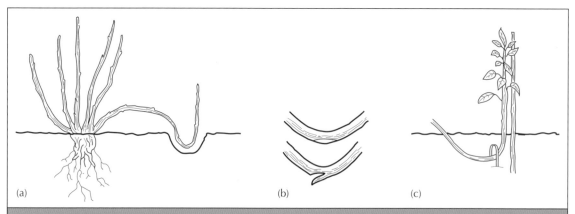

**Fig. 10.4** Propagation by simple layering. Shoots are bent over to the ground and the tip is turned up (a). A cut is made at the base of the bent part of the layered shoot (b), which is then covered with soil and pegged into the ground (c). The tip of the shoot is usually fastened to an upright stake. Rooting occurs at, or near, the site of wounding in the buried part of the stem and the rooted plant is later severed from the parent.

# Air layering

This technique is most commonly used to improve the appearance of ornamental plants that have become leggy with bare stems, as often occurs in the rubber plant, *Ficus elastica*. It is one of oldest forms of propagation, probably originating about 4000 years ago in China.

The technique, which can be successful with both monocotyledons and dicotyledons (Table 10.5), involves the removal of a layer of outer tissue to girdle the stem and interrupt the downward flow of nutrients, especially sugars, which accumulate above the girdled area; sugars normally move in the phloem from the photosynthesising leaves to the roots. Auxins moving down the plant also accumulate above the girdled area, leading to the formation of roots. It is necessary to prevent the girdled area from drying out, usually by covering it with moist sphagnum moss wrapped in polythene. Where possible, a high humidity level should also be maintained around the plant by placing it in a propagating case. When roots have formed, the new plant is severed below the girdled region.

**Table 10.5** Examples of plants that can be propagated by air layering.

| |
|---|
| Cabbage palm (*Cordyline australis*) |
| Daphne (*Daphne* spp.) |
| Dracaena (*Dracaena* spp.) |
| Rubber plant (*Ficus elastica*) |
| Swiss cheese plant (*Monstera deliciosa*) |
| Philodendron (*Philodendron* spp.) |

# Division

Division is a simple technique that can be applied to plants that produce a mass of closely knit shoots or buds, forming a clump that can be split into smaller pieces. Each portion will have shoots and/or buds and roots, and will therefore be capable of independent growth. Although division is most commonly used in the propagation of herbaceous perennials, some woody and semi-woody plants can also be divided (Table 10.6). Most of these plants, if left alone, develop into a crowded patch which often dies out

**Table 10.6** Some common garden plants that can be propagated by division.

**Herbaceous plants with fibrous crowns**
Yarrow (*Achillea* spp.), Michaelmas daisy (*Aster* spp.), Avens (*Geum* spp.), leopard's bane (*Doronicum* spp.), Gypsophila (*Gypsophila* spp.), day lily (*Hemerocallis* spp.), Shasta daisy (*Leucanthemum* × *superbum*), catchfly (*Lychnis* spp.), bergamot (*Monarda* spp.), coneflower (*Rudbeckia* spp.), scabious (*Scabiosa* spp.), meadow rue (*Thalictrum* spp.), globe flower (*Trollius* spp.), valerian (*Valeriana officinalis*), speedwell (*Veronica* spp.)

**Herbaceous plants with tough, compacted crowns and fleshy buds and roots**
Astilbe (*Astilbe* spp.), hellebore (*Helleborus* spp.), plantain lily (*Hosta* spp.)

**Alpine plants**
Aubretia (*Aubrieta* × *cultorum*), gentian (*Gentiana sino-ornata*), cranesbill (*Geranium cinereum* var. *subcaulescens*), primrose (*Primula*, European species), saxifrage (*Saxifraga* spp., mossy and rosette-forming types), stonecrop (*Sedum cauticola*), creeping thyme (*Thymus serpyllum*)

**Semi-woody herbaceous plants**
Astelia (*Astelia chathamica*), sedge (*Carex* spp.), pampas grass (*Cortaderia selloana*), New Zealand flax (*Phormium* spp.), Adam's needle (*Yucca filamentosa*)

**Woody plants**
Purple chokeberry (*Aronia prunifolia*), buckeye (*Aesculus parviflora*), vine maple (*Acer circinatum*), common quince (*Cydonia oblonga*)

in the centre. In the wild the outer shoots continue the growth of the plant, but in the garden the over-crowded or dying centre is unsightly.

The division of herbaceous plants with fibrous crowns and of alpine plants is done by either pulling the mass of shoots and roots apart by hand or teasing them apart with two forks placed back to back. Herbaceous plants with tough compacted crowns with fleshy roots and buds are difficult to divide by hand and often need to be cut carefully with a sharp knife. The cut surfaces should be dusted with a fungi-cide as the fleshy roots are prone to rotting.

The semi-woody herbaceous plants that can be divided produce sword-like leaves in dense terminal clusters, each with its own root system. Because of their tough, woody nature, they are difficult to divide by hand and are usually divided by cutting with a knife. The small number of woody plants that can be divided produce clumps of stems from suckers arising below ground level.

The best time to divide most subjects is at the start of the dormant season, when the plant dies back. For most alpine plants, division is best done soon after flowering, although autumn-flowering plants should be divided in the spring to minimise over-wintering losses.

## PROPAGATION FROM SPECIALISED STRUCTURES

In addition to shoots, leaves and roots, a number of plants, especially many *geophytes*, have specialised structures that can be used as the starting point for propagation. In many of these *sympodial* branching is common. This is a pattern of growth in which the apical bud withers at the end of the growing season and growth is resumed in the following year by the lateral bud or buds below the old apex. This charac-teristic ensures that units of growth such as daughter bulbs, *offsets*, *cormels* or branched rhizomes are maintained and are capable of independent growth once roots are produced. The natural rate of multi-plication varies with the species. For example, in daffodils (*Narcissus* spp.) only about one and a half new bulbs are produced each year compared with five in tulips (*Tulipa* spp.). In contrast, a plant of *Gladiolus* can produce a hundred or more cormels.

Most of these plants can be propagated by lifting the dormant plants after the leaves have died back, separating out the individual offsets, such as daughter bulbs or cormels, and replanting them. Exceptions are snowdrops (*Galanthus* spp.), which are very sus-ceptible to desiccation and must therefore be lifted while still green.

Bulbs, which are underground buds consisting of fleshy, modified leaves (*scales*) attached to a base plate, lend themselves to a number of special methods of propagation. Those that consist of loosely packed scales, especially lilies (*Lilium* spp.) and some fritil-laries (*Fritillaria* spp.), may be propagated from single outer scales, a process called *scaling*, while bulbs with a tighter structure, such as daffodils (*Narcissus* spp. and cultivars) and hyacinths (*Hyacinthus* spp.), must be cut into pairs of scales, called *twin scaling*.

Small bulbs, and non-scaly bulbs such as *Hippea-strum* spp., can be propagated by dividing them into several pieces, each retaining a portion of the base plate, a procedure called *chipping*. Hyacinths and some other bulbs (e.g. crown imperial, *Fritillaria imperialis*) can be induced to form bulblets by deeply scoring the base plate, or scooping it out, to damage or excise the growing point and remove its apical dominance. Some bulb plants can be induced to form offsets, even where none occur naturally. Onion, for example, can be induced to form bulbils by removing the flower buds from the seed head, leaving a 'ball of bristles'; small bulbs then form between the bristles.

*Rhizomes* are specialised underground stems, which also serve as storage organs. The main shoot dies at the end of the season and is replaced by one or more lateral buds in a typical sympodial branching pattern. These secondary rhizomes, in turn, produce an aerial shoot after a single node, as in Peruvian lily (*Alstroemeria*), or after many nodes, as in lily-of-the-valley (*Convallaria majalis*). Rhizomatous plants (e.g. rhizomatous irises) are propagated by cutting the rhizome into portions, each with one or more growing buds. In the invasive weed, couch grass (*Elymus repens*), cultivation breaks up the rhizomes into fragments, which removes the dominance of the apical bud and allows each piece to regenerate. This is why couch grass is so difficult to eradicate by cultivation.

*Tubers* are of two types, both having a storage function. Root tubers, as typified by *Dahlia*, consist of swollen portions of roots near the base of the stem. The tuber itself is unable to develop shoots as it consists only of fibrous roots and cannot produce buds. When propagating by division, therefore, it is important to see that each piece has at least one dormant bud and one tuber.

Stem tubers are pieces of swollen stem and can arise in different ways. Tubers of potato (*Solanum tuberosum*), for example, develop at the end of underground *stolons* (thin underground stems), whereas the tubers of gloxinia (*Sinningia*) are modified, swollen *hypocotyls* (the tissues linking stems and roots) which increase in size annually. All stem tubers have axillary buds in the axils of modified scale leaves. The method of propagation from stem tubers depends on how the tuber is formed. Potato tubers are collected at the end of the season, stored and re-planted the following year. In *Cyclamen*, tuberous *Begonia* and gloxinia, which are all modified hypocotyls, the tubers can be divided when they are sufficiently large and more than one growing point has formed.

Other specialised structures that are used in propagation include corms, which are solid, bulb-like, underground stems (*Crocus* spp., *Gladiolus* spp.), the pseudobulbs of some orchids, which are thickened, bulb-like stems (*Cattleya* spp., *Cymbidium* spp.), stolons, which are long horizontal stems that form new plants at the nodes and the tip (*Fragaria* × *ananassa*) and the plantlets that develop on the edges or ends of the leaves of some plants (e.g. devil's backbone, *Kalanchoe diagremontiana*). For detailed descriptions of how to propagate from such structures and of other specialised propagation procedures, see the Further reading section at the end of this chapter.

## GRAFTING AND BUDDING

There are many reasons for grafting. As with cuttings, the use of adult scions bypasses the juvenile phase during which flowering and fruiting cannot occur, which means that trees flower and fruit much more rapidly than if raised from seed. A second reason is that plants with complex genetic backgrounds, such as most fruit trees, do not breed true from seed, resulting in wide variation in both their fruit characteristics and growth habit. All tree-fruit cultivars are clones derived from a single original parent and therefore all have the same genetic make-up. Most of

them do not root readily from cuttings, and grafting is a quicker and much more reliable method of obtaining new plants. Budding, as for example for rose cultivars on seedling rootstocks, is also a way of obtaining the rapid multiplication of new plants.

There are other special reasons for grafting. For ornamentals, it is possible to produce plants with enhanced decorative features. A pendulous form can be grafted to an upright rootstock at a suitable height to form an attractive weeping tree (for example, the Kilmarnock willow, *Salix caprea* 'Kilmarnock'). Or plants with different attractive features can be combined, such as *Prunus* × *subhirtella* 'Autumnalis' (winter flowers) grafted on to *Prunus serrula* (coloured bark). Several fruit cultivars can be grafted on to one plant to produce a 'family tree' to ensure pollination in small gardens where there is no room to plant several trees; in this case, the grafts are made once a framework of branches has developed.

One of the most important reasons for grafting is the effect of the rootstock on the growth of the scion. For some plants, rootstocks are available that can tolerate unfavourable soil conditions or soil-borne pests or diseases (for example cucumbers, *Cucumis sativa*, can be grafted to the wilt-resistant *Cucurbita ficifolia*). For many purposes, both in ornamental gardens and in orchards, small trees are desirable and size-controlling rootstocks are available that cause the grafted tree to become partially or extremely dwarfed. Rootstocks for apple range from those that are very vigorous (e.g. MM109) to those that are extremely dwarfing (M9, M27 and EMLA 9), the resultant trees on the dwarfing stocks being only about one-third the size of those on the vigorous stock. Dwarfing rootstocks also result in early fruiting.

The effect of the rootstock on the tree vigour can be explained by the fact that roots supply water, mineral nutrients and hormones to the aerial part of the tree, whereas the resistance to soil-borne problems is a direct function of the root cells. Several experiments have indicated that the amounts of naturally occurring growth regulators supplied to the shoot may be implicated in the size-controlling effects of rootstocks. For example, for apple rootstocks a reduction in auxins and cytokinins and an increase in inhibitors may be involved in the dwarfing effect.

Clonal rootstocks, such as the Malling (M) and East Malling, Long Ashton (EMLA) series, are frequently employed, especially for tree fruits. These have been selected for particular effects such as reduction in vigour, and because they are clones they are highly uniform. They must, of course, be raised from cuttings, often hardwood cuttings taken from stool beds or hedges (see the section on juvenility above). Seedling rootstocks, although not clonal, are used in some cases and have several advantages. In particular, the production of large numbers of seedlings is relatively simple compared with taking cuttings and, moreover, not all plants are easily propagated by cuttings. Although seedlings, which are used as the rootstocks for roses, are likely to be more variable than clonal material, this can be reduced by careful selection of the seed source.

Successful grafting and budding depend on the development of a graft union that is physiologically functional and can transport substances between rootstock and scion. Many organic substances cannot cross by simple diffusion but must be transported by living cells, often in the phloem. Ultimately a vascular connection must develop between the two graft partners. This is well illustrated by the many studies on flowering that have been carried out using grafting techniques. Non-flowering plants can be caused to flower in many cases by grafting a flowering shoot on to them. However, this is only successful when a graft union has been established, demonstrating the need for the active transport of a flower-inducing substance across the union.

## Incompatibility

Grafts are more likely to be successful when the plants are closely related botanically. For example, grafting between plants of the same clone is always possible whereas, very occasionally, even grafting between different clones of the same species fails. At the other end of the scale, successful grafting between plants of two different botanical families is usually considered to be impossible. Successful grafts between plants of different genera in the same family are rare. A good example of the complexity of the problem is that of quince and pear. Quince has long been used as a dwarfing rootstock for some pear varieties, but the reverse combination, quince grafted on to a pear rootstock, is never successful. When different species of the same genus are grafted together experimentally, the results are quite confusing. Some are successful, some are not, and some, like the quince/pear graft, are successful only in one direction.

When plants cannot produce a satisfactory and stable graft union, they are said to show *incompatibility*. The graft union may fail completely or may initially appear to be successful but fail later, sometimes after several years, at the point of union. In one type of incompatibility the problem can be overcome by inserting a piece of another plant (an *interstock*) between the two incompatible plants, the incompatibility seeming to depend on local factors at the site of the failed graft union. In other cases incompatibility cannot be overcome in this way and it seems that some substance which can be transported across an interstock is the cause of the failure to establish a union. In at least one case, sweet orange on sour orange stock, a virus present in the scion is not tolerated by the cells of the stock.

Incompatibility is clearly related to genetic differences between the graft partners, but the underlying physiological mechanisms are not fully understood. Various types of structural abnormality can develop at the graft union, suggesting some kind of adverse physiological response between the two graft partners. Hormones from the wounded regions are assumed to be involved in the regeneration process, leading to a connection between the living cells of the graft partners, called a *symplastic* connection. It has been suggested that incompatibility may result when substances antagonistic to the action of hormones are produced. However, some kind of cell–cell recognition response involving signalling molecules has also been proposed.

## Grafting techniques

The origins of grafting can be traced back to ancient times and it was a well-established and popular practice in the days of the Roman Empire. It is not surprising, therefore, that several different techniques have been developed over the years. The goal of grafting is to fit together two pieces of living tissue in such a way that they will unite. Irrespective of the technique used, there are a number of basic requirements for successful grafting. Clearly, the two plants must be related and genetically compatible. They must also be physically compatible in the sense that the cambial regions (see Chapter 2) of the stock and scion must be capable of being brought into direct contact, as it is the cambial cells that will divide to generate a successful union. This means that usually only Gymnosperms and dicotyledons can be grafted successfully, since they have a continuous vascular cambium between the xylem and phloem. Monocotyledons have scattered vascular bundles which are difficult to connect, resulting in a low percentage take. Obviously the two cut surfaces must be held together by tying or taping.

The physiological status of the graft partners is also important; usually the best results are obtained when the scion buds are dormant and the rootstock is just coming into growth. Grafts with scions in leafy growth are rarely successful. It is essential to prevent drying out of the cut surfaces by placing the graft in a humid atmosphere, by covering the union with a water-impermeable wax, or by using special plastic ties. Subsequent management involves removing shoots from the rootstock and/or staking if the scion begins to grow vigorously.

A few popular types of graft are shown in Fig. 10.5 to illustrate the variety of techniques that can be used.

## MICROPROPAGATION

Micropropagation, although not a technique that can yet be used by most growers and gardeners, is becoming increasingly important. Indeed, micropropagated plants of some orchids, for example, are now appearing for retail sale still growing in small culture flasks on their nutrient medium, although such plants require careful handling to establish them as free-growing individuals.

Micropropagation is carried out for several reasons. One of the most important is the elimination of viruses. Another is the production of large numbers of new clonal individuals from small amounts of starting material as, for example, the rapid multiplication of a new cultivar. It is also used for some difficult-to-root subjects, and in some instances micropropagation can be successful for plants that it

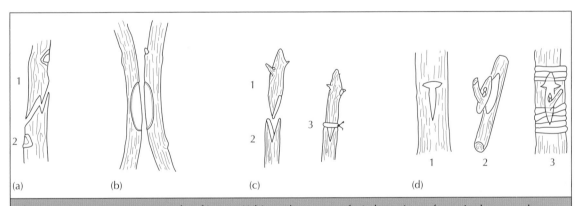

**Fig. 10.5** Some common types of grafting. (a) Whip and tongue graft. A short piece of stem is chosen as the scion and a tongue is cut into it as shown (1). The upper part of the rooted stock plant is removed with a slanting cut and a similar tongue is made in the cut surface at the top of the stock (2). The scion and stock are fitted together with the tongues interlocking and tied. (b) Approach graft. This is usually carried out with two potted plants which can be closely approached together. Similar cuts are made in stock and scion and the cut surfaces are brought together and tied. Once union has taken place, the top of the rootstock *above* the union and the base of the scion *below* the union are removed. (c) Cleft and saddle grafts. To form a cleft graft the base of the scion is cut on two sides to form a wedge (1). The top of the rooted stock plant is then removed and its stem either split or a v-shaped cut made to accommodate the cut end of the scion (2). The two are finally fitted together and tied (3). A saddle graft results from an inversion of this procedure, the top of the stock being cut on two sides to form a wedge, with a v-shaped cut being made in the base of the scion to fit over the wedge like a saddle. (d) Budding. A T-shaped cut is made in the stem of the rootstock (1). The scion consists of a single bud removed with a small piece of bark – the bud shield (2). The bud is then inserted into the cut made in the stock and pushed under the two flaps of bark until the bud shield is completely covered (3) and the graft is tied. In all cases, close contact between the cambium of stock and scion is essential for a successful graft union.

has not been possible to propagate by other means. It is also of interest because it is being used in the conservation of rare or uncommon plants and as a means of facilitating the exchange of plant material between countries. Immature embryos can also be grown in aseptic culture, a technique that has been used to 'rescue' embryos that would not continue to develop in the seed. Micropropagation is also essential for *somatic hybridisation* and for the insertion of new or modified genes into plant cells since, for both techniques (see Chapter 5), the starting points are protoplasts, which must subsequently undergo *organogenesis* in culture media to develop into new plants.

Micropropagation is applied to fruit-tree rootstocks and scions that are difficult to root and is an accepted method of rapid multiplication while maintaining the integrity of the clone. However, fruit trees propagated in this way show that differences can develop in culture. In particular, it has been observed that there is a progressive increase in rooting potential when the material is repeatedly moved from one culture vessel to new culture medium in another (i.e. they are subcultured). For example, in the apple rootstock M9, which is difficult to propagate conventionally, rooting ability increased with repeated subcultures over a year. Based on the fact that some other plants in culture, such as the giant redwood *Sequoiadendron giganteum*, have developed vegetative characteristics that are markers of juvenility, it is thought that the most likely reason for this is some degree of rejuvenation. It is an advantage for fruit trees as it produces plants that retain the property of easy rooting by conventional means.

Flowering is less affected than would be expected if the plants were truly juvenile and, moreover, flowering has actually been accelerated in some micropropagated plants. An interesting example of this effect has been observed in some species of bamboo (e.g. *Bambusa bambos*) where flowering occurred in culture within 6 months, whereas seedlings of the same species normally take some 30 years to flower. The causes of these phenotypic changes are still debated.

Micropropagation usually begins with a piece of tissue cut from the parent plant. One starting point is to remove a piece from the tip of a shoot (about 1 mm long) consisting of the apical dome plus two or three leaf primordia (see Chapter 2). This technique of *meristem culture* is used when virus elimination is required since it is known that very young, actively dividing cells are often free from many viruses and also certain bacterial diseases. It also results in fewer mutations than when other tissues form the starting point, since it does not require that non-meristem cells become organised into meristems. However, explants taken from any tissue can be used as a starting point for micropropagation, as can protoplasts and embryos.

Whatever its origin, the starting explant must be initially aseptic (uncontaminated by microorganisms as, for example, when taken from inner tissues) or must be chemically surface-sterilised before being placed on some kind of support, usually a gel solidified with agar in a small glass or plastic culture vessel, although protoplasts are initially cultured in a liquid medium. Such small pieces of tissue do not photosynthesise so the basal medium contains sugar as well as mineral nutrients and some vitamin supplements. Even though sugars are supplied and photosynthesis does not occur, some light is still necessary for normal development of the emerging shoots, which would otherwise be etiolated, and fluorescent lamps (warm white or daylight) are usually employed.

One of the early findings was that the way in which new plants developed could be altered by the addition of plant hormones to the culture medium. A high cytokinin/auxin ratio favours the initiation of shoots while a high auxin/cytokinin ratio favours the initiation of roots (Fig. 10.1). This knowledge from basic research still underpins the manipulation of root and shoot formation in micropropagation, whether plants are being produced for sale or for experimental purposes. Initially a high level of cytokinin is used to regenerate multiple shoots, which are then divided and subcultured. The auxin level is increased later to allow root initiation to occur.

Once individual plants have been established in aseptic culture, they must be moved out of the culture vessels into soil. This has to be done gradually in order to minimise the shock of the transition from the high humidity of the culture environment to the open greenhouse. The roots that develop in agar gel are somewhat abnormal to begin with, and for this reason shoot tissues are sometimes removed from the culture vessel and rooted in a normal rooting medium as *micro-cuttings*.

## CONCLUSION

Vegetative propagation is an aspect of horticulture where the results of using traditional practical procedures, some of which are very ancient, may be significantly enhanced as a result of simple modifications based on modern scientific knowledge. Moreover, current research on the subject is generating new procedures with the capacity to greatly expand the range of garden plants which may be propagated vegetatively, increase the efficiency and success of propagation of those species and cultivars that can already be propagated successfully, and make possible the application of new approaches to the breeding of garden plants and the conservation of endangered lines. It is an excellent example of the blending of traditional and new knowledge in the advancement of practical gardening.

## FURTHER READING

Bleasdale, J.K.A. (1973) *Plant Physiology in Relation to Horticulture*. Macmillan Press, London.

Brickell, C. (ed.) (1992) *The Royal Horticultural Society Encyclopedia of Gardening*, pp. 540–1. Dorling Kindersley, London.

Gardiner, J. (1997) *Propagation from Cuttings*. Wisley Handbook. Cassell, London.

Goodwin, D. (1999) Bending the rules (layering). *The Garden* **124**, 92–4.

Hartmann, H., Kester, D.E., Davies, Jr, F.T. & Geneve, R.L. (eds) (1997) *Plant Propagation. Principles and Practices*, 6th edn. Prentice Hall International, Saddle River, NJ.

Hide, D. (1997a) Taking the straight and narrow (leafless hardwood cuttings). *The Garden* **122**, 38–49.

Hide, D. (1997b) A fresh look at propagation composts. *The Garden* **122**, 268–70.

Hide, D. (1998) Simple but effective (leafy hardwood cuttings). *The Garden* **123**, 731–3.

Hide, D. (1999) Splitting up (division). *The Garden* **124**, 522–3.

Hide, D. (2005) At the root of it all (root cuttings). *The Garden* **130**, 166–9.

Lumsden, P.J., Nicholas, J.R. & Davies, W.J. (eds) (1994) *Physiology and Development of Plants in Culture*. Kluwer Academic Publishers, Dordrecht.

McMillan Browse, P. (1997) *Plant Propagation*. Mitchell Beazley Publishers, London.

Rees, A.R. (1992) *Ornamental Bulbs, Corms and Tubers*. Commonwealth Agricultural Bureau International, Oxford.

Thompson, P. (2005) *Creative Propagation*, 2nd edn. Timber Press, Cambridge.

Toogood, A. (ed.) (2003) *Royal Horticultural Society Propagating Plants*. Dorling Kindersley, London.

Vince-Prue, D. (1975) *Photoperiodism in Plants*. McGraw Hill, Maidenhead.

# 11

# Shape and Size

## SUMMARY

First, factors within the plant that regulate growth and development are considered, notably hormones (gibberellins, auxins, cytokinins and brassinosteroids), nitric oxide and dwarfing genes. Next, the ways in which growth may be regulated by the application of chemicals is described. There follows an account of environmental factors that affect plant growth, including light quality and direction, day length and temperature. Various mechanical procedures used for shaping plants are then described, including pruning, bonsai production, topiary, stooling and pollarding, and shaking and brushing.

## INTRODUCTION

The shapes of plants add structure, backbone and interest to the garden, especially in winter when colour is lacking. Size also matters when choosing plants and the trend towards smaller garden plots, especially in towns and cities, means that plants which take up less room and do not require staking are increasingly desirable. The use of containers also requires plants of a particular shape and size, notably those that are short with a compact habit.

There is great natural variation in the shapes and sizes of plants (see Chapter 1). For example, shape varies from upright to spreading while size varies from huge forest trees to the smallest alpine plants. Plants also have a remarkable ability to respond to external conditions, while their high developmental plasticity allows the gardener to model them into a variety of shapes and sizes. Moreover, their genetic constitution may be modified by plant breeders to control the amount of, and/or sensitivity to, the hormones that determine height and shape (see

Chapter 5). This chapter will examine some of the many factors that may modify the shapes and sizes of plants, including those, such as pruning and the application of chemicals, that are under the direct control of gardeners.

## ENDOGENOUS REGULATION

The basic elements for plant growth (water, light, $CO_2$ and minerals) determine how fast and how large a plant can grow within its genetic constraints. Basic morphology, growth habit and size, however, are ultimately determined by the plant's genetic code. Understanding how genes control the pattern of growth may lead to the development of plants that are better tailored to the needs of gardeners. This goal has in recent years been brought closer by the increasing availability of mutations that modify growth and development, and these have been especially valuable in elucidating the roles of plant hormones in controlling size and shape.

## Plant hormones: gibberellins

*Gibberellins*, which belong to the terpene group of secondary plant products (see Chapters 2 and 12), are widespread throughout the plant kingdom. They have many effects on plants, for example in flowering (see Chapter 13) and germination (see Chapter 9), but the control of stem growth is a major function. They will be referred to collectively as gibberellins or, in the case of individual gibberellins, as $GA_1$, $GA_2$, etc.

Gibberellins were first discovered in Japan, where they were associated with a fungal disease of rice called the *bakanae* or foolish seedling disease, in which the infected plants grew excessively tall. More recently, mutant studies have shown that the production of gibberellins results in long spindly plants, while loss of the ability to synthesise them or the capacity to respond to them results in dwarf plants. Of the many known gibberellins, the most important endogenous regulator of stem elongation appears to be $GA_1$. However, only $GA_3$ (gibberellic acid) and $GA_{4+7}$ are readily available to growers, as Berelex and Regulex, respectively. The majority of higher plants respond to the application of $GA_3$ but some conifers do not and, for these, $GA_{4+7}$ is more effective.

## Plant hormones: auxins and cytokinins

The naturally occurring *auxin, indole-3-acetic acid (IAA)* has a major role in controlling apical dominance, which is the basis of all pruning treatments. The apical bud (see Chapter 2) in many species prevents or slows the growth of the lateral buds below it. If it is removed by pruning or, under natural conditions, is damaged by wind or animals, the lateral buds are released from its dominance and start to grow. The uppermost lateral bud then grows towards the vertical and becomes the new leading shoot which, in turn, begins to exert dominance over the lower laterals. The presence of the shoot apex also makes the lower branches grow out towards the horizontal, while the apical bud itself grows vertically. It is not known whether this effect results from a difference in response to gravity, or is dependent on some particular angular relationship between the stem and the lateral branches. It is known, however, that auxins appear to be involved.

In releasing lateral buds from apical dominance, it has been known for some time that removing the youngest visible leaves is as effective as removing the whole of the apical bud, and it is evident that one or more of the inhibitory factors are formed in them. However, the precise nature of the inhibitory factor(s) is still in some doubt. If the apical bud is removed and replaced with auxin, the inhibition of lateral bud growth is re-imposed, indicating that apical dominance is due to auxin being transported downwards through the plant (see Chapter 10) and accumulating in the lateral buds. This does not seem to be the whole story, however, since the amount of auxin required to re-impose dominance is some thousand times greater than the amount that is actually present in the apical bud. Moreover, when auxin labelled with radioactive carbon ($^{14}$C) is applied to a decapitated shoot, it can be shown to travel down the plant as expected, but does not move into the inhibited lateral buds. When auxin is applied direct to these buds, they begin to grow out as if dominance had been released.

Such evidence indicates that auxin moving from the apex into the lateral buds is not the immediate cause of apical dominance. One suggestion is that the dominant, actively growing terminal shoot is a *sink* into which nutrients and hormones are diverted, resulting in the inhibition of lateral bud outgrowth. The involvement of auxins and also *cytokinins* (see Chapters 2 and 10) in the control of apical dominance is supported by the results of genetic-modification experiments. When tobacco plants were genetically modified to produce higher than normal amounts of cytokinins or lower than normal amounts of auxin, in each case they branched excessively when compared with plants that had not been modified. Thus the auxin/cytokinin ratio could be the factor that determines the degree of apical dominance and the extent of branching (see Chapter 10 for the effects of the auxin/cytokinin ratio on the development of roots and shoots).

## Plant hormones: brassinosteroids

Brassinosteroids also belong to the terpene group of secondary plant products and evidence from studies of mutants has implicated them as factors that can modify growth and development. For example, mutant plants of thale cress (*Arabidopsis thaliana*) that have lost the ability to synthesise or perceive brassinosteroids exhibited a number of phenotypic characters, notably reduced apical dominance and dwarfism.

## Nitric oxide

The gas nitric oxide is involved in many plant signalling responses. Although it is a highly toxic chemical that damages living tissues, the controlled production of this substance appears to be an essential first step in switching on cell expansion in roots and shoots.

## Genetic dwarfs

A number of genetic dwarfs are known to under-produce or be insensitive to gibberellins. Mention has already been made of the fact that plants have been genetically engineered to investigate the role of endogenous factors, such as hormones, in controlling shape and size. However, genetic mutants with desirable garden features have mostly arisen from selection and conventional plant-breeding methods. For example, the Green Revolution in farming was underpinned by the development of semi-dwarf cultivars of wheat, with shorter stems and consequently a greater proportion of harvestable product. These are known to have a low content of gibberellins, or to be insensitive to them, and it is probable that many dwarf garden plants are similar, although other hormones may also be involved.

## CHEMICAL REGULATION

Gibberellins have been used commercially to promote stem growth in a small number of cases, such as in producing standard *Fuchsia* plants and increasing flower stalk length in gerberas (*Gerbera jamesonii*), but they are not widely used. Much more important are the several groups of synthetic chemicals that have been found to reduce stem length. Known generally as *growth retardants*, these are among the most widely used chemical growth regulators in horticulture, especially in the production of ornamental pot plants, such as poinsettias (*Euphorbia pulcherrima*) and *Chrysanthemum* cultivars. In addition to reducing height, growth retardants have been found to accelerate flower bud production in some woody plants, especially in azaleas (*Rhododendron* spp.), to increase resistance to some stresses such as drought and to increase the colour of leaves. Depending on which compound is used, growth retardants may be applied either as a soil drench to each pot, or as a foliar spray.

Several widely different groups of chemicals have been found to have plant-growth-retardant activity, although no one category is effective on all plants. The choice, therefore, depends on the plant being treated. Chlormequat (Cycocel) is used commercially on poinsettias to reduce their height and on azaleas to increase the number of flower buds. Daminozide (B-nine) is used on chrysanthemums and azaleas, but is less effective than chlormequat on poinsettias, whereas ancymidol (A-rest) is effective on a wide range of plants, including chrysanthemums, poinsettias and many annuals. Paclobutrazol (Bonsai) is effective at low concentrations and is relatively persistent, as well as controlling elongation in a wide spectrum of plants.

Although growth retardants vary considerably in chemical structure, research indicates that they all act in a similar way and achieve their effects by reducing the amounts of gibberellins in the plant. In many species, the inhibition of growth by retardants may be completely overcome by the application of gibberellins, suggesting that their major effect is to reduce gibberellin synthesis. However, individual growth retardants block different steps in the complex biosynthetic pathway leading to gibberellin production. The fact that growth retardants depress the biosynthesis of gibberellins means that treated plants are effectively *chemical dwarfs* and so are comparable with the genetic dwarfs discussed above that under-produce or are insensitive to gibberellins. The essential difference, however, is that the genetic dwarfs remain dwarf throughout their lives, while the dwarfing effects of a chemical treatment is ultimately lost. This means that chemically dwarf plants bought from a commercial source do not maintain their dwarf habit indefinitely if planted out in the garden, although the effect of the chemical may be quite persistent in some cases.

The application of growth retardants requires a lot of labour (for example more than 15 sprays of chlormequat are not unusual for poinsettias). Growth retardants are also expensive, and it is easy to make a potentially disastrous mistake in the timing of application or in the concentration of the chemical used. Moreover, they are regarded as undesirable, especially where the crop is edible. Growers are therefore trying to find alternatives and methods being considered, or in some cases used, are control of temperature and the restriction of water supply in vegetable seedlings.

## ENVIRONMENTAL EFFECTS

## Light

The most dramatic effects of the environment on shape and size are those of light and day length. The effects of light have already been described in Chapter 8 and only a brief resumé is needed here. In the absence of light plants show *etiolation*; following exposure to light, leaves expand, stem elongation is reduced and plants become green (Fig. 8.4). For these responses, light is a source of information, not of energy as in photosynthesis, so only small amounts of light are required. The main light-absorbing pigments controlling these responses are *phytochrome* and cryptochrome.

The main influence of light on plant shape depends on both the ratio of red (R) to far-red (FR) wavelengths in the light and the total amount of light. Many species growing in heavy shade from a plant canopy, which filters out light in the red region of the spectrum, resulting in a low R/FR ratio, are tall and often have undesirably weak stems. Other types of shading from walls or nearby structures reduce the total amount of light without affecting its spectral composition, resulting in similar increases in height, plus the development of thinner, larger leaves. A further response to a low R/FR ratio is the suppression of axillary growth so that the plants tend to grow without branching; this effect is seen in forestry practice where trees are deliberately planted close together so that the trunks develop without excessive branching. Increasing the spacing between plants can, therefore, be considered as a means of increasing branching and reducing overall height. However, there is a considerable variation in the degree of

response and many shade-tolerant plants are largely insensitive to spacing.

Gardeners need to be aware of the particular importance of light in controlling *hypocotyl* elongation in seedlings. Many amateurs, of necessity, sow their seeds under the low light conditions of winter in greenhouses without artificial light or on window sills, while seed trays are often shaded to maintain a high humidity. Seedlings should, if possible, be transferred to better light conditions as soon as they have emerged, to avoid excessive elongation of the hypocotyls.

Other than the effects of shading, the effects of light on plant shape are relevant to horticulture in a number of ways. One of the most obvious is that etiolated, dark-grown plants have less fibre and so are less tough than those grown in the light. The exclusion of light in blanching, to achieve this, is a long-established practice as, for example, in the production of chicory, sea kale and forced rhubarb. The latter is interesting because, although the crop is grown in dark sheds, some light is needed for harvesting. Based on an understanding of human vision and a knowledge of plant responses to light, it is clear that green light would be the best colour to use, since our eyes are most sensitive to green wavelengths (Fig. 12.2) and this is the region which has least effect on the responses controlled by phytochrome and cryptochrome. In fact, green safe-lights have been used for many years in laboratories where research on the effect of light to prevent etiolation is carried out.

For plants growing entirely in artificial light, as in micropropagation, or where supplementary light is given during winter to increase photosynthesis, the light spectrum to which they are exposed may be very different from that of sunlight. Therefore, the effect of the light on plant shape must be taken into consideration when choosing lamps for these purposes. The use of artificial light is discussed further in Chapter 14.

### Day length

As discussed in more detail in Chapter 13, day length may also profoundly modify the pattern of growth by inducing flowering (often accompanied by stem elongation in rosette plants), dormancy in woody plants (formation of bud scales instead of foliage leaves) and the development of storage organs (scales in onion bulbs instead of foliage leaves). The planting date can also affect plant height as it may determine the duration of growth before terminal flower buds form. In *Chrysanthemum* cultivars, for example, later planting will mean shorter plants because autumnal short days induce flower-bud formation earlier. Use is made of this in commercial all-year-round production, where the time at which plants are transferred to short-day conditions is determined by the length of stem desired.

## Light direction

Finally, the direction from which light comes can also change plant habit. It is well known that, particularly in young seedlings, plants grow towards light coming from one direction, a process known as *phototropism*. This can result in lopsided plants and it is important to ensure, as far as possible, that light is uniform from all directions. Placing a reflecting surface behind pots on a window sill, for example, may help to reduce bending towards the outside light. A less well-known response is found in the seedlings of some tropical climbers, such as the Swiss cheese plant (*Monstera deliciosa*), which grow away from the light. In nature this would be towards the relative darkness of the nearest tree.

## Temperature

There are also some interesting responses to temperature. A higher temperature during the day leads to longer stems, while a lower day temperature results in a more compact and sturdier growth habit. For some time it was widely assumed that these effects were due to the difference between day and night temperatures. A day temperature lower than the night temperature would result in shorter plants, while a day temperature higher than the night temperature would result in taller plants. However, more recent work has shown that the effect on elongation is due to the absolute day and night temperatures, rather than to the difference between them. There was a strong positive effect on increasing height with an increase in day temperature in both tomato (*Lycopersicon esculentum*) and *Chrysanthemum* cultivars, while night temperature had little or no effect (Fig. 11.1). In many cases, for example *Chrysanthemum* cultivars, poinsettia (*Euphorbia pulcherrima*), tomato and *Fuchsia* spp., the effect of lowering the day temperature was greatest in the early part of the day. In both tomato and poinsettia, stem length was decreased by dropping the day temperature from 20 to 12°C for 4 hours at the beginning of each day, while lowering the temperature later in the day had much less effect. Unfortunately, this is not always the case and some plants (e.g. *Petunia* spp.) are more sensitive later in the day when it is more difficult to lower the temperature.

Gibberellin-deficient mutants do not show a response to a drop in temperature, and so it is suggested that the dwarfing effect of lowering the temperature may be associated with a decrease in

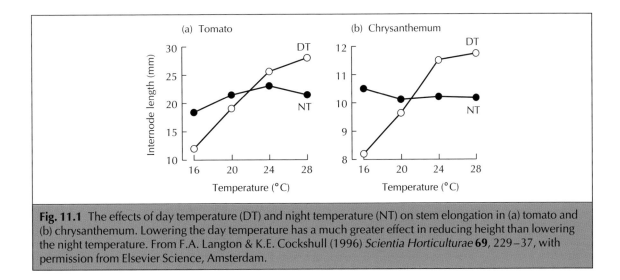

**Fig. 11.1** The effects of day temperature (DT) and night temperature (NT) on stem elongation in (a) tomato and (b) chrysanthemum. Lowering the day temperature has a much greater effect in reducing height than lowering the night temperature. From F.A. Langton & K.E. Cockshull (1996) *Scientia Horticulturae* **69**, 229–37, with permission from Elsevier Science, Amsterdam.

the production of gibberellin by the plant. Although not practicable out of doors, lowering the temperature for part of each day, especially in the early part of the day, is feasible in the glasshouse and has been suggested as an alternative to costly chemical sprays to achieve dwarfing.

## MECHANICAL TREATMENTS

The modification of growth patterns by mechanical manipulation has long been practised by gardeners who discovered by trial and error that certain treatments led to profound effects on the shape of the plant. With an increased understanding of the roles and movement of plant hormones and the operation of inter-organ signalling, it is now possible to understand better why some of these treatments result in the observed effects.

## Pruning

Among horticultural practices, pruning is one of the oldest and most widely used ways of achieving the desired shape and size of a plant, as well as being used to encourage maximum flowering and fruiting. The basis of all pruning treatments is apical dominance: when the apical bud is removed, the lateral buds are released from dominance and begin to grow out, while the uppermost bud begins to grow towards the vertical and becomes the new leading shoot. Before undertaking any pruning treatment, however, it should be recognised that this treatment always causes stress and is not necessarily the best way of achieving the desired results. It is better to choose the right plant for the right place and so avoid the need for remedial pruning. For example, no amount of pruning will make a Leyland cypress (×*Cupressocyparis leylandii*) suitable for a tiny garden and the right solution is to grow alternatives that are genetically small, or have been grafted on to dwarfing rootstocks (see Chapter 10).

One of the problems about pruning to restrict size is that plants have a constant root/shoot ratio and re-direct growth to maintain it. Consequently, hard pruning actually stimulates vegetative growth and increases the vigour of the pruned shoot. Thus hard pruning does not necessarily reduce size, although constant cutting back will ultimately weaken the plant. The general recommendation, therefore, is to prune vigorous shoots lightly and less vigorous shoots more drastically in order to achieve more balanced growth. It is better to thin out shoots by removing some of them entirely, rather than to cut them back. In order to achieve a shapely plant, cuts should be made to a bud that, as far as possible, faces in the desired direction.

Pruning is not only undertaken to control plant shape. In order to obtain maximum flowering, pruning must be considered in relation to the flowering habit of the plant in question. For plants that flower late in the year on the current season's wood, cutting back is normally carried out in the early spring to allow maximum growth before flowering; for plants that flower earlier in the season on the previous season's growth pruning normally takes place after flowering. Finally, pruning may be used to let light and air into the centre of a tree or shrub by removing weak or crossing branches, thereby encouraging strong growth and discouraging the development of diseases.

Detailed recommendations for the pruning of individual plants are widely available in the horticultural literature.

### Pruning established trees and shrubs

This often requires the removal of large branches and the recommendations for doing this have changed in recent years. Originally cuts were made flush with the main trunk to avoid leaving a stub that could die back and be the entry port for pathogenic organisms. However, it is now known that cutting flush with the trunk tends to result in large wounds that develop dead areas, rather than producing protective wound callus. Today, the recommendation is to carry out target pruning, which cuts through that part of the limb where cell division in the reaction zones immediately begins to form a defensive callus that spreads from all around the perimeter of the cut.

To carry out target pruning requires the identification of both the bark ridge, which is a conspicuous line of folded bark at the upper point of the union between branch and trunk, and the branch collar, which is a swollen area where the underside of the branch meets the parent stem. The ideal cut is one that slopes from just outside the bark ridge to a point where the branch meets the branch collar, without penetrating either area. The practice of applying tree paints to seal the wound against possible infection

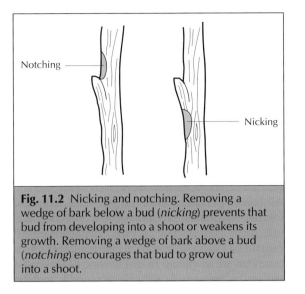

Notching

Nicking

**Fig. 11.2** Nicking and notching. Removing a wedge of bark below a bud (*nicking*) prevents that bud from developing into a shoot or weakens its growth. Removing a wedge of bark above a bud (*notching*) encourages that bud to grow out into a shoot.

often inhibits the cut from healing and is not now recommended.

### Pruning for shape: training young fruit trees as espaliers, fans or cordons

It is well known that vertical branches grow more vigorously than more horizontal ones and this fact has long been the basis of training fruit trees into special shapes, such as espaliers and cordons. When branches are tied horizontally, apical dominance is weakened and the lateral buds begin to grow out, a necessary pre-requisite for the formation of fruiting spurs. The practices of nicking and notching (Fig. 11.2) during the training of fruit trees also has its origin in apical dominance. In notching, removing a small wedge of tissue above a bud isolates it from the effect of the apex and increases the accumulation of materials from below; the result is to weaken the apical dominance and allow the bud to develop. Notching is particularly useful where it is desirable to encourage bud growth from low down on the plant. The opposite treatment, nicking, involves removal of a piece of tissue immediately below the bud, which has the effect of increasing apical dominance and weakening bud growth. It is useful for fan-training, where the apical dominance effect results in the development of branches at a wider angle to the main trunk.

### Pruning for size; root pruning

The relative constancy of the root/shoot ratio means that one method by which shoot vigour can be controlled reliably is root pruning. Originally this was a common practice to reduce the vigour of orchard trees, but has largely been rendered obsolete by the introduction of dwarfing rootstocks. Root pruning is still, however, used to promote flowering in young trees and to restrain their growth.

In the open ground, root pruning should be carried out over two seasons, removing half of the root in each year. A trench 30–40 cm deep is dug out about 120 cm from, and half-way round, the trunk during the winter and large roots cut through. The soil is then replaced and the tree mulched in the following spring. When pruning container-grown plants that have become pot-bound, the outer roots should be pruned back by approximately two-thirds and the plants re-potted. In order to maintain the balance of growth, it is also necessary to remove about one-third of the top growth.

### Bonsai

Perhaps the ultimate restriction of tree size is found in bonsai, which depends for its success on root pruning and/or restriction, the bending of shoots from the vertical by twisting them round wires and careful shoot pruning to achieve the desired final form.

### Topiary

Pruning to create special shapes such as globes and pyramids is becoming increasingly popular. Slow-growing plants such as box (*Buxus* spp.) and yew (*Taxus baccata*) are ideal and, once shaped, require only an annual trim. Faster-growing species such as *Ilex crenata* are also used widely but need clipping more often.

### Stooling and pollarding

*Stooling* is the name given to the technique of cutting shrubs to ground level to encourage the development of new shoots. This may be done for seasonal interest as, for example, the production of coloured young shoots in shrubs such as *Cornus alba* 'Sibirica' and *Salix alba* subsp. *vitellina*. The stems are shortened to 1–2 cm from the stump in early spring. Stooling is also practiced to obtain suitable wood for grafting (see Chapter 10).

Pollarding is the name given to the practice of pruning the plant back to a stump (usually about 60 cm high) in early spring. It is often used with cider gum (*Eucalyptus gunnii*) to encourage the development of the bluish juvenile foliage that is considered more attractive than the dark green adult leaves (see under Juvenility in Chapter 10).

Coppicing, a form of stooling, and pollarding were once widely used in woodland management to encourage the growth of long, straight branches on appropriate trees such as hazel (*Corylus avellana*), ash (*Fraxinus excelsior*) and oak (*Quercus* spp.) for use in building and furniture-making. These procedures are still used occasionally to provide wood for making items such as hurdles for garden use and some specialist furniture. Coppicing may also be practised in woodland areas of large gardens to let in light, thereby encouraging the growth and flowering of herbaceous species such as bluebell (*Hyacinthoides non-scripta*).

## Shaking and brushing

Manipulation of size by pruning is an old established horticultural practice. More recently, a very different procedure has resulted from the observation that exposure to air movements may affect plant height. Although the effect, called a *thigmo-seismic effect*, is not very noticeable out of doors because of the prevalence of wind, shaking or brushing plants may considerably reduce growth in greenhouses. For example, shaking *Chrysanthemum* cultivar plants for just 30 seconds twice a day was found to decrease stem length by as much as 30%. Since plants are continually being handled during pruning, staking, disbudding and other treatments, thigmo-seismic effects should not be ignored, although not all plants are equally sensitive. For example, marigold (*Tagetes* spp.) and sunflower (*Helianthus annuus*) are relatively sensitive to light brushing, while *Zinnia* spp. and China aster (*Callistephus chinensis*) are not. Salvia (*Salvia splendens*) and marvel of Peru (*Mirabilis jalapa*) have been found to be particularly sensitive, with stems growing to only half the length of those in untreated plants. A relatively small amount of brushing may produce a noticeable reduction in size of young seedlings and further treatments increase the risk of damage.

The effect of brushing to decrease height is an interesting technique because it is environmentally friendly, involving no chemicals, is effective to decrease height in most of the species tested so far and is within the reach of the amateur grower. One important application of the effects of shaking on plants is the recommendation that stakes for newly planted trees should be relatively short, in order to allow for some flexing of the young trunks, resulting in thicker trunks and sturdier growth.

## CONCLUSION

This chapter has shown that the size and shape of plants is in part determined by a diversity of internal factors, notably genetic make-up, the movement of hormones, the position of terminal and axillary buds and the root/shoot ratio. The effects of these are modified and/or modulated by a number of external factors, notably light, day length, temperature and wind. With a knowledge of these factors and how they interact, the gardener may intervene to produce plants that are of the desired shape and size for a site or purpose by choosing the most appropriate cultivar, pruning in the right way at the right time, staking correctly or, for plants grown in the greenhouse, controlling the temperature and light regimes.

## FURTHER READING

Brickell, C. (1979) Pruning. In *The RHS Encyclopedia of Practical Gardening*. Reed International Books, London.

Brickell, C. & Joyce, D. (1996) *Pruning and Training*. The Royal Horticultural Society. Dorling Kindersley, London.

Cleveley, A. (2004) The kindest cut. *The Garden* **129**, 46–9.

Hanks, G. & Langton, A. (1998) Growing to length. *The Horticulturist* **7**, 25–9.

Kirkham, T. (2004) *The Pruning of Trees, Shrubs and Conifers*, 2nd edn. Timber Press, Cambridge.

Salisbury, F.B. & Ross, C.W. (1992) *Plant Physiology*, 4th edn. Wadsworth Publishing Company, Belmont, CA.

# 12

## Colour, Scent and Sound in the Garden

### SUMMARY

The pigments responsible for the colour of flowers and leaves, their perception, and their importance as attractants for pollinators are described. The causes of variegation are outlined, and particular attention given to 'Tulipomania'. The environmental factors that affect the colour of plants, and the use of colour in the garden, are reviewed. Next the perception of scent, taste and flavour by humans is outlined and the different classes of compounds responsible for scent in plants described. The importance of aromatic compounds to plants is considered and the use of scent in the garden described. Finally, the nature and perception of sound in the garden are surveyed.

## INTRODUCTION

Flowers are often highly coloured to attract insects and pollinators and are the main but not the only source of colour in the garden. Others include coloured or variegated leaves and coloured fruits, stems and bark. The coloured or variegated leaves of some tropical plants which surround inconspicuous flowers may, like flowers, function as attractants for pollinators.

Pollinators may also be attracted to plants by scents that are sometimes pleasant and sometimes unpleasant, even disgusting, to the human nose. Pleasant scents are an added attraction to the garden experience and are particularly important in gardens for visually impaired people. Not all scents or aromatic substances produced by plants are attractants for pollinating insects and many have other important functions, such as deterring herbivores or pathogens.

Both colours and scents are caused by a large group of compounds known as secondary products, some of which also have a wide range of other functions in plants, including protection against pathogens

and/or herbivores. This chapter will consider the identity of the secondary products responsible for the colours and scents and the ways in which they are perceived by animals and humans.

Sound is a frequently neglected component of the total garden experience and, like scent, is especially important to those with impaired vision. It may be produced by a great diversity of garden plants, especially through their interaction with wind, and by water, birds and insects, particularly bees. Birds and insects may be enticed into the garden by the deliberate use of plants that provide appropriate food.

## COLOUR IN PLANTS

### The perception of colour

To some extent, the sensation of colour varies with the individual. There is also a strong aesthetic component. For example, harmonious combinations such as blue and purple are attractive to some people, while others prefer contrasting colours, such as red and green. The sensation of colour depends on the wavelength of light that reaches the eye. To human beings, light is that part of the electromagnetic radiation spectrum capable of stimulating the human eye, thereby making sight possible. Light comprises only a very small portion of the total electromagnetic spectrum and its limits are dependent on the absorption characteristics of *visual purple (rhodopsin)*, the light-sensitive pigment in the rod cells of the eye that is responsible for vision. Pigments in the cone cells allow discrimination between different wavelengths of light and give rise to the sensation of colour. Due to differences in these pigments, not all people see colour in the same way and some individuals are colour 'blind' in that they are unable to distinguish between certain colours. Lack of discrimination between green and red is the most common form of colour blindness. The wavelength limits of human vision also differ slightly between individuals. Below about 380 nm, there is no sensation of vision. As the wavelength increases above this, a faint stimulus is produced which creates in the brain the sensation of the colour violet. The colour perception then changes with increasing wavelength through blue, green, yellow, orange and red until at between 700 and 750 nm the light is once more invisible (Fig. 12.1). When all wavelengths are present in the light, the eye perceives the sensation 'white', while the absence of visible light is seen as 'black'. Sensitivity to light at any given wavelength also varies and, for human vision, maximum sensitivity occurs at about 550 nm in the green region of the spectrum (Fig. 12.2). The overall visual effect thus depends on both the sensitivity to light at the wavelengths reaching the eye and on the colour sensation that they evoke.

The limits of light in human terms lie between approximately 380 and 750 nm (it varies somewhat between individuals). The near infrared region at wavelengths beyond 700 nm has a heating effect on biological tissue, while wavelengths shorter than 320 nm in the ultraviolet produce undesirable chemical reactions and are damaging to cells. The mutagenic effect of exposure to the sun's short-wave radiation (less than 290 nm) in causing skin cancer is now well recognised.

The vision of most animals is different from that of humans and, from the point of view of the garden, the most important differences are found in animals that pollinate plants, such as bees and butterflies,

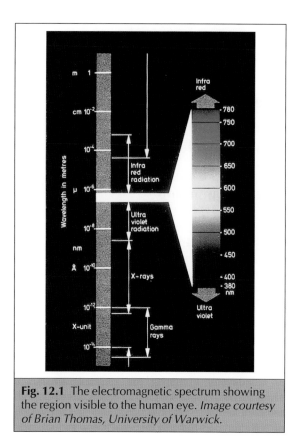

**Fig. 12.1** The electromagnetic spectrum showing the region visible to the human eye. *Image courtesy of Brian Thomas, University of Warwick.*

which can see further into the ultraviolet part of the spectrum. The way in which light is 'seen' by plants is also different. For example, the far-red region between 700 and 760 nm (which is scarcely visible to humans) is important for plants because it is absorbed by the pigment phytochrome, which is involved in the regulation of many important responses to light (see Chapters 8 and 13).

## Plant pigments

In order to produce a photochemical effect (whether vision or any other light-dependent process, such as photosynthesis), light must first be absorbed by a pigment. These are molecules that contain a *chromophoric group* which is responsible for their colour. They do not absorb light equally throughout the visible spectrum and consequently, when they are present in sufficient quantity, result in the colour of the perceived object. This is white when all visible

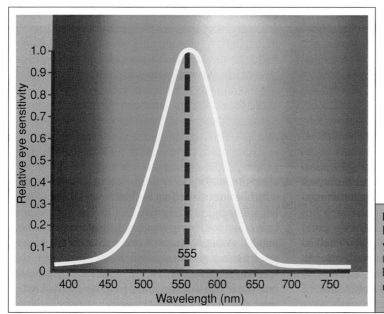

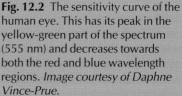

**Fig. 12.2** The sensitivity curve of the human eye. This has its peak in the yellow-green part of the spectrum (555 nm) and decreases towards both the red and blue wavelength regions. *Image courtesy of Daphne Vince-Prue.*

wavelengths are reflected and none are absorbed. Three main groups of pigment are responsible for the colours of leaves, flowers and other organs of the plant, namely *chlorophylls*, *carotenoids* and *anthocyanins*.

## Chlorophylls

The most abundant pigments in plants are the chlorophylls, which absorb light mainly in the red and blue parts of the spectrum to give the green colour of leaves (see Chapter 2). Although chlorophylls are the major light-absorbing pigments for photosynthesis, the chloroplasts also contain accessory pigments which absorb light and transfer energy to the photosynthetic system. Most of these are carotenoids (see Chapter 2). The abundance of chlorophyll means that the red, yellow and purple colours associated with other pigments are not usually evident in leaves until the chlorophyll is broken down. The best example of this is autumn colour, which is discussed in more detail below.

## Carotenoids

There are two groups of carotenoid pigments. The *carotenes*, first isolated from carrots, hence their name, consist of only carbon and hydrogen atoms and are mainly orange in colour. The *xanthophylls*, which have an oxygen atom in addition to carbon

and hydrogen, give the yellow colours of petals and fruits. The carotenoid pigments may be located in the chloroplasts as accessory pigments for photosynthesis, or they may be found in special pigment cells called *chromatophores*.

## Flavonoids

The anthocyanins, derived from the Greek words *anthos* (flower) and *kyanos* (dark blue), belong to a group of chemicals called flavonoids, of which more than 2000 have been identified. Most are soluble in water and accumulate in the vacuoles of cells, although they are synthesised elsewhere. They are often confined to the epidermal or outer layer of cells and are particularly important in determining the colours of flowers, but also occur in stems, leaves and fruits.

Anthocyanins consist of a coloured molecule, called an anthocyanidin, with a sugar group attached. The most common anthocyanidin, cyanidin, is purplish red and was first isolated from cornflowers (*Centaurea cyanus*). Pelargonidin, first isolated from a red *Pelargonium* spp., is orange-red, while delphinidin (from *Delphinium* spp.) is bluish purple. However, the colour may vary, depending on other factors. Anthocyanins are more red at low, more acid, cell pH, and more blue under alkaline conditions. An example of this is found in the flowers of morning glory (*Ipomoea tricolor*): the buds are reddish purple, but as they open the petals become bright blue. At the same time, the pH of the vacuoles in the flower cells increases from 6.6 to 7.7. Confirmation that the increase in pH is the cause of the change in colour comes from a purple mutant that fails to increase the pH of the vacuoles and therefore does not develop the blue colour.

Some anthocyanins can bond with metals to produce colours that are distinctly different from those of the free pigments. Probably the best-known example of this is found in *Hydrangea* spp.: the free pigment results in red or pink flowers, but when it is associated with aluminium it becomes blue. The fact that hydrangeas are more blue on acid soils may seem to contradict the statement that a low cell pH results in the pigment becoming red. The point is, however, that aluminium is only available to the plant in an acid soil (see Chapters 6 and 7) and it is the bonding to aluminium that changes the colour towards blueness. The pH of the cells is unlikely to be affected by the acidity of the soil.

Although similar in colour to anthocyanins, the betalains are unrelated nitrogen-containing secondary products. The two groups of compounds do not occur together in the same family. The betalain, *betanin*, gives the colour to red beets.

Flavones and flavonols also belong to the flavonoid group. They range in colour from yellowish to almost white, but even the colourless forms absorb in the ultraviolet and affect the 'colour' that is visible to bees.

In addition to their function as attractants for pollinators, both flavonoids and carotenoids have a protective action against the tissue-damaging effects of ultraviolet wavelengths, since they occur in the outer cell layers of leaves and absorb ultraviolet wavelengths while transmitting wavelengths that are active in photosynthesis. With the current progressive loss of the ozone layer and the resultant increase in the ultraviolet content of sunlight reaching the Earth, this function is likely to be of greater importance in the future.

### Light-sensing pigments

The light-sensing pigments phytochrome and cryptochrome, which are discussed in detail in Chapters 8, 9 and 13, are present in cells at such low concentrations that they do not result in any colouring of plant tissues.

## Pigments as attractants

In order to attract pollinators, plants need to stimulate the sense of vision and/or smell of the animal. The most common pollinators are insects, which have both a highly developed sense of smell and excellent vision. However, their sensitivity to colours differs from our own. Bees, for example, which are discussed further under Sound, below, are much less sensitive to red light and are more sensitive to blue light than humans and can see further into the ultraviolet. Undoubtedly bees see colours differently from humans and some wavelengths in the ultraviolet, which are invisible to humans, are visible to them. This becomes apparent when certain flowers are photographed using ultraviolet-sensitive film, revealing nectar guides that are invisible when the flower is viewed direct. Tests have revealed that some butterflies have distinct colour preferences: tortoiseshells, for example, prefer yellow whereas swallowtails are attracted more to blue and purple. Flowers pollinated by animals that feed at dusk, such as moths and fruit bats, are normally pale in colour, or white, so that they are more easily visible in poor light. Some flowers are pollinated by birds, which have acute vision but no sense of smell. Their sensitivity to different wavelengths is more like ours than that of insects and they are attracted to bright colours, often in distinctive combinations such as the orange and purple of bird-of-paradise flowers (*Strelitzia reginae*).

## VARIEGATION

Plants with variegated leaves that are streaked, marbled or patterned with colours other than green are prized by many gardeners. Variegation is also found in some flowers, but to a lesser extent than in leaves. The causes of variegation include *chimeras*, *transposons* (see Chapter 5), *pattern genes* and, in a few cases, *virus infection* (see Chapter 15).

## Chimeras

Chimeras are made up of genetically different cells. When the genes of these cells cause obvious differences such as, for example, a change in colour, then the different cell types become visible. Chimeras develop because flowering plants have a layered structure (see Chapter 5). Because of this, the outer layers of one partner in a graft can grow over the inner layers of the other, producing a plant made of cell layers of two distinct genetic types (for example, +*Laburnocytisus adamii*). Chimeras may also result when a mutation occurs in the growing point, causing part or all of a layer of cells to be genetically different from other layers in the same plant. This type of chimera is usually stable and, when the mutations affect the production of chlorophyll, can result in good garden plants with attractively variegated leaves, such as *Hosta* spp. and *Euonymus* spp. (see Chapter 5 for other examples). In evolutionary terms a chimera does not confer any selective advantage, so many may have arisen and died out. Gardeners, however, have selected them for their ornamental value, maintaining the variegation by taking shoot cuttings because they do not breed true from seed. However, they cannot be propagated by taking root or leaf cuttings, which disrupt the stable layered structure of the original growing point.

In some variegated chimeras and variegation resulting from other types of mutation, the variegated form is not stable and reversion to the all-green form may take place in one or more shoots. Because the variegated shoots have no chlorophyll in the white or yellow part of the leaf, they are less vigorous than the all-green shoots and, if left alone, the latter will eventually swamp out the variegation. Thus it is important to prune out any all-green shoots as soon as they are seen. Variegated shoots that lose their green colouration are unable to photosynthesise and normally grow very weakly.

## Transposons

Another cause of variegation in flowers and leaves is the occurrence of transposons (see Chapter 5). These can move around within the chromosome and occasionally 'jump' into a gene and inactivate it. If they affect a gene for flower or leaf colour, a plant with random spots, stripes or sectors may result. One of the best-known garden examples of transposons giving rise to an ornamental plant is that of *Rosa mundi* (*Rosa gallica* 'Versicolor') which has flowers with red streaks on a white background (see Chapter 5 for details of how this is caused). Transposons also affect the patterns of foliage if they disturb the production of chlorophyll. It is not uncommon for pale sectors to occur at random in leaves as, for example, in garden peas (*Pisum sativum*).

## Pattern genes

The variegation patterns in some leaves are controlled by pattern genes and are a normal genetic characteristic of the species. In this case, the plants will breed true from seed. For example, in species of the tropical genus *Maranta* the green leaves are regularly patterned with dark blotches, while in *Actinidia kolomikta* the leaves are tipped with pink and white. Another form of seed-transmitted variegation is seen in Our Lady's milk thistle (*Silybum marianum*) where pockets of air along the veins reflect the light and give a silvery effect.

## Viruses

Although not now of any great importance with respect to the ornamental value of plants, infection with viruses can result in some types of variegation (see Chapters 15 and 16). Two viruses are mainly responsible, and both have limited host ranges so are not a threat to other plants. Infection with vein-clearing virus is the reason for pale leaf veins in *Vinca major* 'Reticulata' and similar forms of ground elder (*Aegopodium podagraria*) and sweet violet (*Viola odorata*). Abutilon mosaic virus only affects species of *Abutilon*, producing a mosaic pattern of yellow/green on the leaves. The forms infected with this virus are *Abutilon pictum* 'Thompsonii', *Abutilon megapotamicum* 'Variegatum' and *Abutilon* × *milleri* 'Variegatum'. The virus has also been introduced by grafting to the hybrids 'Ashford Red' and 'Kentish Belle'.

### Tulipomania

The Netherlands was the setting for one of the strangest episodes concerning a virus-induced variegation in plants. Although it happened to some extent in both England and France, so-called Tulipomania reached its peak in the Netherlands between 1634 and 1637. The tulips that were introduced into Europe from Turkey were single colours, and once the initial stocks had been bulked up the bulbs were

**Fig. 12.3** The *Tulipa* cultivar 'Semper Augustus', in which the 'breaking' is due to a virus infection. *Photograph courtesy of the Royal Horticultural Society, Lindley Library.*

relatively cheap. However, such a plain and relatively valueless bulb could suddenly produce flowers that were marked in contrasting colours. For example, a plain red flower might emerge the following spring with petals marked with intricate patterns of white and deep red. It was this 'breaking' of the colour that produced the most sought-after tulips of the time and resulted in bulbs being sold for extraordinary prices. A single bulb of the variety 'Semper Augustus', which has a white flower patterned with a red flame from the base to the tip of the petals (Fig. 12.3), was already selling for 1000 florins in 1623 and by 1633 was valued at 5500 florins, finally reaching an estimated value of 10 000 florins at the height of Tulipomania. The average annual income at the time was about 150 florins, and the highest price ever quoted for a single bulb of 'Semper Augustus', 13 000 florins, was more than the cost of the most expensive houses in Amsterdam. Tulips of this kind thus became the ultimate status symbol of the time. Since the cause of breaking was a virus, it is not surprising that broken tulips tended to grow poorly and produced few offsets. The scarcity of the latter increased the value still further, since the virus is not transmitted to the seeds, and offsets were the only means of producing plants true to the original variety.

At the time, the cause of breaking was a mystery. It occurred randomly and only in a small number of plants, perhaps one in a hundred, and it was probably this element of chance that drove the spread of Tulipomania in the Netherlands. A plain, relatively valueless bulb might emerge one season miraculously feathered and flamed in contrasting colours and worth many hundreds of times its original cost. Since the cause was not known, the effects could not be controlled, although many attempts were made to induce breaking by treating the soil in various ways. It was not until the late 1920s that breaking was shown to be caused by a virus that partly suppresses the colour of the outer cell layers in the flower, allowing the underlying colour to show through. The latter is always white or yellow and the darker colour, red, purple or brown, looks as if it is painted on the petals from the base. This contrasting colour is always sharply defined, resulting in intricate flame- or feather-like patterns on the petals. Although they were the most sought-after plants of the seventeenth century, the fashion for broken tulips later declined and today growers try to eliminate the virus infection that causes 'breaking', although some enthusiasts continue to maintain the old broken forms.

## ENVIRONMENTAL FACTORS INFLUENCING COLOUR

The intensity of colour developed by flowers and leaves is influenced to a considerable degree by environmental factors such as light and temperature. High light intensity strengthens colours, particularly where these depend on anthocyanins, which are produced as a by-product of photosynthesis. A good example of this is when the reddest apples occur on the outside of the tree. High temperatures have the opposite effect. For example, the bronze colours of *Chrysanthemum* spp. tend to become yellow at high temperatures because the amount of red anthocyanin pigment is reduced, while the yellow carotenoids are unaffected. Similarly, many pink flowers in which the colour is largely due to anthocyanins may become almost white at high temperature. That this is exacerbated by low light is observed in some house plants, which may be deep pink when brought into the home but rapidly fade to almost white if kept in conditions of low light and warmth.

## Seasonal changes

One of the most spectacular environmentally induced seasonal changes in colour occurs during autumn, when leaves can range from the yellows and oranges of birch (*Betula* spp.) and mountain ash (*Sorbus* spp.), through the vivid red of sugar maples (*Acer* spp.) and oaks (*Quercus* spp.), to the purples of some Japanese maples (*Acer palmatum*). The short days of autumn induce the cessation of growth followed by the onset of dormancy (see Chapter 13). In deciduous trees this is accompanied by leaf fall, which is preceded by changes in leaf colour as the chlorophylls begin to break down, revealing the colours of the other pigments present in the leaf. The yellow/orange carotenoids also start to decompose rapidly in autumn but anthocyanins and other flavonoids accumulate, leading to the development of yellow, bronze and red tones. This increase in anthocyanins occurs because, in autumn, the leaves contain more sugar than usual; photosynthesis continues, but at the lower prevailing temperatures the movement of sugars out of the leaves is reduced, resulting in an excess of sugars in the cells. These sugars are then converted to anthocyanins. Since photosynthesis is increased by high light intensity, bright autumn days result in a good display of autumn colour.

Although autumn colour is an annual occurrence, the intensity of the display varies widely from year to year and in different locations. Some of this variation is genetic since it has been found that the location of origin can have an effect on the intensity of autumn colour. For example, when plants of native northeastern azaleas (*Rhododendron* spp.) in the USA were raised from seed collected from different regions it was found that the intensity of leaf colour was related to the origin of the seed. This emphasises the importance of selecting strains for good autumn colour, which is usually most successful when individual plants with good colouring are propagated vegetatively.

The environment also plays a significant part in the timing and intensity of autumn colour. Lower temperatures usually increase the intensity of leaf colour. For example, leaves of *Acer palmatum* 'Bloodgood' developed much brighter colours at a night temperature of 14–18°C than when grown at 26°C. Plants growing in the shade often colour poorly. A good example of this effect is displayed by *Fothergilla major*, where leaves in full sun turn a vibrant red but those in the shade only become pale yellow. Nutrition may be another factor. For example, high levels of nitrogen in the soil may decrease the amount of colour because the sugars, which would normally form pigments, combine with nitrogen to form proteins.

Autumn colour is unpredictable because it is influenced by many factors, several of which are beyond the control of the gardener. Those which are include the selection of the best cultivars, growing plants under the best light conditions and not overfeeding with nitrogen. Other than this, the gardener can only hope for bright autumn days with cool nights which, together, will increase the free sugars in the leaves and so lead to a brilliant autumn display.

## THE USE OF COLOUR IN THE GARDEN

Although the scientific basis of colour in plants is well understood, the use of colour in the garden ultimately depends on the likes and dislikes of the

**Fig. 12.4** A colour chart that can be fanned into a wheel shows primary and secondary colours as well as intermediate shades. *Photograph courtesy of the Royal Horticultural Society, Lindley Library.*

gardener. For many, it is harmonious combinations of colour that appeal and these are generally more restful to the eye than contrasting colours. Harmonious colour combinations are those which are next to each other on a colour wheel, which consists of the primary colours red, yellow and blue, the secondary colours orange, green and purple and a large number of intermediate shades (Fig. 12.4). Contrasting colours are those that lie opposite to each other on the wheel, such as blue and orange, or green and red. There are sound biological reasons for the contrasting effect. If, for example, you stare at a red dot on a white background and then look away at the white area, you will see a faint green circle. The reason for this is that the cones in the retina of the eye become fatigued as they are repeatedly stimulated by the red light reaching them. As you look away from the red spot, the cones send a signal to the brain that the eye is looking at a red-deficient area, which the brain interprets as green because green and red are 'complementary' in terms of the signals that they send to the brain. When red and green plants are grown together in the garden, the messages from the eye continuously reinforce each other as the eye moves between red and green, with the result that both colours appear brighter.

The way colour is experienced depends on the attributes of the light falling on the plant. For example, the direction from which light falls on a border changes depending on the time of day. With the sun behind the viewer, the colour seen is reflected from the plant. When sunlight comes from the other direction, however, the colour seen has been transmitted through the plant. Viewing a plant with transmitted light increases the contrast between light and dark and results in a more dramatic effect; reflected light, by contrast, may leave the border looking flat. The perceived colour also varies widely depending on whether the plant is in the sun or in the shade.

Another factor is that the colour composition of sunlight changes throughout the course of the day, especially early in the morning and late in the evening, and this has a strong influence on the way in which colours are perceived. In the early morning and evening the colours appear more intense, while at midday the colour can appear washed out and leaves start to appear paler and less intensely green.

Many plants change colour as they age, or if they become infected by a pathogen, which may be a problem when choosing colours for the border. For example, *Allium* 'Beau Regard' is mauve when in flower, but the persistent seed heads are an intense straw colour. The petals of *Rosa* 'Nevada' are creamy white when young, but age to pink with intense pinky red spots where infection by an opportunist fungal pathogen has occurred (see Chapter 15).

Colour may be used to deceive the eye. Yellow has the effect of demanding attention and so draws the eye along a border or through a planting. The doyenne of colour-gardening, Gertrude Jekyll, planted bright yellow marsh marigolds (*Caltha palustris*) in the narrow canal at Hestercombe in order to move the eye along the line of the canal. Colour may also be used to distort the length and width of paths. Red tulips scattered in the borders on either side of a path makes it seem wider and shorter, whereas the pastel blue of catmint (*Nepeta racemosa*) makes a short path appear longer.

New cultivars with novel colours or combinations of colours are constantly being produced by plant breeders, since the biochemical pathways responsible for the synthesis of pigments are under the control of genes and therefore heritable (see Chapter 5). Moreover, genetic modification techniques are now being developed to insert the genetic factors controlling the synthesis of particular pigments into species that do not normally exhibit particular colours, for example to produce blue roses. The resulting cultivars are not, however, currently available to gardeners in the UK.

Ultimately the use of colour in the garden is a matter of taste, but a knowledge of the factors that control the amount and composition of the pigments formed by the plant, as well as an understanding of how colour is actually perceived by the eye, can help the gardener in the choice of plants to achieve the desired result.

## SCENT AND FLAVOUR

Sensing the scent of plants and tasting their edible parts are two separate physiological processes, but the impulses they induce combine in the human brain to produce the flavours associated with particular foods. Scent contributes about 80% and taste about 20% to the overall flavour of a plant.

Scents are sensed by specialised receptor cells at the top of the nasal canal. Molecules reaching these cells must not only diffuse through air, but also the mucous layer that surrounds the receptors. Thus, they usually have a molecular mass of less than 400 Daltons (Da), a neutral pH and a high vapour pressure. The receptor cells are elongate, with hairs at their tips that float in the surrounding mucus. The hairs have receptor sites to which scent molecules bind for sufficient time to cause a reaction in the cell. There are different sites for different scents, and once a molecule has bound to a site it induces a change in the membrane resulting in the movement of calcium molecules in or out of the cell. This in turn establishes an electrical current that stimulates nerves attached to the receptor cells, and these carry the impulses to the olfactory lobe of the brain, just above the nasal cavity. Different signals travel to different areas of the lobe, thus allowing discrimination of the scent signals.

Taste is perceived in the taste buds of the tongue and mouth as four principal components: sweet, sour, salt and bitter. Each taste bud is made up of approximately 100 long cells clustered together, their tips protruding from openings in the layer of small cells coating the mouth. The different tastes are detected in groups of taste buds located in different parts of the tongue and mouth and each induces a different physicochemical reaction. Saltiness, only rarely associated with plant tissues, is detected when sodium flows into the taste cells through ion channels in the membrane, resulting in a build up of this chemical.

Sour flavours induce the ion channels to close, preventing the efflux of potassium and therefore causing a build up of this chemical. Sweet molecules are detected by receptors on the protruding surface of the cells of the taste buds. Only substances with particular molecular configurations, such as those of the natural chemicals sucrose (sugar) and fructose (fruit sugar) are able to bind to the receptors to produce the signal that denotes sweetness. Bitter tastes are also detected by the receptors, but these are not well understood. All these changes induce electrical currents which are transmitted to nerve endings at the base of the taste cells and ultimately to the brain. Finally, certain flavours are detected by pain receptors: for example, 'hot' peppers and some 'hot' spices activate the same neurons in the mouth as high-temperature liquids and solids. Further aspects of the taste and flavour of food plants are dealt with in Chapter 17.

Most of the chemicals responsible for the scent of flowers, leaves and other plant parts are from a class of secondary products known as essential oils. These are mainly mixtures of volatile mono- and sesquiterpenes, unsaturated hydrocarbons (i.e. with double bonds) with a common building block of $C_{10}H_{16}$. Some are sweet-smelling, whereas others are more pungent. The chief monoterpene ingredients of peppermint oil, produced by peppermint leaves (*Mentha × piperita*), for example, are menthol and menthone. Eucalyptus oils derive their scent from the monoterpene 1:8 cineole. Many essential oils are highly complex mixtures: orange peel, for example, contains 71 volatile compounds, most being monoterpenes. Other essential oils include phenolic compounds such as thymol from *Thymus* spp., aromatic aldehydes such as vanillin from *Vanilla* spp. and open-chain primary alcohols such as geraniol from *Geranium* spp.

Most essential oils are produced in special cells, multicellular glands or vessels that may occur in leaves or flowers or other plant parts. They may be held, following synthesis, in pockets dotted over the plant surface or they may be secreted direct on to the surface. Sometimes essential oils combine with resins, forming highly aromatic oleoresins. Crushing and rubbing the leaves of scented plants releases the essential oils from the cells or glands containing them. Essential oils are volatile, so they evaporate more easily under warm conditions, thereby intensifying the scent when the plants are crushed. When aromatic plants are dried, for example to make pot pourri, the aromatic oils continue to release their scent until they have evaporated completely.

Most scents, especially of flowers and leaves, enhance the overall garden experience, being important adjuncts to shape, colour and sound (see below). In the case of herbs, they may also have important culinary uses, while scents from flowers and leaves may be extracted by solution or distillation and used as important ingredients of perfumes, as in the case of geraniol. Essential oils from plants may even be responsible for the colour of mountains. The bluish colour of the Blue Mountains of eastern Australia and the Blue Ridge Mountains of Virginia in the USA, for example, is probably the result of the atmospheric scattering of blue light by tiny particles derived from aromatic terpenes (an aromatic compound is defined chemically as one that contains a benzene ring in its molecules).

The biological functions of essential oils are diverse, and only poorly understood in most cases. In some instances they may simply be waste products that are sequestered in leaves and other organs to prevent their interfering with normal metabolic processes. However, most components of essential oils are synthesised by specialised biochemical pathways, suggesting that their production has evolved for specific purposes in response to particular selection pressures. Many clearly act as attractants for pollinating insects, which have a highly developed sense of smell, and are important in combination with shape and colour in this respect. Night-flowering plants in particular rely heavily on scent to attract pollinators because, although their white or pale-coloured flowers show up well in the dark, they are not always completely effective as attractants. It is for this reason that white-flowered species often have a stronger scent than species with coloured flowers. Winter-flowering plants such as ivy (*Hedera* spp.), which often have insignificant flowers, also rely on their powerful scent to attract the pollinating insects that are so scarce at that time of year.

Some of the aromatic compounds produced by plants as insect attractants are not detectable by human olfactory receptors, notably *pheromones*, which are chemicals produced by insects to induce a specific behavioural response, usually mating, in individuals of the same species. Pheromones may, however, be useful to the gardener in biological control procedures for garden pests (see Chapters 15 and

16). In some cases pheromones are produced by plants to attract pollinators, as in the case of bee orchids (*Ophrys apifera*), which employ pheromones to attract bees to attempt to mate with the bee-shaped flowers, thereby transferring pollen.

Not all of the scents that attract pollinators are pleasant for humans. For example, plants pollinated by flies and midges, such as lords and ladies (*Arum maculatum*) and other *Arum* spp., frequently emit a smell similar to that of rotting meat.

Essential oils and other aromatic compounds may also be important to plants in deterring large herbivores, especially in those species that evolved in arid regions where fresh vegetation is scarce, whereas others may be of significance in providing protection against parasitic and herbivorous insects. For example, pyrethroid is an insecticidal monoterpene ester that occurs in the leaves of *Chrysanthemum* cultivars. Because it is not toxic to mammals, only to insects, this compound and its synthetic derivatives are widely used in garden insecticides (see Chapters 15 and 16). Lactones, which are sesquiterpenes found in *Salvia* spp., are known to be strong feeding deterrents to both insects and mammals and, like many other feeding deterrents, they taste bitter to humans. One of the most pungent and best known of the terpenoid essential oils is turpentine, which is formed in specialised cells of *Pinus* spp., and is highly toxic to tree-killing bark beetles.

There are some instances of a single species containing both attractant and repellent essential oils. For example, ponderosa pine (*Pinus ponderosa*) contains both limonene, a terpenoid bark-beetle repellent related to turpentine, and pinene, another terpene which is an insect attractant.

Other aromatic deterrents to herbivores and fungal pathogens include the phenolic substances. All phenolic compounds have an aromatic ring structure with attached substituent groups such as hydroxyl, carboxyl and methoxyl groups or other, non-aromatic, ring structures. Many are synthesised via the shikimic acid pathway, which involves the phenolic acid shikimic acid. This begins with compounds derived from the Calvin cycle of photosynthesis and from the respiratory pathways (see Chapter 2). Ultimately, the shikimic acid pathway leads to the synthesis of *lignin*, the tough, complex polymer that is the major component of wood. Many of the phenolic intermediate compounds in the shikimic pathway are highly toxic to insects and fungi, and appear to be involved in the resistance of plants to attack by such organisms. Indeed, the synthesis of some fungitoxic phenolic compounds, the phytoalexins, is not triggered until infection occurs.

Many phenolic compounds become oxidised to dark, fungitoxic substances when exposed to air, as when cells are wounded or crushed. A well-known example is the browning of the flesh of apples or potatoes when they are cut.

Most of the phenolic compounds that can be detected by the human nose or taste buds are bitter, but some, such as coumarin, are sweet-smelling. More than 500 coumarins exist in nature, although only a few are found in any particular plant family. Coumarin itself is a volatile compound formed from a non-volatile glucose derivative upon senescence or injury, notably in clover (*Trifolium* spp.), where it is responsible for the characteristic aroma of new mown hay. Further phenolic substances, the furanocoumarins, are not toxic until activated by light in the UVA region between 320 and 400 nm. These are abundant in members of the Umbelliferae, including celery (*Apium graveolens*) and parsnip (*Pastinacea sativa*).

Other toxic compounds responsible for scent in plants are the cyanogenic glycosides. These, which occur for example in cherry laurel (*Prunus laurocerasus*), are separated in the cells from enzymes which act upon them to release hydrogen cyanide gas (HCN). Enzymes and substrate only come together when the leaves are crushed, thus releasing their characteristic, but highly toxic, scent of bitter almonds. Leaves of this plant were once used in the 'killing bottles' of entomologists. Poisonous and irritant plants are considered further in Chapter 15.

A final biological function for some essential oils and other aromatic substances may be to prevent the growth of a plant's competitors, a process known as allelopathy. The essential oils are washed into the soil, preventing the germination of the seeds or the survival of the seedlings of competitors which might compete for water, nutrients and light. A related phenomenon involving the coumarin scopoletin is the inhibition of germination of a plant's own seeds. This compound occurs widely in seed coats (e.g. of tomato, *Lycopersicon esculentum*), and prevents germination until it is washed out. Presumably, the ecological significance of this is that if there is sufficient water present to remove the scopoletin, there will also be sufficient water to enable the seed to germinate and for the seedling to become established.

Whatever the biological function of scents, sweet-smelling plants and flowers have for centuries been prized by gardeners and used by them to enhance the pleasure to be derived from their creations. It is, therefore, unfortunate that in the pursuit of bigger, more complex and more colourful flowers during much of the twentieth century, scents, or to be more precise the genes controlling the synthesis of the chemicals that cause them, have often been overlooked by plant breeders, or even sacrificed in favour of what were perceived as more desirable characters. Fortunately, this trend is now being reversed and increasing numbers of scented modern cultivars are appearing in catalogues.

Not all new scented plants will be equally attractive to all gardeners, for the enjoyment of aroma depends on the degree to which an individual's sense of smell is developed, the sensitivity to particular kinds of scents and the previous experiences associated with them. Even aesthetic considerations may also come into play. Whatever the caveats, however, to the vast majority of people the scent of flowers and leaves is an indispensable component of the pleasure to be derived from a garden.

## SOUND

## The nature, perception and appreciation of sound

In simple physical terms, sound is a form of energy generated when an object vibrates, creating movement in the molecules of an elastic medium. The elastic medium most relevant to the gardener is air. Sound moves through air as longitudinal waves in which a region of high pressure travels at 344 ms$^{-1}$ (at 20°C), the so-called speed of sound. Light travels very much faster than this, which is why we very often see a distant event, such as a hammer striking a stake, before we actually hear the sound it makes. Humans hear sound when the moving molecules of the sound waves in air reach the tympanic membrane, or eardrum, causing it to vibrate and send pulses of electrochemical energy to the brain.

The human ear is receptive to only a relatively limited range of frequencies between 20 and 20 000

Hertz (Hz; or cycles per second). Sounds with frequencies below 20 Hz are said to be infrasonic and those above 20 000 Hz are ultrasonic. They may be detected by pets and other mammals in the garden, but are not detected by the human ear. This difference has been used to advantage in the design of devices which emit high-frequency sound to deter unwanted mammals such as moles or other people's cats without disturbing either the human owner of the garden or visiting birds.

The intensity of sound, the loudness, is measured in decibels (dB). Quiet garden sounds, such as rustling leaves, are usually 20 dB or less, while the sound of city traffic is in the region of 70 to 100 dB.

The pitch of a sound, not important to all gardeners, but certainly of importance to those who also enjoy music, is the property that characterises its highness or lowness to a hearer. It is related to, but not identical to, frequency. Below about 1000 Hz the pitch is slightly higher than the frequency, while above 1000 Hz the position is reversed. The loudness of a sound also affects its pitch. Up to 1000 Hz an increase in loudness causes a decrease in pitch; from 1000 to 3000 Hz the pitch is independent of loudness; and above 3000 Hz an increase in loudness causes a rise in pitch. This may be important in determining the acceptability of the sound of certain kinds of garden machinery.

The pleasure or otherwise that a gardener may derive from different sounds depends on a diversity of factors. Most importantly, different people vary in their ability to hear sound of different frequencies and intensities. There are few general rules that can be applied here, except that for most people the ability to detect sound in the higher part of the frequency range declines with age. This information may be of use when deciding which sounds, if any, the gardener wishes to incorporate into, or exclude from, a garden. Other factors are largely aesthetic, but not entirely so, for some relate to human psychology. For example, the emotion associated with a particular kind of sound, or the listener's background, experience and age, may be important. In addition, the location of the source of a sound may be significant. For example, if a sound comes from within one's own garden it may be acceptable, whereas if it comes from someone else's it may be perceived as invading one's 'territory', and therefore be unacceptable. Thus, as a general rule, it is important to keep as low as possible the decibel level of all sound from mechanical garden tools, for although the sound they emit may be acceptable to the user, this may not be the case for neighbours. This can easily be achieved by taking sound emission into account when purchasing equipment. Baffles, such as fences and hedges, may also be used to disrupt sound waves and thereby minimise their impact on others, or to protect oneself from unwanted sounds created by others.

## Sounds for pleasure

There are certain sounds that almost all gardeners would agree are attractive: moving water, the soft rustling or swishing as wind passes through the leaves and inflorescences of grasses or the leaves and branches of trees, the hum of flying insects and the calls and songs of birds, for example. Careful planning, based on scientific knowledge, can ensure that the appropriate sources of sound are incorporated into the garden.

### Water

The sound of moving water from a water feature can be an important means of creating, by association, a sense of being cool and relaxed. As with so many sounds in the garden, however, sensitivity to the pitch and intensity of the sound emanating from moving water varies from person to person. As a general rule, the more turbulent the water flow relative to volume, the more likely it is that the sound generated will be unsettling rather than peaceful. Careful thought and experimentation concerning the volume of water used, the means of inducing turbulence and therefore generating sound waves (waterfalls, rocks and types of fountain for example) and the speed of flow as determined by incline and/or size and type of pump used, should precede the installation of any water feature. Getting the combination wrong can result in the sound of water having the reverse effect of that intended.

Care must also be taken with the selection and siting of pumps. Most submersible, electric pumps create little sound, but those that of necessity must be sited on the surface require careful sound insulation and screening with plants or artificial baffles to disrupt or deflect the sound waves, otherwise the sound of the machinery will modify or even mar the sound of the water itself.

## Plants as sources of sound

Plantings may themselves be used in a variety of ways to create soothing or pleasurable sounds. For example, ornamental grasses and their relatives, especially the more upright forms such as *Miscanthus*, *Molinia* or *Stipa* spp., bamboos (Poaceae) and sedges (*Carex* spp.), may be planted in positions where they catch the breeze. Gentler sounds may be created with quaking grasses (*Briza* spp., especially *Briza media*), whose diffusely branched inflorescences, which appear in May, are tipped with pendant spikelets resembling puffy oats. These rattle and rustle in summer breezes, especially as they ripen and dry out, when the air-filled tissues amplify the sound. Indeed, the sound emitted by all grasses, but especially those with broader leaves, changes as they dry out and the tissues fill with air.

Trees, if carefully chosen and sited, may also be used to generate pleasant sounds as breezes pass through them and set up vibrations in the air. Most broad-leaved species may be used in this way, while the movement of the leaves of trees with a versatile petiolar anatomy (see Chapter 2), such as common birch (*Betula pendula*), adult leaves of cider gum (*Eucalyptus gunnii*) and especially aspen (*Populus tremula*), generate a characteristic fluttering sound (Fig. 12.5). The clattering of the palmately divided, hard sclerophyll leaves (see Chapter 1) of palms such as the windmill palm (*Trachycarpus fortunii*), created as they strike one another, is evocative of the tropics.

Other plant sounds that may give pleasure in the garden and may be encouraged by careful planting are those generated by fallen leaves. For example, the fallen leaves of deciduous trees rustle in the wind or when disturbed by birds searching for food, the sound again amplified by the air-filled dead tissues. The dead, air-filled fallen leaves of hard-leaved, sclerophyll (see Chapter 1) plants such as holly (*Ilex* spp.), cherry laurel (*Prunus laurocerasus*) and *Magnolia* spp. emit a particularly distinctive, clattering sound when disturbed.

Another plant-generated sound worthy of special mention is the popping sound created on a hot summer afternoon as the pods of broom (*Cytisus* spp.) and other shrubby members of the pea family (Leguminoseae) disperse their seeds. The anatomy of the pods, with bands of cells that react to desiccation in different ways, causes the two halves of the pod to twist apart violently to eject the seeds, generating a distinctive sound as they do so. The planting of species, rather than hybrids, which may be sterile, and the choice of a sunny site, are essential to success in introducing this sound to the garden.

A final example of plant sound is that generated by seeds in the dried pods of such plants as *Aquilegia* spp. when tapped or moved by the breeze. Once again, the presence of air in the dead tissues helps to amplify the sound.

The list of ways in which sounds from plants may be incorporated into the garden is almost endless. The reader is encouraged to apply the scientific method

**Fig. 12.5** The leaves of aspen (*Populus tremula*), which have a versatile petiolar anatomy, causing them to emit a characteristic fluttering sound when agitated by wind. *Photograph courtesy of Graham Titchmarsh/RHS Herbarium.*

to discover what suits the garden best. Listen to the sounds that plants make in the wild, in other people's gardens and in garden centres, note the direction, intensity and turbulence of the wind in different parts of the garden and be aware of one's own sensitivity to different kinds, pitches and intensities of sound. Then experiment by planting to create particular sound patterns, modifying the planting scheme in the light of the experimental results or further observations.

## Insects

Nothing is more evocative of summer than the gentle, low-pitched hum of bees or the chirruping of grasshoppers in long grass. The sound of the latter is generated as part of the mating ritual and results from the rubbing of one part of the body, the serrated file, over another, the scraper, as when a comb is drawn across the edge of a card. The sound of bees is generated by their flapping flight, only recently elucidated experimentally with tethered insects in wind tunnels in which the movement of the air around the wings is visualised using smoke. By flapping and changing the shape of their wings insects create eddies of air that not only give them lift and forward movement, but also generate sound waves which may be detected by the human ear. The most soothing are those in the lower register such as emanate from bees. At the other extreme, the high-pitched whine of a mosquito in flight can be most irritating to those with ears that are sensitive to it and who, from previous experience, know what it portends.

Bees of all kinds – bumblebees, solitary bees and honey bees – may be encouraged in the garden by minimal (ideally zero) use of pesticides, by the careful choice of plants to provide pollen and nectar and by the provision of artificial nest sites. The benefits from such actions are considerable: pleasing sounds on warm days and excellent pollination, especially of fruit trees and pod-forming vegetables. Additionally, the adventurous gardener may take up the keeping of honey bees, in which case the production of honey and the opportunity of close observation of the life cycle and behaviour of a colonial insect species are additional benefits.

In selecting plants that are attractive to bees, it is important to remember that they may need a supply of both pollen and nectar throughout the spring, summer and autumn. In the case of honey bees, which may emerge and forage on warm days in winter, sources of sugar-rich nectar and pollen are

required at this time too. Other types of bee tend to hibernate during the winter months, only emerging in the spring. Different species of bees and different types of bees in a colony differ in their ability to take pollen and nectar from different types of flower, depending on their body shape and weight, and their tongue length. To encourage the largest number of bee species possible, it is essential to stock the garden with as rich a mix as is practicable, to provide a range of flowering plants at all seasons. In the northern parts of the UK, garden plants are especially important in extending the length of the foraging season of bumblebees beyond the short flowering period of wild species. Also, it is important to remember that highly bred hybrid plants and selections may not be suitable as sources of food for bees: many hybrids produce little or no pollen and/or nectar and double flowers may prevent access to these foods. The buzzing of bees around the flowers of *Eucryphia* spp. in late summer can be almost deafening, as they are so attractive to insects, but many other species and cultivars are also good bee plants for the garden; Chapter 19 lists some examples. An average, well-stocked garden will attract honey bees and six or more of the 17 known species of bumblebees; indeed, in one wildlife garden in the highlands of Scotland 13 species have been recorded.

Finally, it is important to remember that both honey bees and bumblebees will sting if provoked. The specialised barbed sting of the honey bee remains in the skin and the bee, from which it came, dies. The sting of the bumblebee, a modified ovipositor, is withdrawn from the victim and its owner lives on. The venom of bee stings contains proteins to which many people are allergic, some severely so, and it is therefore important to take care when bees are present and not to interfere in any way with their activities. Of course, many of the plants that are attractive to bees and the non-stinging solitary wasps are also attractive to the rather more unwelcome colonial wasps and hornets. Again, caution should be exercised when these are about.

## Birds

Birds provide a diversity of pleasing sounds in the garden. Their calls and songs are used, however, for a diversity of pragmatic, biological purposes, including marking territories, communication, especially to indicate danger, and attracting or protecting a mate.

Attracting birds to the garden is easy, but some sacrifices must be made. Pesticides and slug-pellets must never be used, as insects and their larvae, worms, slugs and snails are major sources of food for garden birds and for small mammals which provide food for owls. It is also important to leave areas of the garden unweeded, indeed uncultivated, to provide cover for birds and a supply of weed and grass seeds and insects and larvae. A pond, shallow at the edges, will provide water and aquatic insects. And it is essential to grow a range of plants that will provide food throughout the winter: buds, seeds, berries and other fruits, insects and their larvae. Evergreen and deciduous trees and shrubs, ivy, old walls and buildings provide nest sites and safe havens.

The common garden birds, whether they are resident or migrants, and their principal sources of food, are discussed in Chapter 19 as are common garden plants that are attractive to birds.

## CONCLUSION

Colour, scent and sound are all components of the well-stocked garden. Which of these give pleasure and which offend, and what combinations of such sensory stimulants are most desirable is, however, a question of personal preference. By combining horticultural and scientific knowledge, and by using the scientific method, the gardener may effectively manipulate the garden environment to create and attract whatever combinations of colours, scents and sounds give the greatest pleasure.

## FURTHER READING

Baines, C. (2000) *How to Make a Wildlife Garden*. Francis Lincoln, London.

Burras, J.K. (1997) Streaks and sectors. *The Garden* **122**, 12–15.

Dourado, A. (1997) Unpredictable pigments. *The Garden* **122**, 717–19.

Hawthorne, L. & Maughan, S. (2001) *RHS Plants for Places*. Dorling Kindersley, London.

Jekyll, G. (1988) *Colour Schemes for the Flower Garden*. Frances Lincoln, London.

Lancaster, R. (2001) *Perfect Plant Perfect Place*. Dorling Kindersley, London.

Lawson, A. (1996) *The Gardener's Book of Colour*. Frances Lincoln, London.

Lee, D. (2007) *Nature's Palette: The Science of Plant Color*. University of Chicago Press, Chicago, IL.

Macdonald, M. (2003) *Bumblebees*. Scottish Natural Heritage, Edinburgh.

Pavord, A. (1999) *The Tulip*. Bloomsbury, London.

Pollock, M. & Griffiths, M. (1998) *Royal Horticultural Society Shorter Dictionary of Gardening*. Macmillan Publishers, London.

Salisbury, F.B. & Ross, C.W. (1992) *Plant Physiology*, 4th edn. Wadsworth Publishing Company, Belmont, CA.

Simpson, B.B. & Conner-Ogarzaly, M.C. (2001) *Economic Botany*, 3rd edn., pp. 192–217. McGraw-Hill, New York, NY.

Taiz, L. & Zeiger, E. (1991) *Plant Physiology*. The Benjamin/Cummins Publishing Company, San Francisco, CA.

# 13

# Climate, Weather and Seasonal Effects

## SUMMARY

The effect of day length on the flowering of plants (photoperiodism) and the possible identity of a floral 'signal' are discussed. The mechanisms for the detection of day length by the leaf are outlined, and the influence of day length on the formation of storage organs, leaf fall and dormancy is reviewed. The triggering of flowering and the breaking of seed and tree dormancy by low temperatures is described. The direct effects of temperature on flowering are outlined. The effects of below-freezing temperatures on plants is discussed. The chapter ends with a comment on the effects of climate change on the growth of plants.

## INTRODUCTION

All gardens change with the seasons. In this chapter the ways in which some of these seasonal changes are brought about are discussed. Many environmental conditions alter during the course of the year but those which have the greatest effect on plants are temperature, rainfall, the intensity of light and the daily duration of light. The ways in which these vary depends on latitude. Fig. 13.1 gives examples of yearly changes in temperature, light and the length of day for two latitudes, one in the tropics at latitude 5°N and one in the northern hemisphere at 50°N, the latitude of London. Most seasonal factors show some year-to-year variation depending on the weather patterns. In contrast, the length of the day is the one factor that changes with season but does not vary from year to year. It should come as no surprise, therefore, to find that many of the seasonal responses of plants (and animals) are controlled by the length of day,

usually abbreviated to day length. As will be shown later, however, such responses may also be modified by other environmental conditions, leading to the differences, such as in the time of flowering, that occur from year to year.

Responses to environmental stimuli first evolved to enable plants to succeed in particular climatic zones or ecological niches. For example, the control of flowering by day length is important in ensuring that flowers open relatively synchronously when appropriate pollinators are in greatest abundance; the control of leaf fall and *dormancy* by day length and temperature ensures that certain species or varieties are able to survive the adverse weather conditions of winter in northern latitudes or summer droughts in Mediterranean regions. Such responses to environmental stimuli have been retained in some cultivated plants and have a profound effect on how plants grow in our gardens and how they should be managed, especially cultivars that had their origins in a climatic zone different from the one in which they are being grown.

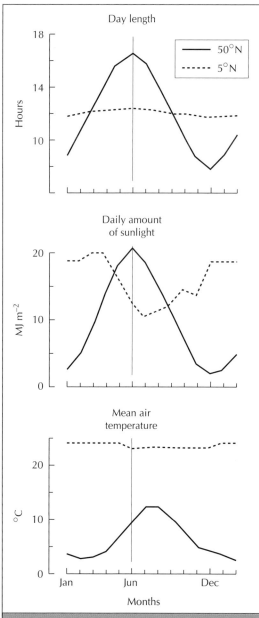

**Fig. 13.1** Seasonal changes in day length, the total amount of light each day and the daily mean air temperature at two latitudes.

# DAY LENGTH

Despite major changes in climate during the Earth's history, day-length patterns have remained remarkably constant, so it is not surprising that plants should have evolved to respond to them in the regulation of their seasonal patterns of growth and development. Perhaps more surprising is that it took such a long time for biologists to realise this, for it was not until the beginning of the twentieth century that the biological effects of day length began to be suspected. At about that time, Julien Tournois of Paris was puzzled as to why hop plants should flower so precociously when sown in greenhouses in winter. He later found that he could also achieve earlier flowering in spring by placing covers over the plants to reduce the hours of daylight. In a paper published in 1914, he concluded that the precocious flowering of hops in winter was caused by the short day lengths or, more likely, by the longer nights that accompanied them. The latter was an inspired guess, since the importance of the duration of darkness was not confirmed until many years later.

Tournois died in the 1914–18 war and it was not until the mid-1920s that the effects of day length were clearly established. Working at the US Department of Agriculture in Washington DC at that time, Wightman Garner and Henry Allard were trying to find out why a variety of tobacco (*Nicotiana tabacum*), 'Maryland Mammoth', continued to grow throughout the summer without flowering, even though it reached a very large size. On the other hand, quite small plants flowered rapidly when they were kept in a greenhouse during the winter. They were also puzzled by the fact that, when the 'Biloxi' variety of soybean (*Glycine max*) was sown at intervals during spring and summer, all the plants tended to flower at the same time. After eliminating temperature and light intensity as possible causes of the differences, they concluded that day length was the only other seasonal factor that might be having an effect. They tested their idea in a very simple way by putting plants of soybean and tobacco into a light-tight shed for 17 hours each night during the summer. These plants flowered, while their counterparts remaining in the open did not. The conclusion was that these varieties of tobacco and soybean would flower only when the daily duration of light was sufficiently short.

**Table 13.1** Some responses of plants to the length of day.

| Response | Promoted by long days | Promoted by short days | Example |
| --- | --- | --- | --- |
| **Flowering** | | | |
| Initiation of flowers | + | | Fuchsia 'Lord Byron' |
| | | + | Poinsettia (Euphorbia pulcherrima) |
| Growth of flowers | + | | Strawberry (Fragaria × ananassa) |
| | | + | Caryopteris × clandonensis |
| **Sex of flowers** | | | |
| Promotion of maleness | | + | Maize (Zea mays) |
| Promotion of femaleness | + | | Begonia × cheimantha |
| **Dormancy in woody plants** | | | |
| Formation of resting buds | | + | Birch (Betula), poplar (Populus) |
| Shedding of apex | | + | False acacia (Robinia pseudoacacia) |
| Leaf fall | | + | Tulip tree (Liriodendron tulipifera) |
| Increased frost resistance | | + | Dogwood (Cornus), spruce (Picea) |
| **Formation of storage organs** | | | |
| Underground stem tubers | | + | Jerusalem artichoke (Helianthus tuberosus) |
| Aerial stem tubers | | + | Begonia grandis subsp. evansiana |
| Root tubers | | + | Dahlia hybrids (Dahlia) |
| Corms | + | | Triteleia laxa |
| Bulbs | + | | Onion (Allium cepa) |
| **Seed germination** | + | | Birch (Betula spp.) |
| | | + | Nemophila insignis |
| **Other changes** | | | |
| Plantlets on leaves | + | | Devil's backbone (Kalanchoe daigremontiana) |
| Runners | + | | Strawberry (Fragaria × ananassa) |
| Rooting capacity | + | | Weigela florida |
| | | + | Holly (Ilex crenata) |
| Leaf size | + | | China aster (Callistephus chinensis) |
| Formation of coloured pigments (anthocyanins) | | + | Flaming Katy (Kalanchoe blossfeldiana) |
| **Increased stem elongation** | | | |
| Rosette plants | + | | Henbane (Hyoscyamus niger) |
| Plants with stems | + | | Fuchsia 'Lord Byron', French bean (Phaseolus vulgaris) |

Since these early experiments, the length of the day (or photoperiod) has been found to control many responses in both plants and animals. Some examples for plants are given in Table 13.1. In addition to flowering, responses that are particularly important for gardeners include the onset of dormancy in many tree species and the formation of underground storage organs such as bulbs and tubers.

The study of how day length regulates flowering and other aspects of growth and development has acquired its own terminology. *Photoperiodism* may be defined descriptively as a response to the length of day that enables an organism (plant or animal) to adapt to seasonal changes in its environment. A more useful definition is that it is a response to the timing of light and darkness during the daily cycle. This definition is important because, as will be shown in Chapter 14, the time when a relatively short exposure to light is given to a plant during a dark period may have a profound effect. Indeed, such *night-break lighting* treatments are used by commercial growers to regulate the flowering time of a number of crops.

With a mechanism sensitive to day length, the seasonal timing of events can be controlled precisely.

Indeed, a day-length difference of only 12 minutes has been found to switch between flowering and vegetative growth in some tropical plants which, contrary to what is sometimes suggested, are often highly sensitive to the relatively small differences in day length that they experience in their native habitats. A point that has to be taken into account is that the day lengths in spring are the same as those in autumn. Confusion between the two seasons in a plant may be avoided by an associated response to temperature or by having a different day-length requirement at different stages of development. For example, autumn-flowering *Chrysanthemum* hybrids form terminal flower buds when the day length is less than about 14.5 hours in either spring or autumn, but their further development into open flowers only occurs in the shorter autumn days of about 13 hours. In contrast, in many varieties of strawberry (*Fragaria × ananassa*), flowers are formed in the short days of autumn but their further development into open flowers is most rapid in long days, encouraging flowering in late spring.

Some plants must experience a sequence of long and short days before they can flower. Where long days must be given before short days, plants will normally flower in the autumn, while the reverse sequence will result in late spring or early summer flowering. Another way is to couple a response to day length with a response to low temperature (*vernalisation*; see below). A requirement for exposure to cold for several weeks will prevent flowering until a winter has been experienced by the plant and so ensures that flowering does not occur until spring or early summer. Biennial plants fall into this category.

What are the advantages to the plant of ensuring that a response occurs at a particular time of year? One advantage is that synchronous flowering within a population of plants in the wild can ensure good cross-fertilisation. Flowering can also be timed to occur when the environment is favourable as, for example, when water is plentiful during a rainy season, or in the summer when there is enough light for photosynthesis to meet the energy demands of seed production. Another important advantage is the ability to avoid the damaging effects of unfavourable conditions such as drought or low temperatures by shedding leaves and/or becoming dormant.

For gardeners, however, responses to day length can be both advantageous and disadvantageous. For example, many plants from lower latitudes require short days for flowering and so do not flower, or flower poorly, during a northern summer (e.g. poinsettia, *Euphorbia pulcherrima*, Table 13.2). Similarly, plants from high latitudes are usually adapted to flowering in the long days of summer when temperature and light conditions are favourable; hence, they will not flower, or flower poorly, in the tropics (e.g. *Salvia × sylvestris* 'Blaukönigin'; Table 13.2). Fortunately much is known about the genetic control of photoperiodic responses, so breeding can often be used to eliminate such problems (see Chapter 5). A major advantage for commercial growers is the possibility of using artificial day lengths in order to time plants for a particular market as, for example, poinsettias for Christmas, or to spread out the cropping period to reduce gluts and peak demands on labour (see Chapter 14).

## Flowering and day length

The time of flowering has been by far the most intensively studied of the many responses to day length and most of our ideas about the underlying mechanisms come from these experiments. Plants in which flowering is responsive to day length are usually classed in one of two main groups (Fig. 13.2 and Table 13.2). These are *short-day plants* (SDPs) which only flower, or flower earlier, when days are shorter than a particular duration (known as the *critical day length*), and *long-day plants* (LDPs) which only flower, or flower sooner, when the days exceed the critical day length. Those plants in which the time of flowering is independent of day length are called *day-neutral plants*. A few plants are known to flower only, or more rapidly, when the days are neither too long nor too short. These are called *intermediate-day plants*.

It is important to understand that the difference between short- and long-day plants does not depend on the actual value of the critical day length. For example, the critical day length for some varieties of marigold (*Calendula officinalis*) is only 6.5 hours; nevertheless they are LDPs and will flower only when the day length exceeds this value. In contrast, the critical day length for some varieties of Japanese morning glory (*Ipomoea nil*) lies between 15 and 16 hours, but these are SDPs and will flower only when the day length is shorter than this value. Thus, in any one day length, both long- and short-day plants may

**Table 13.2** Examples of the critical day length for the formation of flowers*.

**Short-day plants**

A
Chrysanthemum cultivars 11–16
Coleus (*Plectranthus fredericii*) 13–14
Poinsettia (*Euphorbia pulcherrima*) 11–12.5
Strawberry (*Fragaria* × *ananassa*) 11–16
Soybean (*Glycine max* 'Biloxi') 12
Flaming Katy (*Kalanchoë blossfeldiana*) 12

Tobacco (*Nicotiana tabacum* 'Maryland Mammoth') 14
*Perilla*, red-leaved 14
*Perilla*, green-leaved 16
Japanese morning glory (*Ipomoea nil* 'Violet') 15

**Long-day plants**

A
Dill (*Anethum graveolens*) 10–14
Pot marigold (*Calendula officinalis*) 6.5
Coneflower (*Rudbeckia bicolor*) 10
Ice plant (*Sedum spectabile*) 13

Campion (*Silene armeria*) 11–13.5
Mustard (*Sinapis alba*) 14
Spinach (*Spinacea oleracea*) 13

B
Yarrow (*Achillea* 'Moonshine') 16
Japanese anemone (*Anemone hupehensis*) 16
Thrift (*Armeria pseudarmeria*) 16
Milkweed (*Asclepias tuberosa*) 16
Astilbe (*Astilbe chinensis* var. *pumila*) 16
Bellflower (*Campanula carpatica* 'Blue Clips') 16
Tickweed (*Coreopsis verticillata* 'Moonbeam') 14; (*Coreopsis* 'Sunray') 13
Blanket flower (*Gaillardia* 'Kobold') 13
*Gaura lindheimeri* 'Whirling Butterflies' 13
Cranesbill (*Geranium* × *cantabrigiense*) 16; (*Geranium dalmaticum*) 16
*Gypsophila* 'Happy Festival', 16
*Helenium autumnale*, 16
*Hibiscus moscheutos* 'Disco Belle', 16

Plantain lily (*Hosta* spp.) 14
*Leucanthemum* × *superbum* 'White Knight' and 'Snowcap', 16
*Lobelia* × *speciosa* 'Compliment Scarlet', 16
Bergamot (*Monarda didyma* 'Gardenview Scarlet') 16
Evening primrose (*Oenothera fruticosa*) 16; (*Oenothera missouriensis*) 16
*Phlox paniculata* 'Tenor', 'Eva Cullum' 14
*Physostegia virginiana* 'Alba' 16
Coneflower (*Rudbeckia fulgida* var. *sullivantii* 'Goldstrum') 13
Sage (*Salvia* × *sylvestris* 'Blaukönigin') 16
Saxifrage [*Saxifraga* 'Triumph' (× *arendsii*)] 15
Ice plant (*Sedum* 'Herbstfreude') 16
Stokes' aster (*Stokesia laevis* 'Klaus Jellito') 13

* The critical day length (in hours) is given for each plant. Short-day plants (SDPs) flower in day lengths *shorter* than the critical, while long-day plants (LDPs) flower in day lengths *longer* than the critical. However, the critical day length is not always defined in the same way. In group A, the critical day length is that above (LDP) or below (SDP) which some flower parts are formed, although open flowers may not always follow. This is the definition used by many plant physiologists. For the LDP in group B, the critical day length is that at and above which rapid and uniform flowering occurs. This definition is likely to be more useful to gardeners. The critical day lengths in group B were determined as part of a research programme at Michigan State University under the direction of A. Cameron, W. Carson and R. Heins. It is important to emphasise that, irrespective of how it is defined, the critical day length may vary in other cultivars or strains and under different environmental conditions. Moreover, some plants must be given a period at low temperature (see text, under Vernalisation) before they become sensitive to day length.

be able to flower and their classification can only be determined by growing them experimentally in a range of different day lengths (Fig. 13.3). Under natural conditions, flowering in LDPs will be delayed until the critical day length is exceeded during the lengthening days of spring and early summer. Many SDPs will flower in autumn, when the day length becomes shorter than the critical duration. The critical day lengths of some common horticultural plants are given in Table 13.2.

**Fig. 13.2** Flowering responses to day length. Plants were grown either in short days (8 hours, plants on left of each panel) or long days (16 hours, right of each panel). The short-day plant (*Kalanchoe blossfeldiana, left*), flowered rapidly in short days but was markedly delayed in long days. The long-day plant *Antirrhinum majus* (*right*) flowered much more rapidly in long days. *Photograph courtesy of Daphne Vince-Prue.*

Although the changes that lead to flowering take place at the main or lateral shoot tips, it is the leaves that actually detect the day-length signal. This is also true for other responses to day length such as the onset of winter dormancy in woody plants and the formation of underground storage organs. The special role of the leaves was demonstrated as early as 1934 when it was shown that exposing only the leaves of the LDP spinach (*Spinacea oleracea*) to long days resulted in the formation of flowers at the shoot tip. Perhaps the most spectacular result was with the SDP *Perilla* (red-leaved perilla) where leaves were exposed to short days after they had been cut off from the plant (Fig. 13.4). These leaves were able to cause flowering when they were subsequently grafted on to vegetative (i.e. non-flowering) plants growing in long days. It follows from these experiments that a leaf exposed to appropriate day lengths is able to transmit some kind of signal, presumed to be a chemical, that is able to travel to the shoot apex and trigger a flowering response there. We can, therefore, separate the process of the day-length control of flowering into three components:

1. the detection of day length and subsequent changes in the leaf (known as *photoperiodic induction*);
2. the transmission of some kind of signal from the leaf;

3. the changes at the shoot tip that result in flowering (known as *evocation*).

It is highly likely that the events at the shoot tip are not unique to photoperiodism, but also occur where the onset of flowering is independent of day length, as in day-neutral plants.

From the point of view of horticulture there are two major questions to be addressed. First, what is the identity of the signal that is exported from the leaves? This obviously has implications for the possible control of flowering by the use of appropriate chemical treatments. The second question concerns the way in which day length is detected by the leaves, since this becomes important when growers use artificially imposed day lengths to control flowering and other responses to day length.

## Chemical control of flowering

Several approaches have indicated that the signal exported from leaves following photoperiodic induction is a chemical that appears to move mainly in the phloem (the sugar-transporting system; see Chapter 2). Since it moves from a site of production (the leaves) to a site of action (the shoot tip), the floral

(a)

(b)

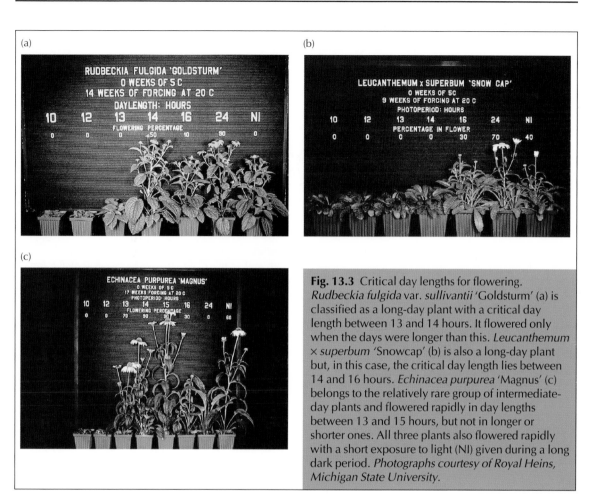

(c)

**Fig. 13.3** Critical day lengths for flowering. *Rudbeckia fulgida* var. *sullivantii* 'Goldsturm' (a) is classified as a long-day plant with a critical day length between 13 and 14 hours. It flowered only when the days were longer than this. *Leucanthemum × superbum* 'Snowcap' (b) is also a long-day plant but, in this case, the critical day length lies between 14 and 16 hours. *Echinacea purpurea* 'Magnus' (c) belongs to the relatively rare group of intermediate-day plants and flowered rapidly in day lengths between 13 and 15 hours, but not in longer or shorter ones. All three plants also flowered rapidly with a short exposure to light (NI) given during a long dark period. *Photographs courtesy of Royal Heins, Michigan State University.*

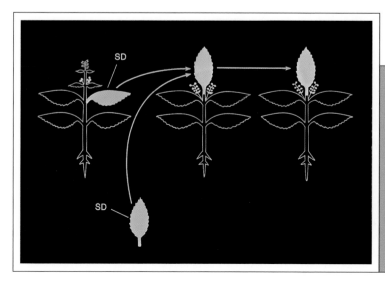

**Fig. 13.4** The detection of day length takes place in the leaf. A leaf of the short-day plant *Perilla* was exposed to short days (SD) either while still attached to the plant (top left) or after removal (bottom). When grafted, this induced leaf (see text) was able to cause flowering in a receptor plant maintained in long days (top middle) and retained the ability to cause flowering when detached and re-grafted to another plant (top right). *Photograph courtesy of Daphne Vince-Prue.*

stimulus has the characteristics of a hormone. The term *florigen* is sometimes used for this hypothetical hormone but *floral stimulus* is probably a better term. Florigen implies that the action of the hormone is uniquely concerned with the formation of flowers, whereas the existence of such a flower-specific hormone has not yet been demonstrated.

One interesting line of experimentation into the nature of the floral stimulus has been to use the horticultural technique of grafting (see Chapter 10). The stimulus will move from a LDP grafted to a SDP and, conversely, from a SDP grafted to a LDP. To take a typical example, the LDP henbane (*Hyoscyamus niger*) will flower in short days when grafted with the SDP tobacco (*Nicotiana tabacum*); this will only occur, however, when the donor plant, tobacco, retains its leaves and is kept in short-day conditions. Thus only when they are themselves induced by exposure to the appropriate day length can the leaves of the tobacco plants bring about flowering in a plant of a different day-length response group.

The fact that the floral stimulus is equivalent in plants of different day-length response groups and different genera is one of the pieces of evidence that led to the suggestion that there may be a chemical that is effective in a wide range of plants and perhaps in all, irrespective of their day-length response. There are, however, two problems with this proposal. One is that the experimental evidence is limited because grafting is only possible between closely related plants (see Chapter 10) where any floral stimulus might be expected to be similar. The second

is that, so far, all attempts to isolate and identify such a universal flowering hormone have been unsuccessful.

Recent genetic and molecular studies have resulted in new ideas about the possible identity of the transmitted floral signal. In the LDP *Arabidopsis thaliana*, activation of a flowering time (*FT*) gene in the leaf vascular tissue induces flowering at the shoot apical meristem. Activation of *FT* occurs in long days but not in short days, as would be expected for this LDP. The CONSTANS (*CO*) gene that activates transcription of *FT* is regulated by the circadian clock in the leaf. It is thought that the autonomous pathway (i.e. independent of day length) to flowering in *A. thaliana* also operates by activation of the *FT* gene in the leaf (Fig. 13.5). It is interesting that *FT* family genes have also been found to be involved in the photoperiodic control of bud dormancy in at least two tree genera (*Populus* and *Picea*).

It is now evident from studies on more than one plant that the activation of *FT* genes in the phloem cells of the leaf leads to flowering at the shoot apical meristem and to the onset of bud dormancy in woody plants. The question remains as to the precise nature of the resultant transmitted signal. Earlier suggestions were that the signal might be the FT mRNA or some unidentified product of the FT protein. However, there is recent evidence that the FT protein itself (i.e. the protein product of the activation of the *FT* gene) is the signal that is transmitted through the phloem to the shoot apex, where flowering or bud dormancy result.

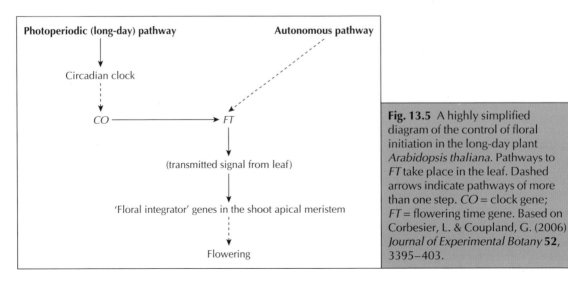

Fig. 13.5 A highly simplified diagram of the control of floral initiation in the long-day plant *Arabidopsis thaliana*. Pathways to *FT* take place in the leaf. Dashed arrows indicate pathways of more than one step. *CO* = clock gene; *FT* = flowering time gene. Based on Corbesier, L. & Coupland, G. (2006) *Journal of Experimental Botany* **52**, 3395–403.

There is also good evidence from grafting experiments that naturally occurring inhibitors of flowering exist. For example, when a day-neutral variety of tobacco was grafted on to the LDP henbane (*Hyoscyamus niger*), flowering in the tobacco shoot was prevented when the grafts were kept in short days and the henbane failed to flower. This implies that an inhibitor of flowering moved from the henbane to the tobacco, since the variety of tobacco used normally flowers in short days. Since the inhibitory effect increased with the number of leaves remaining on the henbane plant, it is clear that, like the floral stimulus, the inhibitor originated in the leaves.

## Plant hormones and flowering

The enormous practical applications that could follow from the discovery of chemical regulators of flowering that are relatively inexpensive and easy to apply have led many research workers to examine the effects on flowering of the known plant hormones.

Although hormones belonging to the *auxin* and *cytokinin* groups may be involved in the changes that are undergone at the shoot tip during the transition to flowering, they have little or no effect when applied to intact plants. In contrast, hormones in the *gibberellin* group have been found to bring about flowering in several species, while the gaseous hormone *ethylene* is highly effective in one particular family, the Bromeliaceae.

### Gibberellins

It has been known for a long time that the application of a solution of gibberellins (see Chapters 3 and 10) can substitute for a particular day length in many plants (Table 13.3). In particular, they cause flowering when applied to many long-day plants that grow as rosettes in short days. The flowering response to either gibberellins or long days is often accompanied by elongation of the flowering stem. The problem of interpretation here is that the normal function of gibberellins is to promote stem elongation, so the flowering response may simply be a consequence of this. There are, however, a few LDPs where gibberellin application can bring about flowering without stem elongation, which suggests that the two responses are separate.

Gibberellin application can also substitute for a low-temperature signal in the vernalisation (see below) of several biennials and in other cold-requiring plants such as tulip bulbs. In addition, striking

**Table 13.3** Gibberellins influence flowering in many different ways.

| | Effect of application of gibberellin* | Plant |
|---|---|---|
| **Long-day plants** | Stimulates flowering in short days<br>Inhibits flowering in long days<br>No effect | Henbane (*Hyoscyamus niger*)<br>*Fuchsia × hybrida*<br>White mustard (*Sinapis alba*) |
| **Short-day plants** | Stimulates flowering in long days<br>Inhibits flowering in short days<br>No effect | *Zinnia elegans*<br>Strawberry (*Fragaria × ananassa*)<br>Soybean (*Glycine max*) |
| **Day-neutral plants** | Stimulates flowering<br>Inhibits flowering<br>No effect | Cypress (*Cupressus* spp.)<br>Apple (*Malus domestica*)<br>Evening primrose (*Oenothera* spp.) |
| **Plants requiring vernalisation**<br>Long-day plants<br>Short-day plants<br>Day-neutral plants | Stimulates flowering in long days<br>No effect<br>Stimulates flowering<br>No effect | Oat, winter strains (*Avena sativa*)<br>*Chrysanthemum* cultivars<br>Cauliflower (*Brassica oleracea*)<br>*Saxifraga rotundifolia* |

* The application of gibberellin can promote, inhibit or have no effect on flowering, even with plants that have the same environmental requirements for flowering. Note that there are many different gibberellins and that these experiments were mostly carried out with $GA_3$.

effects to accelerate flowering in juvenile plants have been seen in several conifers. Thus, in an impressive list of plants the application of gibberellins has been found to substitute for the seasonal signals of day length and low temperature as well as for the internal trigger of 'age'. It is also well established that day length affects certain steps in the biochemical pathways leading to the synthesis of gibberellins and so alters the content and composition of gibberellins within the plant (at least 84 different gibberellins have been detected and several can occur together in the same plant).

Recent work with the LDP *Lolium temulentum* has supported the conclusion that gibberellins can function as floral hormones. Specific gibberellins ($GA_5$ and $GA_6$) have been shown to increase rapidly in the leaf following the transfer to long-day conditions; this is followed by transport to the shoot apex where flowers are produced. There is also evidence that when $GA_5$ is applied to the leaf it is transported to the apex and causes flowering. It is also pertinent that, unlike $GA_1$, $GA_5$ and $GA_6$ have little effect on stem extension. They therefore seem to be specific for flowering in *Lolium*. As with the LDP *Arabidopsis*, expression of the *FT* gene does increase in long days, but this does not happen until after the increase in gibberellins.

Unfortunately, any simple interpretation of the role of gibberellins in flowering is ruled out because not all plants respond in the same way and in many cases the application of gibberellins has no effect or may even be inhibitory. Even in the same plant gibberellins can have opposite effects. Flowering in many temperate grasses requires exposure to short days, followed by long days. Here gibberellins are inhibitory to flowering when applied during the short-day treatment, but promote flowering when applied after the short-day requirement has been completed. However, most of the experiments have used $GA_3$ (which is one of the few gibberellins that are generally available) whereas only $GA_5$ and $GA_6$ were effective in causing flowering in *Lolium*. Some examples of the different flowering responses to the application of $GA_3$ are given in Table 13.3, but few of these have yet been found to be of any practical use in horticulture.

## Ethylene

This gaseous hormone (see Chapter 2) has many different effects on plants, including the modification of flowering in ways that have practical uses. Although the application of pure ethylene gas to plants is difficult, chemicals such as 2-chloroethyl phosphonic acid (CEPA) that release ethylene within the plant are available and can be applied as sprays. Ethylene and ethylene-releasing compounds inhibit flowering in several SDPs, including *Chrysanthemum* cultivars and Japanese morning glory (*Ipomoea nil*). Ethylene also inhibits flowering in sugar cane and CEPA has been used commercially with this crop to increase the yield of sugar by preventing flowering, which leads to cane senescence.

The most striking effect of ethylene in promoting flowering occurs in pineapple (*Ananas comosus*) and other members of the Bromeliaceae, as discovered when it was found that turning pineapple plants on their side resulted in rapid flowering. This is now known to be associated with the accumulation of auxin at the underside of the shoot apex under the influence of gravity, as it has been established that a high concentration of auxin results in the production of ethylene by cells. As far as is known, the promotion of flowering by ethylene applies to all bromeliads and has been exploited commercially by applying ethylene-releasing compounds to stimulate flowering in pineapple and ornamental members of the Bromeliaceae such as *Aechmea* and *Billbergia*. An intriguing variant is the traditional use of smoke from burning green wood to cause flowering in pineapples grown commercially under glass in the Azores. Since ethylene is known to be produced during fires, it is possible that the flowering response is due to ethylene and related gases in the smoke. Ethylene is also the only hormone with a significant effect on flowering in bulbs, a response that has been used commercially in forcing Tazetta narcissus and bulbous iris.

The control of flowering by ethylene in the bromeliads also involves prevention when not desired. A single experimental treatment with an inhibitor of ethylene synthesis in the plant can prevent flowering for several months. Treatment with a chemical that leads to the production of ethylene by the plant will then induce flowering.

Despite its quite dramatic effect on flowering in some species, particularly in the bromeliads, the mechanism through which ethylene causes flowering is unknown. It does not appear to substitute for, nor interact with, day length and because it is a gas it is most unlikely that it is part of any floral stimulus.

This does not, however, exclude chemicals that are required for the synthesis of ethylene by the plant. One or more of these could be exported from the leaves and converted to ethylene gas at the shoot apex where flowering takes place.

## How is day length detected by the leaf?

Under natural conditions the nights get longer as the days get shorter, so it is not possible to say whether plants are responding to the duration of darkness or to the duration of light in each daily cycle, or perhaps to both. To answer this question it was necessary to carry out experiments in which the durations of darkness and light to which plants were exposed were varied independently. When, in 1934, Karl Hamner and James Bonner carried out such experiments with the SDP cocklebur (*Xanthium strumarium*), they found the duration of darkness to be decisive, as had already been suspected by Julien Tournois in 1914. Plants grown in long days would flower, provided that the duration of darkness exceeded a critical value. Conversely, they failed to flower in short days when these were coupled with nights that were shorter than the critical value. Even a single long night was enough to cause flowering in cocklebur, irrespective of the length of the associated light periods. A similar pattern of response was later seen in other SDPs and also in some LDPs (Table 13.4).

We know now that the measurement of the duration of darkness by the plant is controlled by a biological clock of the same kind and with many of the same properties as that which controls other time-dependent processes in plants and animals, such as the daily rhythms of leaf movement in plants and the

'jet-lag' effect in people. This internal clock keeps time by going through a cycle that returns to the same point at approximately 24-hour intervals. Because the cycles do not repeat in exactly 24 hours the clock is usually called a *circadian clock*, which is derived from the Latin for 'about a day'. Since the repeating cycles are longer or shorter than 24 hours, time as measured by the endogenous clock would drift in relation to real time unless the clock is re-set by some external signal. In plants, the times of sunrise and sunset are the daily signals that re-set the clock to local time.

The circadian clock in plants is thought to measure the critical duration of darkness as follows. In plants that are sensitive to day length, each cycle goes through a period when exposure to light triggers a switch between flowering and not-flowering. This occurs at a certain number of hours from the beginning of darkness, although the exact number can vary depending on the species. When the dawn (daylight) arrives before this light-sensitive period occurs, the switch is triggered by the light and the critical night length is not reached. However, when the light-sensitive period occurs before dawn (i.e. during darkness), the switch is not triggered and the critical night length is exceeded. As will be shown in Chapter 14, this is also the basis of the night-break response in which a relatively short exposure to light given at a particular time during the night triggers flowering in LDPs but inhibits flowering in SDPs. The switch between flowering and not-flowering occurs only when the night-break light is given during the light-sensitive period of the circadian clock.

So far, the discussion about the way in which day length is detected has concentrated on the responses of those plants in which the duration of darkness is

**Table 13.4** Long nights rather than short days are important for the control of flowering in plants that are sensitive to day length*.

| Light treatment | | Flowering response | |
|---|---|---|---|
| **Day length** | **Night length** | **Short-day plants** | **Long-day plants** |
| 8 hours | 16 hours | Flowering | Not flowering |
| 16 hours | 16 hours | Flowering | Not flowering |
| 8 hours | 8 hours | Not flowering | Flowering |
| 16 hours | 8 hours | Not flowering | Flowering |

* Short-day plants flower with *long* dark periods, whereas long-day plants flower with *short* dark periods.

decisive. Most SDPs and a few LDPs behave in this way, which is sometimes called *dark-dominant*. However, when scientists began to look at LDPs in more detail, it was found that many have certain characteristic features which differ from those of SDPs. In particular, it seems that the spectral quality of light during the day is important in many LDPs, even when the duration of darkness is favourable for flowering. For this reason, they are sometimes termed *light-dominant*. The differences between light-dominant and dark-dominant responses are not important under natural conditions in sunlight but can be important when artificial light, which may have different properties from natural light, is used to manipulate day length (see Chapter 14).

As with other responses to light, perception of the day-length signal in most SDPs and LDPs depends on the red/far-red-sensitive pigment phytochrome, although more than one kind of phytochrome may be involved (see Chapter 14). However, a few plants (notably long-day species of the Cruciferae) are sensitive to blue light, indicating the participation of another chemical, cryptochrome.

Irrespective of the particular mechanism, plants that are sensitive to day length flower seasonally in response to changes in the natural photoperiod. For plants growing in the open, the transitions between light and dark are not abrupt but occur through a gradually changing intensity of twilight. At what point, then, does a plant begin to respond to darkness in the evening, or to light at dawn? Is the effective length of day influenced by morning or evening clouds, or moonlight, or even street lamps? The answers have been shown to vary with species and also with the kind of light given.

The presence of clouds during twilight and dawn might influence the time at which a threshold intensity between light and darkness is reached, allowing the night to begin earlier or end later. This could lead to small changes in the time of flowering from year to year. In some varieties of Japanese morning glory (*Ipomoea nil*), for example, the effective day length on clear days was found to be longer by 20–30 minutes than on cloudy days. For more sensitive species, however, the presence of clouds has little effect and errors in timing would be very small in such plants. The longer twilights of high latitudes and the short ones in the tropics may also influence how long it takes to go from above to below a threshold value of light. Because of the twilight effect, critical day lengths determined under artificial conditions

when light is switched on and off instantaneously may differ from those actually experienced by plants growing out of doors.

As far as the moon is concerned, present evidence indicates that even light from the full moon is below the threshold necessary to influence flowering. Street lights are, however, a different story. Although there are considerable differences between species, for many plants the threshold light level above which flowering was promoted (LDPs) or inhibited (SDPs) was less than 100 mW m$^{-2}$ (milliwatts of light energy reaching a square metre of the surface; about 50 lux) in experiments where light from tungsten-filament (incandescent) lamps was given continuously throughout the night (Table 13.5). The possible effect of street lamps depends on the light level reaching the plant, the sensitivity of the species to light and the spectral composition of the lamp. Sodium-vapour lamps emit light containing a high proportion of active wavelengths and are thus more likely to have an effect than mercury-vapour lamps, which emit less light in the active part of the spectrum. The present trend of keeping street lights on during the entire night makes it more likely that they could influence flowering or other responses to day length. For example, street lights may be sufficiently bright to delay the onset of dormancy in trees, thus increasing their susceptibility to freezing injury (see below), especially in northern latitudes where autumn frosts begin early.

## The effects of day length on the formation of storage organs

One of the seasonal features of some plants is the formation of resting structures in which food is stored (see Chapter 2). Such storage organs arise by lateral swelling of a number of different tissues including stems (tubers and corms), roots (tuberous roots) and leaf bases (bulbs). Their formation is usually accompanied by the cessation of active growth followed by death of the rest of the plant. Once formed, storage organs often become dormant, during which time they are more resistant to unfavourable environmental conditions such as water stress or the extremes of temperature that occur in winter. In many cases, the formation of storage organs depends on or is accelerated by particular day lengths and, with the exception of bulbs in the genus *Allium*, these are usually favoured by short days (Table 13.6).

**Table 13.5** The minimum level of light necessary to influence flowering is not the same for all plants*.

| Short-day plants | Minimum light level needed to prevent flowering† (mW m$^{-2}$) |
|---|---|
| *Chrysanthemum* cultivars | 90 |
| Flaming Katy (*Kalanchoe blossfeldiana*) | 90 |
| Poinsettia (*Euphorbia pulcherrima*) | 21 |
| Japanese morning glory (*Ipomoea nil*) | 4–40 |
| Soybean (*Glycine max*) | 0.4 |
| **Long-day plants** | **Minimum light level needed to promote flowering† (mW m$^{-2}$)** |
| Turnip rape (*Brassica campestris*) | 4515 |
| Campion (*Silene armeria*) | 31–90 |
| Barley (*Hordeum vulgare*) | 10–20 |
| China aster (*Callistephus chinensis*) | 4–12 |

* Most long-day plants require higher light intensities to stimulate flowering than is required by most short-day plants to inhibit flowering. There is, however, a lot of difference between plants and the values only apply for these particular conditions; that is, when light from tungsten-filament lamps was given continuously throughout the night. They would be different with other light sources or when light is given at different times and/or durations.

† mW m$^{-2}$ = milliwatts of light energy reaching a square metre of the surface.

**Table 13.6** The formation of storage organs is often influenced by day length.

| Plant | Type of storage organ |
|---|---|
| **Favoured by short days** | |
| Peanut (*Apios tuberosa*) | Root tubers |
| *Begonia socotrana* | Aerial stem tubers |
| Tuberous begonia (*Begonia tuberhybrida*) | Underground stem tubers |
| Dahlia (*Dahlia* hybrids) | Root tubers |
| Artichoke (*Helianthus tuberosus*) | Underground stem tubers |
| Potato (*Solanum tuberosum*) | Underground stem tubers |
| **Favoured by long days** | |
| Shallot (*Allium cepa*) | Bulbs |
| Onion (*Allium cepa*) | Bulbs |
| Garlic (*Allium sativum*) | Underground and aerial bulbs |
| *Triteleia laxa* | Corms |

As with other responses to day length, the formation of storage organs is completely dependent on exposure to appropriate day lengths in some plants, but only accelerated in others. In the potato (*Solanum tuberosum*), for example, cultivars differ considerably in the extent to which they respond to day length. In many European and North American cultivars tuber formation occurs in both long and short days, but occurs earlier in the latter. The early dieback of the haulms associated with the formation of tubers often results in a lower yield in short days, despite the fact that the onset of tuber formation is accelerated. This is a good example of the need to match the response of the species or cultivars planted to the day-length conditions in any particular geographical location.

The regulation of tuber formation in potato by short days appears to be very similar to the induction of flowering in many SDPs. The duration of darkness is the controlling factor and the formation of tubers

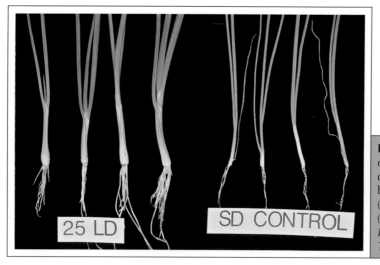

**Fig. 13.6** The formation of bulbs in onions requires exposure to long days. Bulb formation had already begun in plants given 25 long days (left), while those in short days (right) showed no signs of bulbing. *Photograph courtesy of Brian Thomas, University of Warwick.*

will only occur when this exceeds a critical value. Thus tubers develop during the lengthening nights of late summer or early autumn. In contrast, bulb formation in onions (*Allium cepa*) is dependent on exposure to long days (Fig. 13.6). In lower latitudes, such as in Egypt, many varieties commonly grown in northern Europe do not form bulbs and the so-called short-day bulbing types must be grown. However, these are also LDPs, the difference being that they have a shorter critical day length (Table 13.7) and so will form bulbs in the shorter days of lower latitudes. When they are grown at higher latitudes the critical day length may be attained early in the summer, leading to rapid bulb formation in young plants, with a consequent reduction in bulb size. As with tuber formation in potato, therefore, the yield is reduced when day lengths appropriate for bulbing occur too early in the life of the plant.

As with flowering, the formation of storage organs depends on exposing the leaves to the appropriate day lengths. Again, some kind of chemical stimulus must be exported from the leaves to the region where the storage organ develops. One intriguing observation arising from grafting experiments is that the generation of a tuber-forming stimulus is not confined to the leaves of plants which themselves are able to form tubers. In one experiment, shoots of sunflower (*Helianthus annuus*) were grafted on to Jerusalem artichoke plants (*Helianthus tuberosus*). Artichokes normally develop tubers only when their leaves are exposed to short days. Although sunflowers do not form tubers themselves, they were able to cause tubers to develop on the artichoke plants to which they were grafted when the leaves of the sunflower donors were grown in short days. There is some evidence that the signal for flowering and tuber

**Table 13.7** Different varieties of onion have different day-length requirements for bulbing.

| Variety | Percentage of plants with bulbs | | | | |
|---|---|---|---|---|---|
| Day length (hours) . . . | 10 | 12 | 13 | 14 | 16 |
| 'Sweet Spanish' | 0 | 29 | 77 | 93 | 100 |
| 'Yellow Flat Dutch' | 0 | 0 | 15 | 92 | 100 |
| 'Yellow Rijnsburg' | 0 | 0 | 9 | 46 | 100 |
| 'Yellow Zittau' | 0 | 0 | 0 | 40 | 100 |

Although the variety 'Sweet Spanish' forms bulbs on shorter day lengths than 'Yellow Zittau', both only form bulbs when the day length is longer than the critical value. This is about 14 hours for 'Yellow Zittau' but only 12 hours for 'Sweet Spanish' which can, therefore, be grown successfully at lower latitudes.

formation may be the same, or similar, with the nature of the response being dependent on the plant itself. For example, when shoots of tobacco (*Nicotiana tabacum*) varieties with different day-length requirements for flowering were grafted on to potato plants, tubers formed in the latter only when the leaves of the donor tobacco plants were in day lengths that induced flowering in them (Table 13.8).

What is the nature of this tuber-forming stimulus? The fact that it is also produced by plants which are themselves incapable of forming tubers suggests that we are not dealing with a unique tuber-forming substance but rather with a non-specific compound. Consequently, many substances have been tested for their possible ability to bring about the formation of storage organs, especially in economically important crop plants.

Many chemicals, including sugars and known plant hormones, have been found to have some effect on the formation of storage organs in particular species, but the results are often conflicting. It is, however, well established that gibberellins play a significant role as chemical signals for the control by day length of tuber formation in the potato. For example, a dwarf mutant with a low content of gibberellins formed tubers in both long and short days, as did plants treated with a chemical which prevents the production of gibberellins. The likely explanation is that the formation of tubers is prevented in long days by an inhibitor coming from the leaves, which is presumed to be one of the many gibberellins. There appears to be no role for any of the other known plant hormones in potato, and discovering the identity of the tuber-forming signal that is indicated from the grafting experiments remains an elusive goal.

# The role of day length in controlling leaf fall and dormancy

Perhaps the most majestic of all seasonal displays is that of autumn colour (see Chapter 12), which precedes leaf fall and entry into a rest period (dormancy) in deciduous trees of high latitudes. In most cases this is a response to the longer nights and shorter days of autumn and, under their influence, stem and leaf growth stop and winter resting buds develop. The value to the plant is that the seasonal signal of short days gives an advance warning of low temperatures to come and enables plants to increase their resistance to freezing temperatures. Thus the time at which they become dormant and develop cold hardiness may be crucial for survival in a particular location.

The photoperiodic mechanism(s) for the control of dormancy appear to be the same as those controlling flowering. For example, in Norway spruce (*Picea abies*) the control of dormancy by short days in a southern population is clearly dependent only on the duration of darkness, whereas a population from the far north at latitude 64°N depends also on the duration and quality of light during the day.

In some plants, including many evergreens, dormancy involves only metabolic changes without the development of any specialised structures. In others, protective bud scales are formed from overlapping leaf bases or by the development of modified leaves (see Chapter 2). Perhaps the most obvious morphological change is the shedding of leaves in deciduous trees (see Chapter 2). However, although leaf fall is influenced directly by day length in some plants, temperature is also important. In the tulip tree (*Liriodendron tulipifera*) leaves are always shed in

**Table 13.8** The formation of tubers on potato plants is affected by grafting them to flowering or non-flowering plants of tobacco*.

| Tobacco donor | Response | |
|---|---|---|
| | **Grafted plants maintained in short days** | **Grafted plants maintained in long days** |
| *Nicotiana tabacum* 'Mammoth' (short-day plant) | Tubers formed | No tubers formed |
| *Nicotiana sylvestris* (long-day plant) | No tubers formed | Tubers formed |

* Tubers formed on potato plants only when the tobacco donors were flowering ('Mammoth' in short days, *N. sylvestris* in long days), indicating that the signal which causes flowering in tobacco may also cause tuber formation in potato.

short days, but in the false acacia (*Robinia pseudoa-cacia*) leaves are retained when the temperature remains high, whatever the length of the day.

The response to day length is of great importance for the adaptation of trees to the latitude in which they originate. That this is controlled by the plant's genes can be seen when seeds are collected from several different latitudes and grown together in one place. In one experiment, as many species as possible were collected from three different latitudes in Norway and Denmark and grown together in day lengths ranging from 12 to 24 hours. It was found that plants originating from the same geographical area had approximately the same critical day length for the maintenance of stem growth and the prevention of dormancy, irrespective of species (Table 13.9).

The shorter critical day lengths of more southerly populations of the same species means that they continue growing for longer if they are moved further north and may be damaged by low temperatures in autumn. For example, the onset of dormancy and the development of cold resistance in a species of willow (*Salix pentandra*) from 60°N, with a critical day length of about 14 hours, was delayed when plants were grown outdoors in the longer days of Tromsø in Norway (at about 70°N) and they were severely damaged by frost. The critical day length tends to be

**Table 13.9** The critical day length for the onset of winter dormancy varies with the latitude at which a plant originates*.

| Latitude of origin | Approximate critical day length (hours) |
| --- | --- |
| 55°N | 15 |
| 60°N | 17 |
| 65°N | 19.5 |
| 70°N | 22 |

* When seeds were collected from trees growing at different latitudes in Scandinavia, seedlings from the far north ceased growing and became dormant when grown in day lengths shorter than 22 hours, whereas seedlings from further south at latitude 55°N only ceased growing when the day length was shorter than 15 hours.

longer for trees growing at higher altitudes where winter comes early. Consequently they stop growing earlier in the season before the temperature drops (Fig. 13.7).

Stunting of growth is often seen when plants adapted to grow at high latitudes are grown in shorter day lengths further south. For example, plants from the far north at latitude 70°N have a critical day length of about 22 hours, which may never be achieved

**Fig. 13.7** The latitude of origin affects growth, leaf fall and dormancy in birch trees. Seeds were collected from different locations in Sweden (from left to right, latitude 68, 65, 63, 63 (mountain) and 60°N) and grown together in Uppsala (latitude 60°N). Trees on the far left (origin 68°N) are stunted and had already shed their leaves on 25 September when the photograph was taken. In contrast, those on the far right, which originated locally in Uppsala (60°N) are much taller and the leaves were still green. Trees from a high altitude (*second from right*) have a longer critical day length and behave like trees originating in lowland sites further north (*second from left*). *Photograph courtesy of David Clapham, Uppsala Genetic Centre.*

when they are grown further south. Consequently, following the breaking of dormancy by winter cold (see under Temperature below) the shoots cease to elongate and become dormant again after making only a small amount of growth in the spring. This dwarfing effect was very clearly seen in birch trees (*Betula* spp.) grown from seeds collected at 68°N and grown in Uppsala, Sweden, at latitude 60°N (Fig. 13.7).

Winter cold is not the only seasonal stress that can be avoided by a response to day length. In the Mediterranean regions of hot, dry summers and cool wet winters, the shedding of leaves in the summer allows plants to survive a period when water is in short supply. In some cases the shedding of leaves has been shown to be speeded up in long days. Summer dormancy is also associated with regions which have hot, dry summers and may be triggered or accelerated by long days as, for example, in the crown anemone (*Anemone coronaria*).

Bud dormancy is a complex process which, depending on the species, may include the cessation of shoot growth, the development of resting buds with various types of morphological modification, entry into dormancy, increase in frost hardiness, increase in drought resistance and leaf fall. Each of these processes has been shown to be influenced to some extent by day length, but there are considerable differences between species in the degree of control that day length can exert. Although low temperatures are important for the later stages in the development of deep dormancy and greater frost tolerance, short days are the main signal for the first stage; that is, the cessation of shoot growth and some frost tolerance. This day-length-dependent stage goes faster at mild temperatures and can actually be inhibited if the autumn temperature is too low.

As with flowering and the formation of storage organs, it is the leaves that detect the dormancy-inducing day-length signal and it is presumed that some kind of chemical stimulus is produced in them and exported to the buds where the changes leading to dormancy take place. Given that the overall process is so complex, it is perhaps not surprising that several substances appear to be involved. The known plant growth hormones, particularly the gibberellins, have been found to influence many aspects of dormancy and are generally thought to be major controlling factors. As noted above, recent experiments have shown the involvement of *FT* family genes.

## TEMPERATURE

Changes in the average temperature are also seasonal factors, although from year to year the actual temperature is strongly influenced by the immediate weather conditions. Temperature has a major effect on plants since all plant processes are influenced by it to some extent. The growth rate is speeded up at higher temperatures, provided that other factors, such as the amount of light, are not limiting and that the temperature is not so high as to be damaging. Temperature may also alter the plant's response to day length. Temperature as well as day length changes with latitude, and plants are often sensitive to both factors. Thus, although day length may be the overriding factor controlling whether or not a particular response can take place, the magnitude and even the timing of the response may be altered not only by temperature, but also by the amount of light and the supply of water.

In the temperate and cold regions of the world, temperatures below freezing during the winter are damaging, except for plants with adaptations or avoidance mechanisms. Some of these mechanisms (e.g. winter rest and the formation of storage organs) have already been discussed and later in this chapter we will look at the physiology of freezing damage. For some plants, however, exposure to low temperatures during winter is an essential step in their development. Two of these processes, vernalisation and the breaking of winter dormancy, have important implications for gardeners.

### Vernalisation

Not everyone agrees on the precise definition of *vernalisation*, but from the point of view of understanding the underlying mechanism it is probably best restricted to the triggering of flowering that occurs in response to a cold treatment given to seeds or young plants. It is sometimes used to refer to the triggering of germination in some seeds by low temperatures (see Chapter 9) but this is a different process and should not be called vernalisation (although it is often erroneously referred to as seed vernalisation). Vernalisation is usually an 'inductive' process, in the sense that after the cold treatment is

**Fig. 13.8** Vernalisation in the long-day plant *Coreopsis* 'Sunray'. In the absence of a cold treatment (vernalisation) plants failed to flower and continued to grow as rosettes. Stem elongation and flowering occurred rapidly after exposure to 5°C, but only in long days (LD). The cold treatment began when the seedlings had developed 12 leaves. SD, short days. *Photograph courtesy of Royal Heins, Michigan State University.*

complete no physical changes can be detected in the plant, and the formation of flowers begins only after plants have been returned to higher temperatures, and in many cases to particular day lengths. Exceptions are Brussels sprouts and some bulbs, where the flower initials are actually formed during the cold or cool treatment.

The requirement for vernalisation is most commonly found in LDPs. These may be annuals (e.g. winter cereals), biennials (e.g. carrot, *Daucus carota*; beet, *Beta vulgaris*) or perennials (e.g. perennial rye-grass, *Lolium perenne*). Without exposure to an adequate period of cold, these plants show delayed flowering or flowering may fail completely, and they may grow as rosettes (Fig. 13.8). The optimum temperature for vernalisation lies between about 1 and 7°C, with the effective temperature ranging from just below freezing to about 10°C. In experiments where different parts of a plant were given localised cooling treatments, it was found that cold was sensed by the shoot tip and its effect seemed to be largely independent of the temperature experienced by the rest of the plant.

Most studies indicate that the vernalisation effect is restricted to the cells that directly experience cold. Once vernalisation is complete it is perpetuated through subsequent cell divisions, and all of the cells arising from those that were originally exposed to cold are also vernalised. In annual and biennial plants the requirement for vernalisation is re-established during the process of reproduction. Perennial plants, in contrast, have to be vernalised again each winter

if they are to flower. In *Chrysanthemum* cultivars, plants become 'de-vernalised' during the preceding summer, while in some perennial grasses the vernalisation effect is not perpetuated indefinitely through cell divisions and the tillers formed late in the summer are not vernalised.

The nature of the cellular changes that lead to the semi-permanent vernalised state are unknown. Because vernalisation leads to stem elongation (bolting) in plants such as carrot (*Daucus carota*), which grow as rosettes in the absence of a cold treatment, it is thought that gibberellins may be involved. Indeed, the application of gibberellins to plants often substitutes for low temperature and causes bolting and flowering (e.g. foxglove, *Digitalis purpurea*). However, applying gibberellins does not cause flowering in all plants, even though the stems may elongate, for example in Canterbury bell (*Campanula medium*).

### Seed and plant vernalisation

Many annual plants, such as the winter forms of rye (*Secale cereale*) and oat (*Avena sativa*), can be vernalised by treating imbibed or germinating seeds. Although some biennial plants can also be vernalised as seeds, many biennial root crops only become sensitive to cold after they have reached a certain size. The duration of this *juvenile phase* during which they cannot be vernalised can vary considerably between varieties and has important implications for horticulture. A long juvenile phase may allow earlier sowing and so enable gardeners to take advantage of a longer growing season without the problem of bolting in the first year, which would result in failure to produce a harvestable storage root. This is particularly important where the spring temperatures are low and within the range that can cause vernalisation. For example, when sugar beet (*Beta vulgaris*) was sown in Scotland in mid-March, some 40% of the plants bolted, but when sown in mid-April most did not. In contrast, hardly any plants bolted from the earlier sowing of a non-bolting type.

The duration over which a cold treatment is required is also important, and non-bolting types may require a longer exposure to low temperature in order to vernalise them, as well as, or perhaps instead of, a long juvenile phase. If flowering is required for the production of seeds or in plants grown for their ornamental value as cut flowers, pot plants or garden plants, the need for a long exposure to low temperature can prevent successful cultivation in regions with mild winters. Prior vernalisation is also essential if cold-requiring plants are to be grown for flowering under heated glass.

### De-vernalisation

A further complication is that periods of higher temperatures (above about 15°C) may reverse the vernalising effect of a previous exposure to cold. This *de-vernalisation* by higher temperatures, notably high daytime temperatures, progressively decreases as vernalisation proceeds and is finally lost once the vernalisation process is complete, usually after several weeks of cold.

### Vernalisation and day length

It is perhaps not remarkable that vernalisation is linked with photoperiodism, since this enables plants to respond more precisely to seasonal changes. Only a few plants are known to require exposure to short days after vernalisation (e.g. some *Chrysanthemum* cultivars). The majority of plants requiring vernalisation also require subsequent exposure to long days in order to induce flowering (Fig. 13.8). This combination of responses ensures that flowering occurs during the longer days of summer when conditions are more favourable for the energy demands of seed production. It prevents the flowering of small seedlings in the previous summer and allows the plant to build up food reserves, often in the form of storage organs, for seed production in the following year (e.g. in biennial crop plants such as carrot).

Other interactions between day length and vernalisation are known. Exposure to short days, in particular, can often substitute either partly or entirely for cold in some LDPs, for example winter strains of wheat (*Triticum aestivum*) and rye, and Canterbury bell (*C. medium*).

## Breaking winter dormancy

Once they have become fully dormant, most tree species from the temperate zone will not resume growth until dormancy has been broken by exposure to low temperature. In most cases day length has no effect once dormancy has been completely broken by cold, although long days may accelerate bud-break in trees that have received some low temperature but are still in a state of partial rest. This is seen in beech (*Fagus sylvatica*), where exposing dormant cuttings to artificial long days in November and December had no effect on bud-burst, whereas such treatments became increasingly effective during February and March. This effect of day length on the time of bud-burst could be an effective strategy for breeding plants (see Chapter 5) that would be more tolerant of early spring frosts. Plants with a strong requirement for long days to accelerate bud-burst would leaf out later and so be less likely to be damaged.

The most effective temperatures for the breaking of winter dormancy are similar to those for vernalisation, and range from just above freezing to about 10°C. The cold requirement varies considerably between species and varieties and the failure to break dormancy can be a considerable problem in regions with mild winters. This is particularly true for temperate-zone fruit trees, such as apple (*Malus* spp.), where the duration of winter cold may be insufficient to break dormancy completely, resulting in erratic or delayed bud growth. This affects the timing of crop spraying programmes and, in the worst cases, reduces leaf expansion to such a degree that growth is affected. Consequently, it is commercial practice in some areas to promote dormancy-breaking by the use of various chemical sprays, including mineral oils, often together with dinitro-orthocresol. Their action in dormancy breaking is not fully understood, although the effect of mineral oils has been associated with the imposition of anaerobic conditions within the dormant buds, the coating of oil creating an airproof seal. In some cases, cultivars with shorter chilling requirements are available and are more suitable for warmer climates. These could become increasingly important in temperate regions if winters become warmer due to climate change.

## Direct effects of temperature on flowering

As we have seen, there are numerous interactions between day length and temperature in their effects on flowering. However, there are also many plants in which flowering is directly influenced by temperature, with often quite specific requirements. Some cultivars of *Chrysanthemum*, for example, require a minimum temperature to ensure uniform and regular bud formation. Other plants, for example calceolaria (*Calceolaria* Herbeohybrida Group) and cineraria (*Pericallis* × *hybrida*), require temperatures below a certain critical value in order to flower.

Probably the most detailed information regarding specific temperature requirements for flowering relates to bulbs and is widely used to prepare them for forcing. In the major bulb crops, notably cultivars of hyacinth (*Hyacinthus* spp.), tulip (*Tulipa* spp.) and daffodil (*Narcissus* spp.) the production of normal flowers requires first a period of high temperature for the initiation of flower buds, followed by a period of cool temperatures (8–12°C) to trigger stem elongation. However, other bulbs (e.g. Dutch irises, *Iris* × *hollandica*) initiate flowers only at relatively cool temperatures. Several different mechanisms are likely to be involved in these diverse responses to temperature and most of them are not well understood.

## Damage by temperatures below freezing

A major problem for plants of the temperate zones is exposure to temperatures below freezing, and many plants have evolved mechanisms that enable them to survive these adverse conditions. This may simply be avoidance by, for example, shedding of leaves or die back and overwintering by underground organs, or it may involve changes in the cell that increase the resistance to damage.

Ice crystals first begin to form in the spaces between cells and in the cell walls, where the concentration of dissolved substances is lower than inside

the cells themselves, and the freezing point is therefore higher. Water from within the cells then diffuses out and condenses on the growing ice masses. As water diffuses out, the increasing concentration within the cells further lowers the freezing point and so prevents the formation of ice crystals in the living protoplasm. The loss of water increasingly dehydrates the cells, and this in itself can be damaging to the living tissues.

When thawing occurs, particularly if this is gradual, the ice crystals outside the cells melt and water goes back into the cells. However, in non-hardy plants damage to membranes and other cellular components may already have occurred, so water does not re-enter the cells completely and normal metabolism cannot be resumed. Damage is particularly likely to occur with fast thawing, with ice crystals melting before the cells have regained their ability to absorb water sufficiently rapidly. For this reason, direct exposure to early sunlight, such as against an east-facing wall, often leads to frost damage, whereas those plants in positions where thawing occurs more slowly may escape such damage.

The ability of plants to develop frost resistance is quite remarkable. Some woody plants have been found to be able to withstand temperatures as low as that of liquid nitrogen (−196°C) when in the dormant, winter-hardy state and yet, when actively growing, the same plants can be killed at temperatures of −3°C, only just below freezing. We can see this readily in the damage to young expanding leaves that is often observed when spring frosts occur soon after bud-break. In woody plants winter hardiness is, in part, induced by the short days of autumn, but it is also dependent on subsequent exposure to low temperatures. In herbaceous plants, frost-hardiness typically develops during exposure to relatively low temperatures (usually about 5°C) and the gradual 'hardening-off' of plants raised under glass is an important practice, especially if they are likely to be exposed to temperatures below freezing. Hardening-off is, however, also important for plants that have no tolerance to frost but can be damaged by chilling temperatures.

In some cases frost resistance is based on tolerance to severe dehydration following the formation of ice outside the cells. This has many features in common with the ability to tolerate water and salt stress, and all three processes appear to involve the plant hormone abscisic acid (see Chapters 2 and 8 for the role of abscisic acid in water stress). The experimental application of abscisic acid to plants leads to the accumulation of proteins, called dehydrins. These also accumulate in plants in response to any environmental influence that has a dehydration component, such as freezing, salinity and drought. However, the ice crystals themselves can damage the cells, especially when they are large. This potential source of damage is reduced in frost-hardy plants by the presence of other proteins that bind to the surface of the ice-crystals and prevent them from growing larger. Finally, the properties of the cell membranes may be altered in a way that enables them to function at low temperatures.

## WATER AND LIGHT

The daily amount of light (as well as its duration) and the availability of water also vary seasonally and, like temperature, are modified by the immediate weather conditions. Both have significant effects on plant growth and are considered elsewhere. Light is discussed in relation to the limitations to growth imposed by low winter light on plants growing under glass in Chapter 14, and in relation to the effect of high light intensities on increased water loss from leaves in Chapter 10. Adaptations to water stress are considered in relation to the selection of plants for growing in dry conditions (Chapter 8,) and the avoidance or amelioration of water stress by shedding leaves and the development of summer dormancy (this chapter).

## CLIMATE CHANGE

The greenhouse gases causing global warming (see Chapter 18) are likely to have a number of effects on plants. The climate changes we are most likely to see include an increase in carbon dioxide ($CO_2$), hotter, drier summers and milder, wetter winters with an increased frequency of storms.

What do these changes mean for garden plants? Based on the experience of growers who enrich their greenhouses with extra $CO_2$, many but not all plants would be expected to grow faster due to the increased rate of photosynthesis at higher $CO_2$ concentrations (see Chapter 14). Warmer summers and winters with less frost should mean that gardeners would be able to grow a wider range of plants in the open. Even slightly higher winter temperatures would reduce frost damage to marginally hardy plants and extend their range further north. However, the precocious growth encouraged by earlier springs would be vulnerable to sudden cold snaps.

In contrast to the positive effects, there are many negative ones that would make gardening more difficult in the future. Although warmer winters would allow a wider range of plants to be grown, problems might arise in plants that require a sufficient period of winter cold to break dormancy or initiate flower buds. Storm damage is also likely to be more frequent.

Potentially, one of the most serious effects of climate change in the garden would be on soils. An increase in winter rainfall would increase the leaching of nutrients from the soil, especially nitrates, and be likely to cause more soil compaction and erosion. Waterlogging of soils is predicted to increase in frequency, leading to reduced root growth and in some cases root death.

Summer droughts and low water reserves would mean that irrigation in gardens could be reduced or banned. Even short periods of dry weather coupled with a shortage of water can weaken many plants and may even kill them. Lawns in particular are liable to become brown in dry conditions and may not recover if drought is prolonged. Gardeners could plant drought-tolerant species, including lawn grasses, but many of these are not tolerant of waterlogged soils so would be difficult to maintain during winter.

Climate change is discussed in further detail in Chapter 18.

## CONCLUSION

The growth, development and flowering of plants is influenced by a diversity of seasonal factors, including day length, temperature, water and light. Much is known about the mechanisms underlying these influences, and understanding these can help the gardener to manage plants growing out of doors, to site them appropriately and to predict how they will behave during different seasons.

## FURTHER READING

Atherton, J.G. (ed.) (1987) *Manipulation of Flowering*. Butterworths, London.

Bernier, G. (ed.) (2005) The florigen quest: are we beginning to see the end of the route? *Flowering Newsletter* **40**, 4–59.

Bisgrove, R. (2002) Weathering climate change. *The Horticulturist* **11**, 2–5.

Burroughs, W. (1988) Degrees of damage. *The Garden* **123**, 249–51.

Burroughs, W. (1999) Winter frieze. *The Garden* **124**, 22–5.

Clapham, D.H., Dormling, I., Ekberg, I., Qamaruddin, M. & Vince-Prue, D. (1998) Dormancy: night timekeeping and day timekeeping for the photoperiodic control of budset in Norway spruce. In *Biological Rhythms and Photoperiodism in Plants*, P.J. Lumsden & A.J. Millar (eds), pp.195–209. Bios Scientific Publishers, Oxford.

Gates, P. (1999) Burgeoning beginnings. *The Garden* **124**, 28–33.

Gates, P. & Ardle, J. (2002) Climate change coming soon to a garden near you. *The Garden* **127**, 912–17.

Halevy, A.H. (ed.) (1985, 1989) *Handbook of Flowering*, vols I–VI. CRC Press, Boca Raton, FL.

Jackson, S.D. (1999) Multiple signaling pathways control tuber induction in potato. *Plant Physiology* **119**, 1–8.

Jackson, S.D. & Thomas, B. (1997) Photoreceptors and signals in the photoperiodic control of development. *Plant Cell and Environment* **20**, 790–5.

King, R.W. & Evans, L.T. (2003) Gibberellins and flowering of grasses and cereals; prizing open the lid of the black box. *Annual Review of Plant Biology* **54**, 307–28.

Salisbury, F.B. & Ross, C.W. (1992) *Plant Physiology*, 4th edn. Wadsworth Publishing Company, Belmont, CA.

Vince-Prue, D. (1990) The control of flowering by daylength. *The Plantsman* **11**, 209–24.

Vince-Prue, D. & Cockshull, K.E. (1981) Photoperiodism and crop production. In *Physiological Processes Limiting Plant Productivity*, C.B. Johnson (ed.), pp. 175–7. Butterworths, Oxford.

Vince-Prue, D. & Thomas, B. (1997) *Photoperiodism in Plants*. Academic Press, London.

# 14

# Gardening in the Greenhouse

## SUMMARY

The ways in which a greenhouse modifies the plant's environment are considered, and the main physiological effects of these changes described. General guidelines for managing a greenhouse are suggested, based on an understanding of how plants respond to the major factors that can be controlled in this environment, particularly temperature, light, carbon dioxide, water, air movement, humidity, nutrition and day length.

## INTRODUCTION: THE GREENHOUSE ENVIRONMENT

Plants in a greenhouse are growing under conditions that are quite different from those in the open air. There are two main reasons for this, the properties of glass itself and the fact that the growing space is enclosed.

### Light

When sunlight passes through glass, the amount of ultraviolet (UV) light at wavelengths shorter than about 380 nm (nanometres) is decreased progressively until, below about 310 nm, there is no UV light. Consequently, the greenhouse is a low-UV environment. This affects the way plants grow, especially their height.

Both plastics and glass reduce the amount of light reaching the plants, compared with those growing in the open. Clean glass transmits about 86% of the incident light, whereas dirty glass significantly reduces this and can cause more than 20% of available light to be lost. The importance of this is emphasised by the fact that for many crop plants, such as tomato

(*Lycopersicon esculentum*), the results of experiments indicate that a 1% loss of light results in a 1% loss in yield. The amount of light transmitted through glass also depends on the angle of the sun. At low solar angles some 30% of the incoming light is reflected and only about 58% passes into the greenhouse. Glazing bars and any nearby shading structures also contribute to the overall reduction in the available light. Because of the strict relationship between light and photosynthesis (see Chapters 2 and 8) it is essential to maximise the available light by paying attention to factors such as the siting of the greenhouse, clean glass and minimal structural interference with light transmission.

### Temperature

Sunlight transmitted through glass is absorbed by the internal surfaces of the greenhouse and re-radiated at longer wavelengths as radiant heat. Since glass does not allow these wavelengths to pass through it, the re-radiated heat is trapped within the greenhouse and warms the internal atmosphere significantly, particularly on sunny days. Mostly designed to extend growing seasons by trapping solar heat during cold winter weather, greenhouses overheat most plants in summer. Plastics do not have the same transmission characteristics as glass: some are more transparent to radiant heat and so may not trap the re-radiated heat as effectively.

Many commercial growers and some gardeners utilise flexible film plastics, attached to a supporting frame (for example *polytunnels*) instead of greenhouses. In the past, film plastics have had a poor reputation. Heat retention was poor and it was possible to have a *temperature inversion*, with frost inside a polytunnel yet none outside. Modern film plastics have, however, been modified substantially, with better heat retention and less condensation, and they often outperform glass in the amount of light they transmit. Although film plastics have to be replaced at intervals, modern UV-inhibited plastics should last for four seasons.

## Ventilation

The fact that plants are growing in an enclosed environment is also important. When there is no ventilation, there is little exchange of carbon dioxide ($CO_2$) between the external atmosphere and that of the greenhouse; this results in a reduction in atmospheric $CO_2$ levels inside the greenhouse due to uptake into the plants by photosynthesis, particularly in bright light. In a closed greenhouse the $CO_2$ level at midday can drop to a level well below that necessary to maintain maximum growth rates (Fig. 14.1). An enclosed environment also reduces air movement, and plants in a greenhouse are growing under much less turbulent conditions than those growing outside.

## SITING THE GREENHOUSE

A major consideration in siting a greenhouse is the nearness of potential shading objects such as trees and buildings. To minimise light loss, a rule of thumb is that the greenhouse should be no closer than four times the height of a nearby shading structure. It is also important to minimise heat loss, which is increased by air movement over the glass. It is, therefore, highly desirable to site a greenhouse where it is protected from cold northerly and easterly winds, remembering that wind-breaks also cast shade. Lean-to structures are the least demanding of heat input because the house wall provides complete insulation on one side. Twin-walled polycarbonate (a type of rigid plastic) provides better heat retention than single glass, but has reduced light transmission. Another way to save heat is to line the greenhouse with bubble polythene, although this also results in the loss of available light. A final point concerns orientation. In small gardens there is usually little choice, but it has been demonstrated that, for a single free-standing greenhouse, an east–west orientation transmits more light in the winter months, when light is the main limiting factor for growth, than does a north–south orientation.

## EFFECTS OF THE GREENHOUSE ENVIRONMENT

Due to the transmission properties of the cladding and to the fact that the greenhouse is an enclosed space, the internal conditions are different from those

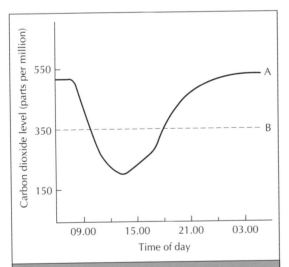

**Fig. 14.1** Changes in the concentration of carbon dioxide in a closed greenhouse during the course of the day. Rapid photosynthesis during the period of high light intensity between noon and 15.00 hours leads to a drop in the concentration of carbon dioxide in a closed greenhouse (A) to well below that in the outside air (B). Note that when this experiment was made the atmospheric $CO_2$ concentration was lower than it is now.

outside. Light, temperature and $CO_2$ concentration are the major environmental factors affecting the rate of photosynthesis, which in turn is a major factor limiting the rate of growth (see Chapter 2). Light also modifies the general appearance of plants. Under low-light conditions, stems tend to be longer and their strength is reduced, whereas leaves are thinner and often larger than in plants growing under better light conditions.

## Light

Photosynthesis is a complex process, which involves reactions that are directly dependent on light, as well as dark reactions that are dependent on $CO_2$ and temperature (see Chapters 2 and 8). Consequently, the responses to these environmental factors are also complex. When a single leaf is exposed to different amounts of light at normal atmospheric levels of $CO_2$ (about 380 parts per million, and rising due to climate change), the rate of photosynthesis increases with increasing light levels until, above a certain value, there is no further increase (Fig. 8.8). This value is known as the *light saturation point* and varies with the species and the previous light history of the plant. Plants with C-4 photosynthesis (see Chapter 8) usually have much higher light-saturation values than plants with C-3 photosynthesis (Fig. 8.8).

Species native to shady habitats have low rates of photosynthesis in bright light and the light-saturation value is also lower than in other plants. Many do not tolerate full sunlight and their leaves can be scorched under these conditions. However, under very low-light conditions they usually photosynthesise at higher rates than do other species. Many ornamental house plants, such as African violet (*Saintpaulia ionantha*), fall into this category and are the plants of choice when light is limiting.

At the other extreme are 'sun' plants, which can only grow satisfactorily in bright light. However, there are many sun plants that, when grown in the shade, can adapt (acclimate) and develop so-called shade leaves; these are thinner and more expanded than leaves of the same species developing in full sunlight and they have more light-harvesting chlorophyll (see Chapter 2) per unit weight of leaf. These characteristics increase the ability of the leaf to capture light, enabling the plant to grow in shady conditions. The growth rate, however, will normally

be slower and undesirable morphological changes such as long, weak stems may still occur. This ability to acclimate to low light may be particularly important in large plants and in closely planted crops, where new leaves on the lower branches develop under shadier conditions than those on the upper ones.

Plants can also acclimate to an increase in available light, developing smaller, thicker leaves with a higher light-saturation value. At high light intensities such leaves are able to photosynthesise at a faster rate than those developing in shade, and so enable the plants to utilise the better light more efficiently. It has been shown experimentally that the plants detect in some way the total amount of light received each day (the *light integral*) and not the brightness (*irradiance*) at any one time; this means that a given amount of light for a certain length of time results in the same degree of acclimation as twice the light intensity for half the time. In young expanding leaves, these changes can occur within a few days when plants are transferred from low to high light levels, although leaves previously grown in shady conditions can be severely damaged and bleached if they are exposed suddenly to bright sunlight.

## Carbon dioxide

Provided that light is not limiting, the rate of photosynthesis increases with an increase in the amount of $CO_2$ in the atmosphere until a certain level is reached. The actual value at which saturation occurs depends on the amount of light; when there is more light, saturation occurs at a higher concentration of $CO_2$. Plants with C-3 photosynthesis (see Chapter 2) show a strong response to $CO_2$ and, provided that light is not limiting, net photosynthesis in most C-3 plants continues to increase until quite high levels of $CO_2$ are reached (Fig. 14.2). These can be considerably above normal atmospheric concentrations. As greenhouse crops frequently lack enough $CO_2$ for maximal growth, some commercial growers add $CO_2$ to the greenhouse atmosphere. Levels are not usually allowed to exceed about 1000–1200 parts per million (i.e. more than three times the normal concentration in the air), as higher concentrations cause stomata to close and can be toxic. They are also uneconomical, especially when the ventilators are open. However, different plants respond in unpredictable ways to enrichment with $CO_2$. Tomatoes (*Lycopersicon*

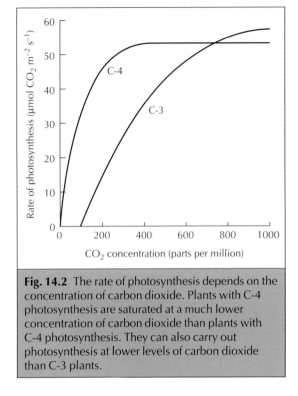

**Fig. 14.2** The rate of photosynthesis depends on the concentration of carbon dioxide. Plants with C-4 photosynthesis are saturated at a much lower concentration of carbon dioxide than plants with C-4 photosynthesis. They can also carry out photosynthesis at lower levels of carbon dioxide than C-3 plants.

*esculentum*), for example, show a high response whereas cherry seedlings (*Prunus* spp.) do not.

It should also be noted that C-3 and C-4 plants (see Chapter 8) respond to increased $CO_2$ concentration in different ways. Photosynthesis in C-4 plants is generally saturated at about 400 parts per million (i.e. just above the normal content in the air), even at high light levels when the demand for carbon dioxide is greatest (Fig. 14.2). At the other end of the scale, C-4 plants are able to continue to carry out photosynthesis at very much lower levels of $CO_2$ than is possible for C-3 plants.

## Temperature

The third factor that influences the rate of photosynthesis is temperature and, as one might expect, the effect of temperature depends on the conditions under which the plant is growing as well as on the species. In general, the optimum temperatures for photosynthesis are similar to the daytime temperatures at which the plants usually grow in their

natural habitats. Although there are exceptions, optimum temperatures are generally higher for C-4 than for C-3 plants.

Temperature influences the rate of carbon fixation in the 'dark' reactions of photosynthesis, and consequently the true rate of photosynthesis usually increases with increase in temperature until the enzymes begin to become denatured. However, the loss of $CO_2$ through respiration also increases with temperature. The net amount of photosynthesis (i.e. carbon fixed by photosynthesis minus carbon lost by respiration) is therefore not promoted by increased temperature nearly as much as one might expect, especially in C-3 plants. In contrast, C-4 plants exhibit a much stronger response to temperature and have a higher optimum value. This is because, in addition to normal dark respiration, C-3 plants carry out a process known as *photorespiration* in the light. Like dark respiration, the rate of photorespiration (i.e. loss of carbon) increases with increasing temperature, so that the net rate of photosynthesis is decreased. C-4 plants do not carry out photorespiration, so this source of carbon loss is absent and net photosynthesis increases with increasing temperature to a much higher value than in C-3 plants.

## Air movement

The lack of air movement in the enclosed space of the greenhouse affects plants in a number of ways, and enhancing the rate of air movement is recognised as being one of the most important ways of improving the uniformity of climatic control, both within the greenhouse and between seasons. In still air, the atmosphere immediately around the leaf may be depleted of $CO_2$, especially in bright sunlight. Even slight air movement can increase photosynthesis by displacing this $CO_2$-depleted air (Table 14.1). Owing to the high moisture content of the still air around the leaf, cooling by transpiration is reduced and leaf temperatures can increase substantially. This increase in leaf temperature will in turn increase water loss because of the temperature differential between leaf and air (see Chapters 8 and 10). Although this will increase transpiration and so cool the leaf, the overall effect of high temperatures in still air is to increase the rate of water loss from the leaf. This often leads to temporary wilting, even when the water supply is

**Table 14.1** Wind velocity affects the rate of photosynthesis.

| Wind velocity (cm per second) | Rate of photosynthesis (cm³ of CO₂ fixed per cm² of leaf, per hour) |
|---|---|
| 10 | 79 |
| 16 | 88 |
| 42 | 101 |
| 100 | 109 |
| 300 | 114 |
| 1000 | 118 |

plentiful. Still air, especially at high humidities, may also increase the incidence of some fungal diseases (see Chapters 15 and 16).

## MANAGING THE GREENHOUSE ENVIRONMENT

So far in this chapter we have considered how the environment in a greenhouse differs from that outside, and the effects of these differences on plants. Managing the greenhouse is largely concerned with minimising the undesirable aspects of the enclosed environment, as well as maximising the desirable ones.

### Temperature

To take the desirable aspects first, the most obvious advantage of growing under a glass or plastic structure is the ability to maintain a minimum temperature. In order to conserve energy, this should be as low as possible with regard to the stage of growth of the plant and the season. High temperatures are undesirable in winter because the loss of carbon by respiration is increased, whereas the rate of photosynthesis is limited by light. No more than protection from freezing temperatures is often sufficient during periods of low light, or when plants are in a dormant or semi-dormant state. However, there are several plants that can be damaged by exposure to low temperature even when this is above freezing, and it is important to

know which plants fall into this category. Common examples include tomato (*Lycopersicon esculentum*), cucumber (*Cucumis sativus*) and pepper (*Capsicum annuum*) as well as ornamental plants such as African violet (*Saintpaulia ionantha*).

The optimum temperature for any plant depends not only on the species, but also on the stage of growth and the amount of light and $CO_2$. There is also the problem that, in most greenhouses, there is a marked variation across the greenhouse, both in light and temperature. This can be ameliorated to some extent by moving plants around and can also be used to some advantage by growing plants with different requirements in different parts of the greenhouse. To do this effectively requires the gardener to make reasonably accurate measurements of both light (see below) and temperature.

### Water

A second and perhaps less obvious desirable feature is the fact that the water supply is under the gardener's control. Correct water management is, therefore, one of the most important and sometimes most difficult aspects of greenhouse culture. This is less tricky for plants growing in soil, since they have a larger reservoir of water to access and there is less danger of over-watering. However, for plants growing in containers, both over-watering and under-watering can be a problem.

A major difference between the soil and containers is that the soil has a continuous system of pores, allowing excess water to drain out, whereas a pot has a base where the pore system is discontinuous. The effect of this is that the growing medium in the pot retains water and does not drain as freely. In winter, temperatures and light are low, so that the rates of water loss by transpiration are also very low. Under these conditions it is very easy to over-water pots and allow the growing medium to become waterlogged. Because the overall growth rate is slow, due to low rates of photosynthesis, root growth is restricted; this also limits water uptake and accentuates the problem. So in winter care must be taken to restrict the supply of water.

The summer situation is obviously quite different; high temperatures lead to heating of the leaf and increase the rate of water loss, especially in modules and small pots, both of which can dry out extremely quickly. Plants are also growing much more rapidly under the better light conditions and so the demand for water is much greater. The high temperatures of the interior mean that the greenhouse environment in summer is a dry one, with a low moisture content. The gradient of water vapour between the inside of the leaf and the surrounding air is increased both by the high leaf temperature and the fact that the content of water vapour in the surrounding air is low. This results in rapid water loss by transpiration.

Several management practices can ameliorate the situation. Lowering the leaf temperature is a major consideration and can be achieved in several ways. Adequate ventilation is probably the most important factor in reducing the greenhouse temperature. Not only does ventilation allow exchange between the warmer air inside and cooler air outside, but the increased air movement lowers the leaf temperature and in this way reduces the rate of water loss. Cooling the interior surfaces by damping down also helps, especially where the floor consists of concrete or paving. Some commercial growers use evaporative cooling systems, known as *fan and pad cooling*, in which the outside air is drawn across wet pads into the greenhouse. This is a very effective way of both lowering the temperature and increasing the rate of air movement, but is outside the range of most gardeners, who must rely on simple ventilation, which can be increased by fans. Evaporative cooling can cool greenhouses to about 5–6°C below those cooled by open-vent natural ventilation.

Another way of reducing the heat load on the leaf is to shade the greenhouse, either by painting the glass with a shading material or by the use of blinds. However, shading has the disadvantage of decreasing the amount of light reaching the leaf. Even in summer, the amount of light reaching the lower, more shaded leaves on the plant is less than that needed for maximum rates of photosynthesis, so shading necessarily results in some reduction in growth. This is exacerbated by permanent shading, which remains in place even on cloudy days. Blinds are more satisfactory as they can be lowered in bright light to reduce the heat load and raised to allow maximum light when the sun goes in. Where permanent shading is used, it is essential to ensure that it is removed as soon as possible at the end of the summer.

## Ventilation

Ventilation not only lowers the air and leaf temperature, but also increases the air movement inside the greenhouse. This enhances photosynthesis by replacing $CO_2$-depleted air in the atmosphere around the leaf (Table 14.1). Ventilation also increases the concentration of $CO_2$ in the greenhouse by exchange with the incoming fresh air. In a closed greenhouse the $CO_2$ concentration is reduced by photosynthesis, especially in bright light, and as discussed above this often limits the rate of photosynthesis in summer. When ventilation is reduced, as in winter, $CO_2$ can fall below the normal atmospheric level. Air movement can also reduce plant height in the same way as shaking (see Chapter 11), and the effect is more marked under glass where air movement is normally restricted. Finally, ventilation is an important management practice in preventing the incidence of some fungal diseases, which can become a major problem in stagnant air with a high relative humidity (see Chapters 15 and 16).

## Growing media

Many soils possess positive features that are helpful in supporting plant growth. They contain:

1. a large reserve of nutrients;
2. living organisms that cycle nutrients and assist in creating soil structure;
3. a range of pore sizes that allows excess water to drain, gases to flow and water to be stored;
4. colloids (e.g. clay and peat) that allow pH to be 'buffered' and change only slowly.

An ideal potting compost would contain all these features, but achieving this in a controlled and cost-effective way is very difficult. When plants are grown in containers, their roots are confined in a smaller volume than when grown in the garden and this means that the demands made on the potting medium for water and nutrients are more intense. Using garden soil in containers gives poor results unless the physical and chemical properties are substantially enhanced and steps taken to eliminate pathogens.

The development of the John Innes composts was a substantial step forward because it produced a medium that had good chemical and physical properties for plant growth. A key feature of the compost was that it contained loam (i.e. soil with properties that ensure a good supply of water and sufficient clay content to provide a buffered source of nutrients and regulation of pH). The advantages of a loam-based compost are that the nutrient supply is good and minor element deficiencies are not common. The disadvantages are that good-quality loam is in short supply, the material must be steam-sterilised to kill potential pathogens, and the composts are heavy and difficult to handle.

These substantial disadvantages have led to the development of many loamless composts, usually based on organic materials. Peat has been the most commonly used organic material, but the growing concern about the environmental damage caused by peat extraction (see Chapter 18) has led to trials with other materials. The use of peat-free media is now recommended by the Royal Horticultural Society but growers should be aware of the fact that many peat-free products handle differently from peat. Particular attention should be given to water requirements and feeding. Peat-free composts can dry out on top but be moist below, while fertiliser requirements may vary.

Loamless composts are generally less variable in composition than loam-based composts, do not require sterilisation and are cheaper and lighter to handle. Their disadvantage is that nutrient supplies are often lower and less well buffered, resulting in a greater dependence on supplementary feeding. Some commercial producers grow their crops entirely on inorganic substrates, such as rock wool. In this case all the essential nutrients are supplied by liquid feeding, and complex computer control is usually employed, together with frequent nutritional analyses, to ensure that correct nutrition is maintained at all stages of the plant's growth.

## Light

Light is most often the factor that limits the growth of plants under glass, especially at higher latitudes in winter, when there is much less available light for photosynthesis because of shorter days, low solar angle and many more overcast days; cloud cover can reduce light to less than 20% of bright sunlight. In the UK, the total light received on a winter day may be only 10% of that received on a summer one (Fig. 14.3). Even if the upper leaves are at light saturation in the summer, photosynthesis in the shaded lower leaves will still be limited by the available light. Although a single leaf may be light-saturated at about 90 W m$^{-2}$ (watts per square metre), up to twice that amount can be required to achieve light saturation in a crop. Obviously the closer together

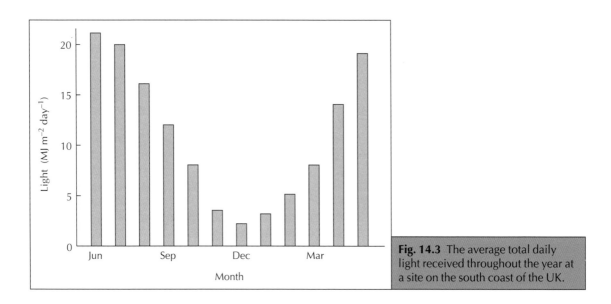

**Fig. 14.3** The average total daily light received throughout the year at a site on the south coast of the UK.

the plants, the greater the shading effect, so one way of increasing the available light for individual leaves is to increase the spacing between the plants. This is particularly important in winter, when light is severely limiting. Some management techniques can also help, such as cleaning the glass and lowering the temperature.

## SUPPLEMENTARY ARTIFICIAL LIGHTING

The problem of low light during the winter months, usually from November to March in northern Europe, can be ameliorated to some extent by the use of artificial light to supplement natural daylight. In some facilities, such as specialist propagation units, plants are grown in insulated windowless structures and natural daylight is entirely replaced. However, artificial light is more commonly used to 'top-up' winter daylight in a greenhouse by adding to the natural light during the day, or giving extra light during the night, or both. This is especially useful for seedlings in winter and early spring, when large numbers of plants can be lit in a small area. Extra light at this stage gets the seedlings off to a good start and the ambient light levels are generally improving by the time the plants get bigger and need to be spaced out. The additional light results in a faster growth rate and better-quality, sturdier plants.

The primary effect of the additional light is to increase the rate of photosynthesis, so the major requirement is to give sufficient extra light to make a difference. This can be expensive and, in order to make the best and most economic use of artificial light, several points need to be taken into account, as follows.

1. How should light be measured in relation to plant responses?
2. What type of lamp gives the best results?
3. How much light is necessary for a satisfactory response and how is this best achieved?

In order to answer these questions, it is necessary to have some understanding of the basic properties of light and the way it interacts with plants.

## Light measurement

Measuring the amount of light available for plant growth is not straightforward. Many measuring devices are calibrated in *photometric units* and measure light in terms of the sensitivity of the human eye to different wavelengths (see Fig. 12.2). The light is expressed as *illuminance* and the units are *lux* (an earlier unit that is sometimes still used is the foot candle; 1 ft candle = 10.76 lux). The human eye is most sensitive to green light and is relatively insensitive to red and blue, which are the most important wavelengths for plants. Consequently, light measurements expressed in lux can be very misleading where plants are concerned. Light can also be measured, using a radiometric detector such as a thermopile, in terms of radiant energy per unit area (usually as watts per square metre; $W\ m^{-2}$). Once again this can be misleading because most detectors of radiant energy measure radiation not only in the visible wavelengths that are active in photosynthesis, but also in the ultraviolet and infrared regions beyond the limits of plant responses. One advantage of detectors that measure in lux is that they measure only visible light, which is in the range that is also detected by plants. Illuminance values (lux) can be converted to irradiance ($W\ m^{-2}$) by using the appropriate conversion factors given in Table 14.2.

To be effective in photosynthesis, or any other light-dependent process in plants, light must first

**Table 14.2** Factors used for converting illuminance (lux) to irradiance (milliwatts per square metre, $mW\ m^{-2}$).

| Type of light | Conversion factor* |
|---|---|
| Natural sunlight | 4.0 |
| Tungsten-filament lamp (100 W) | 4.2 |
| High-pressure sodium lamp (SON/T) | 2.4 |
| Tubular fluorescent lamps | |
|   Warm white† | 2.8 |
|   Deluxe warm white† | 3.6 |
|   Daylight† | 3.7 |

* Multiply by this factor to convert measurements in lux into $mW\ m^{-2}$.
† The precise conversion factors for the various colours of tubular fluorescent lamps may vary with the manufacturer, but the values given here are adequate for use in practice.

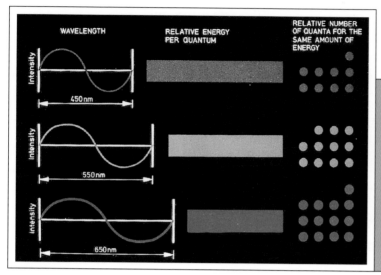

**Fig. 14.4** The relationship between the wavelength of light, the energy per quantum and the number of quanta per unit energy. Each quantum of blue light (at a wavelength of 450 nm) has a shorter wavelength (*left*) and more energy (*centre*) than a quantum of red light (at 660 nm). Consequently, for the same amount of energy there are fewer quanta in blue than in red light. *Image courtesy of Brian Thomas, University of Warwick.*

be absorbed by a pigment. For photosynthesis, the chlorophylls and some associated carotenoids are the light-absorbing pigments and they absorb at wavelengths throughout the visible spectrum, although not equally at all wavelengths (see Chapter 2). Absorption is least in the green part of the visible spectrum and greatest in the red part. Moreover, it is important to remember that light is absorbed by pigments in the form of discrete units of energy called *quanta* (or *photons* for energy within the visible spectrum), and that the amount of energy in each quantum is inversely proportional to the wavelength of the light. This means that, for the same amount of measured energy, there will be fewer quanta available for photosynthesis in blue light at an average wavelength of 450 nm than in red light at 650 nm (Fig. 14.4).

The most useful measurement of the amount of light available for photosynthesis is the number of quanta (photons) per unit area (the *photon flux density*) in the visible spectrum, between 400 and 700 nm. This is sometimes called *photosynthetically active radiation* (PAR). However, when only sunlight is being considered, the irradiance (expressed as $W\,m^{-2}$) between 400 and 700 nm is also a useful way of comparing the amounts of light available for photosynthesis at different times and in different places; as, for example, to look at the shading effects in different parts of a greenhouse, or to determine how much light is being transmitted through the glass.

## The choice of lamp

Because the effect on photosynthesis and other photochemical reactions is dependent on the number of absorbed photons for the same amount of measured energy, photosynthesis will be stimulated more by red wavelengths (more quanta per unit of energy) than by blue or green ones with fewer quanta per unit of energy. So other things being equal, one might expect that lamps emitting red light would be the recommended ones for supplementary lighting. However, other things are not equal. First of all, the horticultural market is small and lamp manufacturers are mainly concerned with lighting for human vision, which is most sensitive to green wavelengths and less sensitive to red. The spectral output for most lamps is, therefore, related to this. A few lamps specifically designed for plants are available, but they are considerably more expensive and not considered to offer sufficient advantage to justify their extra cost.

In addition to photosynthesis, light has many other effects on plant growth (see Chapters 8 and 11). As discussed in Chapter 11, stem growth in many species (especially sun plants) responds to the ratio of red to far-red wavelengths (R/FR) in the light; the greater the relative amount of FR light, the more stems elongate. Not all plants are sensitive to the R/FR ratio, however, and in some stem growth appears to be controlled mainly by the amount of blue light. This is well illustrated by the response

**Fig. 14.5** The responses of different lettuce (*Lactuca sativa*) cultivars to light from low-pressure sodium lamps. Some cultivars (B and C) grown in light from low-pressure sodium lamps with added red light (SOX + R) show elongated and abnormal growth, while others (A amd D) grow normally. When blue light is added (SOX + B) all of the cultivars are normal. *Photograph courtesy of Brian Thomas, University of Warwick.*

of lettuce (*Lactuca sativa*) seedlings (Fig. 14.5). When grown under low-pressure sodium lamps, which do not emit any blue light, some cultivars develop an extremely elongated and unsatisfactory habit of growth. Only when blue light is added do the seedlings grow in a normal manner. Other cultivars, however, are clearly dependent on the R/FR ratio of the light and do not require blue light. These cultivars grow normally under low-pressure sodium lamps, which are the equivalent of red light and establish the same amount of the biologically active form of phytochrome. It is clear that plant responses have to be considered individually when choosing a suitable lamp type.

Of the lamps currently available, probably the most satisfactory from the point of view of the wavelengths they emit are the various types of 'white' fluorescent lamps. Most of these have a very low FR emission and emit light throughout the visible spectrum, producing a sturdy habit of growth. Indeed, because the R/FR ratio of most of these lamps is considerably higher than sunlight (Table 14.3), plants growing under them often develop a shorter habit of growth, with dark green leaves (Fig. 8.3). These lamps are the ones most often used where daylight is replaced completely, as in micropropagation and some other specialist uses, and spectral quality is a more critical issue than when some daylight is present to supplement the spectrum.

**Table 14.3** The ratios of red to far-red light in lamps that are commonly used in horticulture.

| Type of light | Ratio of Red to Far-red light[1] |
|---|---|
| Tungsten-filament lamp (incandescent) | 0.7 |
| Tubular fluorescent lamp | |
|    Warm white (colour 29) | 22.7 |
|    Deluxe warm white | 3.6 |
| High-pressure sodium lamp (SON/T) | 3.7 |
| Sunlight | 1.15 |

[1] Sunlight is given for comparison. The red/far-red ratio of tungsten-filament lamps is less than sunlight, resulting in excessive stem elongation in some plants. The red/far-red ratio of most fluorescent lamps is much higher than sunlight, resulting in a compact habit of growth.

In other respects, fluorescent lamps have both advantages and disadvantages. A major advantage, especially in small glasshouses where there is little headroom, is that they emit little heat and have a low surface temperature. This means that they can be suspended quite close to the plants without damaging them. Because of their linear shape, they also have a uniform light distribution. Their disadvantage is that their overall light output is low, lamps being no more than 125 watts. This means that several lamps are usually needed to obtain light levels high enough to stimulate photosynthesis. In turn, this means that they can cast a considerable amount of shade, especially when used with reflectors, thus reducing the amount of natural light reaching the plants.

Fluorescent lamps are the ones most commonly used by amateur gardeners, but for most commercial growers the high-pressure sodium lamp has become the preferred choice. These lamps have a high light output, a long useful life and a spectrum that produces a satisfactory habit of growth when used to supplement daylight. High-pressure sodium lamps are relatively small in size and, for the same light output, cast considerably less shade than fluorescent lamps.

Tungsten-filament (TF) lamps (i.e. ordinary domestic electric light bulbs, or incandescent lamps) are not satisfactory for supplementing daylight. They have a R/FR ratio considerably lower than that of daylight (Table 14.3) and result in an excessively 'leggy' type of growth in many plants (Fig. 8.3). Additionally, their high output in the infrared region of the spectrum results in a considerable heat load on the leaf.

## How much 'extra' light?

The amount of supplementary light needed for satisfactory winter growth in high latitudes varies with the plant. However, in practice light levels between 5 and 19 W m$^{-2}$ produce satisfactory results when used for between 8 and 24 hours each day. It is generally true that plants respond to the total amount of light received each day, so lower light levels from the lamps can be compensated for by increasing the daily duration of extra light. The daily duration of light can, however, have a marked influence on plants, so this must be taken into account when deciding on the precise schedule for supplementary lighting. For example, tomatoes should not be continuously lit for 24 hours each day as this can result in damage to the leaves. Economies can be obtained by switching lamps off during bright periods in late autumn and early spring.

Detailed recommendations for particular installations and geographical locations are outside the scope of this book but are given in several manuals (such as *Lighting for Horticultural Production*, see the Further reading section in this chapter).

## DAY-LENGTH LIGHTING

The physiology of photoperiodism is discussed in Chapter 13 in relation to the effects of changes in natural day length. However, the use of artificial day lengths to manipulate flowering is largely confined to the greenhouse, at least in the UK, and so needs to be considered here. The short days of winter can be extended by using artificial light, while the long days of summer can be decreased by covering the plants each day with a light-excluding material. Both are practised commercially.

Summer-flowering plants such as *Antirrhinum, Calceolaria, Alstroemeria* cultivars and even perpetual-flowering carnations (*Dianthus caryophyllus*), which flower much more profusely in summer, can be brought into bloom in the winter. *Chrysanthemum* cultivars that flower in the short days of autumn can be marketed at any time of year by a combination of covering them in summer, to allow flowering, and giving them long days with artificial light in winter, to delay flowering and obtain a satisfactory length of stem for a cut flower (Fig. 14.6). 'All-year-round' flowering (AYR) of *Chrysanthemum* cultivars is now a major industry, but before about 1950 growers in the UK had great difficulty in holding back flowering to obtain the higher prices of the Christmas period. Other short-day plants for which artificial long days are used commercially to time the crop or to achieve a desirable size include poinsettia (*Euphorbia pulcherrima*), flaming Katy (*Kalanchoë blossfeldiana*) and winter-flowering *Begonia* spp.

### Short-day plants

Covering in summer to obtain summer flowering of short-day plants (SDPs) poses no major difficulties, although the actual duration of covering depends on the plant. SDPs will flower when the days are shorter than a certain critical length, but unfortunately this varies with species and even cultivars (some examples are given in Table 13.2). However, a rule of thumb is that the majority will flower on a 10-hour day, or shorter. Plants differ considerably in their sensitivity to light, so it is important to exclude daylight as much as possible. Domestic lighting is bright enough to delay or prevent flowering in some SDPs, such as poinsettia, especially when the lights

**Fig. 14.6** 'All-year round' production of *Chrysanthemum* cultivars. During the winter months, plants are given artificial long days (night-breaks with tungsten-filament lamps) to prevent premature budding. When the stems are sufficiently long, the lights are switched off and the plants returned to short days to allow bud formation. The short days are either given naturally in winter or obtained by covering the plants with a black cloth each day during summer. *Photograph courtesy of Daphne Vince-Prue.*

are on for several hours each evening. Flowering can be achieved by placing the plants in a dark cupboard for about 14 hours every night, but they must be moved into the light for several hours a day in order for the photoperiodic mechanism to function.

## Night-break lighting

Lighting to provide long days in winter is much more complicated, and to be successful it is necessary to have some understanding of the underlying physiological mechanism. In the so-called *dark-dominant* plants, which include the majority of SDPs, flowering depends on whether or not a critical duration of darkness (the critical night length) is exceeded. Flowering in these plants can be controlled by giving a relatively short exposure to light at a particular time during the night. Such a night-break produces the same effect as a long day and considerably reduces the energy load. If responses are to be manipulated by giving a night-break, it is clearly important to know

when is the optimum time, how long the exposure should be and what kind of light is needed.

## Night-break timing

The time at which a plant is most sensitive to a night-break varies with individual plants. Many practical manuals recommend that the night-break should be given near midnight to break the night up into two periods, each shorter than the critical night length. However, this is not how the night-break mechanism works. Time measurement during darkness begins at dusk and plants become sensitive to night-break light after a set number of hours, determined by the biological clock, have elapsed. If the duration of darkness is increased, the time at which a night-break is most effective still occurs at the same number of hours after dusk, even though the subsequent period of darkness far exceeds the critical night length (Fig. 14.7). In other words, the plant 'counts' time from dusk

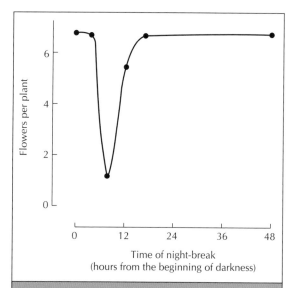

**Fig. 14.7** The time of maximum sensitivity to a night-break in Japanese morning glory (*Ipomoea nil*) when the duration of the dark period is extended to 48 hours. A brief exposure to light (night-break) strongly inhibits flowering when given after 8–9 hours of darkness, even though the subsequent period of darkness (39–40 hours) far exceeds the critical night length (8–9 hours in this plant).

until it reaches the time when it becomes sensitive to a night-break. The subsequent duration of darkness is unimportant. However, if the night-break is very long (from 6–8 hours, for example), the plant recognises this as a 'new day' and counting begins again from the beginning of darkness.

The greatest effect of a night-break occurs at around the middle of a 16-hour night in many plants but this is not always the case and it can be earlier, or later. Consequently it is unwise to assume that the 'best' time to give a night-break is always around midnight. Over the normal growing range, temperature has only a small effect on the time of greatest sensitivity to a night-break, although it may have a considerable effect on other aspects of flowering.

## What kind of light?

The night-break effect on flowering was one of the first responses shown to be under the control of phytochrome, with red light being the most effective. Giving far-red light immediately after the red night-break prevented its effect (Table 14.4), demonstrating that, as well as controlling germination and many growth responses to light (Chapters 8 and 11), the pigment phytochrome also controls the night-break response. There is now evidence that phytochrome B plays an important role in the sensitivity to

**Table 14.4** Giving a night-break with red light inhibits flowering in the short-day plant *Chrysanthemum* 'Honeysweet', whereas exposure to far-red light reverses the effect of red light and promotes flowering. When the sequence of light treatments ended in **red**, flowering was prevented; when the sequence of light treatments ended in **far-red**, flowering occurred.

| Night-break treatment | Response |
|---|---|
| Darkness for 16 hours (no night-break) | Flowering |
| 3 minutes of **red light** | Vegetative |
| 3 minutes of red, followed by 3 minutes of **far-red light** | Flowering |
| 3 minutes of red, 3 minutes of far-red, followed by 3 minutes of **red light** | Vegetative |
| 3 minutes of red, 3 minutes of far-red, 3 minutes of red, followed by 3 minutes of **far-red light** | Flowering |

photoperiod in SDPs, both in the control of flowering and in the tuberisation response in potato.

Lamps with a high R/FR ratio (such as 'white' fluorescent lamps) are the most efficient light sources for night-break lighting because of the sensitivity to red light, although tungsten-filament (TF) lamps are also quite effective and are widely used because they require no control gear and are cheap to install.

## Night-break duration

Although very short exposures to light can be effective (Table 14.4), longer exposures are usually given in practice, especially when TF lamps are used. The recommendation for chrysanthemums in the UK is to use TF lamps for 2 hours during September and March, increasing this to 5 hours in December.

## Cyclic lighting

In large commercial installations the cost of electricity becomes important and this can be reduced by the practice of *cyclic lighting*. Experiments have shown that giving light intermittently can be just as effective as giving light continuously throughout the night-break, provided that the intervals over which light is given are sufficiently short. Only TF lamps can be used in this way, as other lamp types cannot be switched on and off so often. Unfortunately, as with many other aspects of biology, not all plants behave in the same way. Carnations (*D. caryophyllus*) show a very weak response to intermittent lighting even when the dark intervals are extremely short and cyclic lighting offers no obvious advantage for this crop.

## Long-day plants

In contrast to most SDPs, flowering in many long-day plants (LDPs) depends not only on the duration of the dark period but also on the kind of light received during the day. These so-called *light-dominant* plants are less responsive to a short night-break, and many require quite long exposures to light of several hours' duration to achieve a long-day response. Having said this, the same quantity and duration of light given during the middle of the night is usually more effective than the same amount given at the end of the day. So in practical terms, to promote winter flowering in these plants the most efficient schedule is to give light for several hours (usually 4–8), in the middle of the dark period. In some cases, however, for example in carnations, lighting throughout the night from dusk to dawn gives the most satisfactory results.

Light-dominant plants have another important characteristic. Even when the duration of darkness is favourable, the kind of light during the daytime influences the response, and to be effective a long day must contain some FR light. It is now known from genetic evidence that the long-day promotion of flowering by light containing a high proportion of FR light is mediated through phytochrome A, which is particularly associated with the perception of FR light. However, phytochrome B also plays an important role in determining the degree of sensitivity to day length in LDPs.

Sunlight is rich in FR wavelengths (Table 14.3), so under natural conditions only the duration of darkness is important for light-dominant LDPs. However, when using artificial light to extend the length of the natural day (day-extension treatment) it is important to use TF lamps, which have a high content of FR light. Fluorescent lamps are deficient in FR light and are less effective, or even ineffective, in promoting flowering when used as a day-extension treatment for light-dominant LDPs. Unfortunately, TF lamps have a considerably lower R/FR ratio than sunlight (Table 14.3) and when used over several hours each day can lead to the development of 'leggy' plants, with long weak stems.

## Day-length lighting in practice

Only relatively low irradiances are necessary for day-length lighting because light is only acting as a signal for the pigment phytochrome, and is not supplying an energy source for photosynthesis. Although species do differ somewhat in their sensitivity to light, photoperiodic lighting is usually installed at between 0.25 and 0.5 W m$^{-2}$, which is adequate for most plants. Supplementary lighting for photosynthesis can also be used to obtain long days and in this case a much higher irradiance is needed. It is important to take care to avoid overspill to other nearby plants if they are responsive to day length.

From the above discussion and from Chapter 13, it is evident that much remains to be discovered about the mechanism(s) through which plants detect day length, and it is still necessary to experiment with individual species and cultivars, using different durations, timings and lamp types to obtain the best results.

## CONCLUSION

A greenhouse changes the environment of a plant in a variety of ways, in addition to the obvious one of protecting it from low temperatures. Understanding these changes, how they may be modified and the scientific basis of the plant's response to them may enable the gardener to use valuable greenhouse space economically and to the best effect.

## FURTHER READING

Anon (1964) *Commercial Glasshouses. Siting, Types, Construction and Heating.* Bulletin no. 115. Ministry of Agriculture. HMSO, London.

Anon (1987) *Lighting for Horticultural Production. Grow Electric Handbook.* Electricity Council, Stoneleigh.

Canham, A.E. (1966) *Artificial Light in Horticulture.* Centrex Publishing Company, Eindhoven.

Egginton, N. (2003) Chill out. *The Horticulturist* **11**, 6–7.

Hanan, J.L., Holley, W.D. & Goldsberry, K.L. (1978) *Greenhouse Management.* Springer Verlag, New York, NY.

Payne, C. (1997) Horticulture in the next millennium – the contribution of R & D. *The Horticulturist* **6**, 25–8.

Revell, R. & Henbest, R. (1999) A bright future for plastics. *The Horticulturist* **8**, 29–30.

Salisbury, F.B. & Ross, C.W. (1992) *Plant Physiology,* 4th edn. Wadsworth Publishing Company, Belmont, CA.

Thomas, B. (1999) Photoreceptors and photoperiodism – mini review. *Flowering Newsletter* **27**, 10–14.

Thomas, B. & Vince-Prue, D. (1997) *Photoperiodism in Plants.* Academic Press, London.

Vince-Prue, D. (1990) The control of flowering by daylength. *The Plantsman* **11**, 209–24.

Vince-Prue, D. & Cockshull, K.E. (1981) Photoperiodism and crop production. In *Physiological Processes Limiting Plant Productivity,* C.B. Johnson (ed.), pp. 175–7. Butterworths, Oxford.

Vince-Prue, D. & Canham, A.E. (1983) Horticultural significance of photomorphogenesis. In *Photomorphogenesis. Encyclopedia of Plant Physiology, New Series,* vol. 16B, pp. 518–44. Springer-Verlag, New York, NY.

# 15

# The Diversity of Undesirables

## SUMMARY

In this chapter the diversity of macro- and microorganisms that pose a threat to garden plants is described. Also discussed, as appropriate, are the ecological strategies of these organisms, the range of symptoms they cause and their impact on plant growth. The criteria used in deciding whether the application of control measures is necessary are then analysed. Finally, poisonous and irritant plants are reviewed.

## INTRODUCTION

In the imagined Arcadian wilderness before gardening was invented there were no undesirables, only a rich biodiversity. Today's gardeners find this richness excessive and re-label some of it as pests, diseases and weeds. A pest, disease or weed is simply biodiversity being over-assertive, thus limiting or preventing the growth, flowering or fruiting of cultivated plants. It may therefore be necessary to take some corrective action, to either prevent or reduce the problem. Before taking such action, however, it is important to be able to identify the organism or environmental factors involved.

## PESTS

Throughout the year plants in gardens, in greenhouses or on windowsills are at risk from pests, which may be defined for our purposes as any animals that cause harm or damage to plants. Some, like woodlice (Isopoda) and millipedes (Diplopoda), feed mainly on decaying plant material and only occasionally become pests of seedlings and other soft plant growth. Others cause more serious damage to plants of all ages, resulting variously in distorted growth, loss of vigour and a reduction in the yield of flowers and crops. The principal types of pest encountered by gardeners are listed below.

### Nematodes or eelworms

Plant-feeding nematodes (Nematoda) are mostly microscopic, worm-like animals that live in the soil or within plant parts. They have piercing mouthparts that are used to suck sap from host cells. Some of the soil-dwelling nematodes cause stunted root growth and may transmit virus diseases (see below). Leaf and bud nematodes live mainly inside the foliage and cause a distinctive pattern of brown wedges or islands where the infested tissues are separated from the rest of the leaf by the larger veins. Narcissus and phlox nematodes (races of *Ditylenchus dipsaci*) cause distorted and stunted growth which is often followed by the plants' death. Cyst nematodes (*Globodera*, *Heterodera* spp.) and root knot nematodes (*Meloidogyne* spp.) develop inside plant roots, causing poor growth as a result of severe disruption of the uptake of water and nutrients.

## Molluscs: slugs and snails

Molluscs (Mollusca: Gastropoda) are mainly active after dark. They eat holes in, or completely devour foliage, flowers and stems. Some slugs are also a problem underground where they feed on bulbs, potato tubers and root vegetables.

## Mites

Mites (Acarina) suck sap and feed mainly on foliage. The most frequently encountered species is the two-spotted mite (*Tetranychus urticae*), which causes a pale mottled discolouration of upper leaf surfaces. Tarsonemid mites attack the shoot tips of some greenhouse and garden plants, leading to progressive stunting and distortion of growth. Eriophyids or gall mites secrete plant hormones (see Chapter 2) which induce their host plants to produce abnormal growths called galls, in which the mites develop.

## Insects

Insects are by far the most numerous and diverse of plant pests. The more important orders that include pests are the Dermaptera (earwigs), Hemiptera (aphids, whiteflies, scale insects, mealybugs, leafhoppers, froghoppers and capsid bugs), Thysanoptera (thrips), Lepidoptera (butterflies and moths), Diptera (flies), Coleoptera (beetles) and Hymenoptera (sawflies, ants and gall wasps). Between them they exploit every ecological niche provided by plants. Apart from the wholesale consumption of foliage and other plant parts, some insects may specialise as leaf miners or stem borers, or attack seeds or fruits. Plant-feeding insects in the Hemiptera and Thysanoptera are all sap-feeders. As a result, some excrete a sugary substance, honeydew, which soils the foliage and allows the growth of black sooty moulds. Sap-feeding insects may also spread plant virus diseases on their mouthparts when they move to a new host plant. Insects of many orders, but especially in the Hemiptera, Diptera and Hymenoptera, may induce galls on leaves, stems, buds and other plant parts. In the Lepidoptera and Diptera it is always the larval stage that causes the damage; in the other orders damage is often caused by both adult and immature stages.

## Birds and mammals

Birds and small mammals may be welcome in gardens where the presence of wildlife is valued and encouraged (see Chapter 18). The damage they cause may, however, be excessive, especially if population sizes become too large. For example, bullfinches (*Pyrrhula pyrrhula*) take flower buds from fruit trees and bushes and some ornamentals during the winter. Wood pigeons (*Columba palumbus*) often defoliate brassicas and peas, while sparrows (*Passer domesticus*) shred the petals of crocus and primulas. Other birds may damage ripening fruits.

Mice (*Apodemus* and *Mus* spp.), voles (*Clethriomys*, *Microtus* and *Arvicola* spp.), grey squirrels (*Sciurus carolinensis*), rabbits (*Oryctolagus cuniculus*), foxes (*Vulpes vulpes*), deer (especially roe deer, *Capreolus capreolus*) and moles (*Talpa europaea*) are the most important mammalian garden pests. Rabbits and deer will eat most plants; these and squirrels and voles may also kill woody plants by removing bark from the stems. Mice, voles and squirrels also feed on seeds, fruits and bulbs. Moles may create havoc in lawns and seed-beds by tunnelling and thereby producing molehills.

## DISEASES

Disease is a harmful disturbance of normal function, and in plants may be caused by infectious organisms (*pathogens*), non-infectious agencies such as nutrient imbalances, adverse physical conditions such as waterlogging, or pollutants such as ozone, sulphur dioxide or 'acid rain'. The main groups of infectious agents (Table 15.1) will be familiar to gardeners because they also contain the causal agents of animal and human diseases, but the individual organisms that attack animals, such as the influenza and pox viruses, do not attack plants, and vice versa. Also, the relative importance of the groups is not the same for animals and plants. Viruses and Bacteria are both important, but there are no plant-parasitic Protozoa to rival the mammalian malarial parasites in severity, and fungal pathogens like species of *Phytophthora* (Chromista) and *Armillaria* (true Fungi) are generally more numerous and damaging than those attacking animals. Plants are also attacked by

**Table 15.1** The main groups of organisms that include plant pathogens.

| Group | Description | Examples |
|---|---|---|
| Superkingdom: PROKARYOTAE<br>Kingdom: MONERA (Bacteria and Mollicutes) | Bacteria grow mainly as slimy masses of single cells, in or between host cells. May secrete enzymes, toxins and plant hormones which cause the symptoms of disease. Reproduction by cell division and/or spores*, dispersed by wind, water or sometimes vectors†. May be biotrophic or necrotrophic‡. Bacterial diseases usually favoured by wet conditions and widespread, but not as common in the garden as diseases caused by fungi.<br><br>*Mollicutes* (Phytoplasmas and Spiroplasmas) are related to bacteria but have no cell walls, and live within the cells of the host as biotrophs‡. Transmitted by vectors*, notably insects and other arthropods, and by injuries caused by pruning or grafting. Mollicute diseases, once thought to be caused by viruses, are only rarely encountered by gardeners. | *Agrobacterium tumefaciens* (crown gall), *Erwinia* spp. (fire blight, soft rots and wilt)<br><br><br><br><br><br><br><br><br>Yellows, witches' brooms, phyllody (leaflets produced instead of flowers) |
| Superkingdom: EUKARYOTAE<br>Kingdom: FUNGI (true fungi) | Normally grow between and through plant cells as microscopic tube-like structures (hyphae), with haploid nuclei and chitin-containing cell walls. May secrete enzymes, toxins and plant hormones which cause disease symptoms. Reproduce asexually and sexually by mean of non-motile spores*, which may be dispersed by wind, water or occasionally vectors such as insects. Sexually produced spores (resting spores) often have thick walls and considerable powers of survival in the absence of a host. May be biotrophs, necrotrophs or hemibiotrophs‡. Common causes of disease in the garden. | Biotrophs: Uredinales (rusts), Ustilaginales (smuts), Erysiphales (powdery mildews)<br>Hemibiotrophs: *Pyrenopeziza brassicae* (light leaf spot)<br>Necrotrophs: *Penicillium* spp. (blue mould), *Fusarium* and *Verticillium* spp. (vascular wilts), *Colletotrichum* spp. (anthracnose), *Sclerotinia* spp. (soft rots) |
| Kingdom: CHROMISTA (referred to colloquially as 'fungi' and until recently included in the Kingdom FUNGI) | Normally grow through or between plant cells as microscopic, tube-like structures (hyphae), with diploid nuclei and cellulose-containing cell walls. Usually produce motile asexual spores* dispersed in water; otherwise as for FUNGI. Common causes of disease in the garden. | Biotrophs: Peronosporales (downy mildews and white blister fungi)<br>Hemibiotrophs: *Phytophthora infestans* (potato blight)<br>Necrotrophs: *Phytophthora* species (blights), *Pythium* species (rots and damping-off) |
| Kingdom: PROTOZOA (referred to colloquially as 'fungi' and until recently included in the Kingdom FUNGI) | Grow as masses of cytoplasm without cell walls (plasmodia) only in living host cells (obligate parasites). Reproduce by motile asexual spores* and sexual resting spores*. Cause a very limited range of diseases in the garden. | *Plasmodiophora brassicae* (club root)<br><br><br>*(Continued)* |

**Table 15.1** (*Continued*)

| Group | Description | Examples |
|---|---|---|
| Kingdom: PLANTAE (flowering plants) | About 1% (300 000 species) of flowering plants are parasitic on other plants. Reproduce by seeds. Seedlings form a close association with the vascular system of a living host for the exchange of nutrients via complex structures called haustoria [very different from fungal haustoria (see Table 15.2) but perform the same function]. Some only grow in association with living hosts (holoparasites); others live part of their life cycles away from hosts (hemiparasites). Most hemiparasites are components of natural grasslands. Rarely a problem in gardens and often considered desirable. | Holoparasites: *Viscum album* (mistletoe), *Orobanche* spp. (broomrapes)<br><br>Hemiparasites: *Euphrasia officinalis* (eyebright), *Rhinanthus minor* (yellow rattle) |
| Kingdom: VIRUSES | Viruses are sub-microscopic parasites capable of growth and reproduction only inside living host cells (obligate parasites). Lack any cellular structure, consisting only of nucleic acid (RNA or DNA) with a protein coat. Spread by vectors† (usually insects or other arthropods, but also many other organisms including FUNGI, CHROMISTA and nematodes) or by wounding caused by pruning or grafting. Common causes of disease in gardens.<br><br>*Viroids* are simple VIRUSES which lack a protein coat. | RNA viruses: tobamovirus (tobacco mosaic), potyvirus (potato virus Y), tombusvirus (tomato bushy stunt), cucumovirus (cucumber mosaic)<br>DNA viruses: caulimovirus (cauliflower mosaic)<br>Potato spindle tuber |

* Spores are minute propagules of bacteria and fungi, functioning as seeds but without a pre-formed embryo. For further information see text.
† Vectors are carriers (often insects or other arthropods, but sometimes fungi or humans) of diseases (especially virus diseases) from one host to another.
‡ For definitions of biotrophy, necrotrophy and hemibiotrophy, see text.

phytoplasmas (Mollicutes), which have some properties of both Bacteria and Viruses, and by parasitic plants (Angiosperms). The latter never attack animals, except in fiction.

The microorganisms that cause disease are simple in structure and have only a few measurable characteristics (see Table 15.1), compared with the higher organisms for which conventional taxonomic characters were first devised. This makes identifying and classifying them difficult, which is not merely an academic matter, because the name is the key to the scientific literature on a disease-causing organism, and therefore to its control or management.

The organisms in all the groups listed in Table 15.1, with the exception of the Viruses, reproduce by producing *spores*, which are minute propagules

that function as seeds, but do not have a pre-formed embryo (see Chapter 2). Spores may be produced as a result of sexual reproduction, in which case they have great powers of survival in the absence of a host and are called resting spores. Alternatively, they may be produced asexually and act to disperse or spread the pathogen, sometimes over great distances. Asexual spores that are non-motile (unable to move except when carried by wind or water) are called conidia and are produced by true Fungi. Asexual spores that are motile are called zoospores and are able to swim in water with the aid of one or two flagella; they are produced by members of the Chromista and Protozoa. Conidia are usually formed on conidiophores, special branches of the *hyphae* (see Table 15.1) that comprise most Fungi. The conidia of many species, and the zoospores of all species, are surrounded by a thin wall to form a structure called a sporangium, or zoosporangium, respectively. Sporangia and zoosporangia are usually borne on specialised hyphal branches called sporangiophores. Knowing what type of spore or spores a pathogen produces may provide important clues to its control or management, or to forecasting the risk of infection, spread or persistence.

The largest and most diverse group of infectious microorganisms attacking plants are the 'fungi', which are now classified into three groups (Table 15.1): the true Fungi; the Chromista, which includes the very important downy mildews, *Phytophthora* spp. and white blister fungi (*Albugo* spp.); and the Protozoa, which include the slime moulds, especially *Plasmodiophora brassicae*, the cause of club root in the Brassicaceae (cabbage family).

The classification of pathogenic microorganisms in the Fungi, Chromista and Protozoa is based primarily on the sexual structures, as in higher plants. Difficulties arise, however, because some species, notably among the Fungi, rarely form these structures, and some never do. Where sexual structures are not present, various classifications have been based on asexual structures, mostly the conidia and conidiophores.

The problem of accurate naming becomes more acute with the smallest organisms, the Viruses and Bacteria. At present, much reliance is placed on biochemical tests that distinguish species by the enzymes they contain (for Bacteria) and by physical and serological characteristics (for Viruses).

Morphological and biochemical characteristics are ultimately determined by the unique base-pair sequence of the organisms' DNA and RNA (see Chapters 2 and 5), and the sequences of selected parts of the DNA or RNA will probably soon provide the definitive taxonomic criteria for plant pathogens.

Pathogens exploit different strategies to live and reproduce at the expense of their plant hosts. Some examples follow.

## Opportunists

*Botrytis cinerea*, the cause of grey mould of soft fruit, is a typical opportunist pathogen. The spores are always present, because this fungus also lives and grows on dead plant remains. However, it can infect living tissues when they are wounded, weakened by stresses or lacking some of the normal defences. Thus spores infect the styles of fruits like strawberries at flowering, but the fungus is suppressed by chemical defences in the unripe fruit and remains latent until ripening. Then, as the fruit's sugar content rises during ripening and its resistance to infection falls, the fungus becomes active, breaks out and causes mould.

## Necrotrophs: destructive parasites

Necrotrophs invade living tissues, kill them rapidly with cell-destroying enzymes and toxins, extract the nutrients and then disperse, leaving the dead remains to others. *Pythium* spp., the cause of damping-off of seedlings, is a good example of a pathogen that uses this strategy. Such pathogens often have no active life outside their hosts because the only other available food is dead plant and animal material, and they are unable to compete with other non-pathogenic microorganisms, called saprotrophs, which are better adapted to utilising this material. They therefore usually pass the time between attacks on susceptible host plants as dormant resting spores or other survival structures such as sclerotia, masses of hyphae with a hard waterproof rind.

## Biotrophs: non-destructive parasites

Examples of biotrophs are rust fungi (Uredinales) and Viruses. They extract nutrients from living cells without killing them and may subtly alter the metabolism of the host to their own advantage. Viruses are very specialised in this respect. They have no ability to reproduce outside the cells of the plant host and rely on *vectors* (carriers), often insects but sometimes nematodes or humans, to move them from one host to another. Biotrophs cause enormous economic and aesthetic damage to plants by debilitating or disfiguring them, but seldom cause death.

## Hemibiotrophs

Hemibiotrophs are pathogens that grow for part of their life cycle in the living tissues of a host plant, like a biotroph, and then, after killing the tissues with enzymes and toxins, grow as necrotrophs. *Phytophthora infestans*, cause of potato and tomato blight, is a typical example. Initially, following infection, the host leaf remains relatively green. It is only after the appearance of the spores, which break through the surface of the apparently healthy tissues, that the cells begin to die and a necrotic lesion is formed.

## Symptoms caused by the different types of pathogenic microorganism

The disease which all these organisms cause, the 'harmful disturbance of normal function', leads to abnormal growth patterns in plants which are recognised as symptoms. These may appear suddenly: *Phytophthora* leaf blight of tomatoes, for example, frequently devastates outdoor tomatoes in the UK while the unfortunate owners are away for the late summer bank holiday. Others develop slowly as a gradual spread of disease weakens the plant; root infections in trees may cause a slow decline in vigour and dieback of branches over several years. Examples of the great diversity of symptoms caused by plant pathogens are listed in Table 15.2.

## WEEDS

A weed may be defined simply as a plant growing where it is not wanted. In spite of gardeners' best efforts, weeds very successfully interfere with the cultivation and enjoyment of gardens. Their astonishing resilience is no accident, for their life strategies are finely tuned to exploit the opportunities gardeners inadvertently provide. Realistically weeds cannot be eliminated, but they can be managed. The key to this is to understand their strengths and exploit their weaknesses.

Weeds are very diverse and examples may be found among the flowering and non-flowering plants, including mosses, liverworts and algae (see Chapter 1). Perennial and annual life cycles are represented, sometimes in the same species, as for example annual meadow grass (*Poa annua*). Woody and herbaceous plants may be weeds. Sometimes weeds are wild plants that invade the garden: tree seedlings such as sycamore (*Acer pseudoplatanus*), tuberous perennials such as lesser celandine (*Ranunculus ficaria*; Fig. 15.1) or bulbs such as bluebells (*Hyacinthoides non-scripta*) fit this category. Sometimes domesticated plants such as potatoes (*Solanum tuberosum*)

**Fig. 15.1** Mind-your-own-business or baby's tears (*Soleirolia soleirolii*), an example of a plant introduced to the UK as an ornamental which has escaped and become a weed. *Photograph courtesy of the Royal Horticultural Society.*

**Table 15.2** Examples of the diversity of symptoms caused by plant pathogens.

| Symptom | Description |
| --- | --- |
| Rusts | Usually limited lesions in which rusty (brown, black or orange) coloured spores of biotrophic Uredinales (FUNGI) break through the cuticle of living leaves and stems. 'Green islands' of chlorophyll may remain around lesions as infected leaves yellow with age. Hyphae, which grow between host cells, usually form haustoria (specialised feeding branches) that penetrate cells. Some rusts are systemic (grow throughout the plant), and spores may then be produced over large areas of the hosts, which may also grow in a distorted manner. |
| Smuts | Usually limited lesions in which black smut-like spores of the biotrophic Ustilaginales (FUNGI) break through the cuticle of living leaves and stems or burst from reproductive structures (anthers or ovaries). Hyphae may grow in or between the living host cells. Many smuts are systemic and sometimes cause the host to grow in a distorted manner. |
| Powdery mildews | White to grey hyphae and spores of the biotrophic Erysiphales (FUNGI) grow on the surfaces of the leaves and stems of living hosts, connected to epidermal cells by feeding branches called haustoria. Green islands of chlorophyll may remain around infected areas as leaves yellow with age. |
| Downy mildews | Masses of white to grey hyphae and spores of the biotrophic Peronosporales (CHROMISTA) grow through the cuticle or stomata of living leaves and stems. The hyphae usually grow between host cells and form haustoria (see above). Some are systemic and cause the host to grow in a distorted manner. |
| Galls, leaf curls and witches' brooms | Overgrowth, distortion or proliferation of leaves, stems or roots, induced by hormonal imbalance caused by a wide diversity of biotrophic or hemibiotrophic pathogens. Symptoms may be caused by some members of all main groups of MONERA, FUNGI, CHROMISTA and PROTOZOA. |
| Scabs | Limited dry leaf and stem surface lesions with scabby host deposits of corky, lignified and pigmented cells induced by infection by certain hemibiotrophic or necrotrophic bacteria (*Streptomyces* spp.) and FUNGI (mainly Ascomycota). |
| Necrotic spots | Limited necrotic lesions formed on leaves, stems or flowers of hosts by wide range of necrotrophic or hemibiotrophic bacteria, FUNGI and CHROMISTA. Toxins and enzymes secreted by the pathogens may contribute to symptoms. The character of lesions is also affected by host responses to infection, including deposition of dark phenolic substances. |
| Cankers, die-backs, stem and twig blights | Stem lesions and twig and shoot death caused by some necrotrophic bacteria and FUNGI (mainly Ascomycota) that infect and destroy stem and twig tissues of woody hosts. |
| Vascular wilts | Wilt of foliage and stems followed by death of infected plant, caused by invasion of the vascular system and release of toxins and enzymes by certain necrotrophic FUNGI and some bacteria. |
| Blights | Killing of host leaves and shoots by toxins and/or enzymes produced by a wide diversity of necrotrophic or hemibiotrophic bacteria, FUNGI and CHROMISTA. |
| Rots | Destruction of stems, leaves, flowers or roots by enzymes of a wide diversity of necrotrophic and hemibiotrophic bacteria, FUNGI and CHROMISTA. Enzymes break down the cell walls of the infected tissues and cause a rot which may be wet or dry, soft or hard, depending on the species of pathogen and the types of enzymes produced. |
| Hypersensitive (necrotic) flecks | Rapid death, lignification and dark pigmentation of one or small group of leaf or stem cells of resistant host following challenge by spores of strains of bacteria, FUNGI, CHROMISTA or VIRUSES to which the host is resistant. |
| Yellows | Loss of chlorophyll in leaves infected by some VIRUSES and mollicutes. |
| Mottles and mosaics | Patchy yellowing of leaves caused by some VIRUSES. |
| Colour breaks | Streaking of flowers or unusual colour patterns on flowers and fruit infected by some VIRUSES. |
| Phyllody | Replacement of flowers or florets with leaflike structures, caused by some mollicutes. |

or horseradish (*Armoracia rusticana*) become invasive and cause problems as weeds. Ponds and other aquatic environments are home to weeds such as duckweed (*Lemna* spp.) or algae (see Chapter 1).

## Annual weeds

Annual weeds complete their life cycle, seed to seed, in a single season and are the great survivors of cultivated ground. Fat hen (*Chenopodium album*), groundsel (*Senecio vulgaris*), shepherd's purse (*Capsella bursa-pastoris*) and annual nettle (*Urtica urens*) are common garden examples. Although not deliberately domesticated, many annual weeds almost appear to be so, as they are highly dependent on human activities for their success.

Massive fecundity characterises annual weeds: they produce huge numbers of seeds which are also very mobile, constituting a ubiquitous 'seed rain'. Groundsel, for instance, has hairy parachutes that carry its seeds some distance. Other short-range spreading techniques include explosive seed capsules. The Himalayan balsam (*Impatiens glandulifera*) and hairy bittercress (*Cardamine hirsuta*) are species that exploit this strategy. Persistence of dormant seeds in the soil, as a 'seed bank' of viable propagules which may be released from their dormancy by triggers such as exposure to light during cultivation, is another feature of annual weeds (see Chapter 9).

## Perennial weeds

Perennial weeds colonise cultivated ground to some extent, and in gardens are a particular menace in plantings of perennials. Examples are couch grass (*Elymus repens*), ground elder (*Aegopodium podagraria*), horsetail (*Equisetum* spp.), brambles (*Rubus* spp.) and creeping buttercup (*Ranunculus repens*). Perennial weeds produce seeds but also often exhibit successful vegetative propagation strategies: for example, some *Oxalis* spp. and *Ranunculus* spp. (Fig. 15.1) produce numerous bulbils or root tubers, and bindweeds (*Convolvulus* spp.) form rhizomes, which break up into highly viable fragments. Trees are frequently a major problem for gardeners, especially those with wind-blown seeds. The seeds of birch (*Betula pendula*) and willow (*Salix* spp.) are

long-range drifters, while the winged seeds of ash (*Fraxinus excelsior*) and sycamore are a nuisance over shorter distances.

## Weed ecology

Most plants show no weedy characteristics, and weedy species are a distinct minority in the plant world. Those that are or become weeds have lifestyles that are astonishingly well suited to succeed in cultivated areas. Some, like couch grass, survive intense soil disturbance by cultivation, others have life-forms such as the rosettes of common daisy (*Bellis perennis*) or the creeping habit of clover (*Trifolium* spp.) and the dead nettles (*Lamium* spp.) that enable them to exist easily in mown herbage or shelter in the competitive environment of borders and around the bases of woody plants. Even trampled areas have their weeds: annual meadow grass is adapted to exploit this situation. The adaptability of weeds in exploiting unpromising environments is impressive. Some, like groundsel and annual nettle, have very short life cycles ideally suited to exploiting bare soil.

Garden plants are selected for food value, flavour, attractive flowers or foliage, and shape and form rather than their ability to survive and compete. Their competitive potential has been further eroded by breeding for features such as dwarf, leafless and variegated forms, improved palatability and removal of seed dormancy. These characteristics leave them ill-suited to compete with weeds. There is, nevertheless, currently some enthusiasm among gardeners for creating self-sustaining communities of the more competitive cultivated plants that leave no room for weeds.

It is now possible to genetically modify crops so that they possess features that weeds do not have, such as resistance to broad-spectrum herbicides. In this situation a weed's natural adaptation to the crop environment brings no benefit. A herbicide-resistant crop can be treated at almost any time, leaving weeds heavily exposed to damage. However, there may be significant drawbacks to this approach. Herbicide-tolerant weeds may be selected by overuse of a particular herbicide and/or genes for herbicide resistance may cross into related weed species, or the genetically modified crop itself may become a weed in subsequent crops. Proponents of genetic modification suggest strategies for ensuring that such problems do not arise, for example by limiting how frequently genetically modified crops are grown, or by inserting additional genes to prevent reproduction of genetically modified cultivars. Currently there is considerable public resistance to the growing of genetically modified plants, and it remains to be seen whether gardeners will ever be able to use them as an aid to weed control.

## Weed origins

Despite the ancient origins of gardening and agriculture, weeds have probably not adapted to gardening, but already possessed their competitive and invasive characteristics before cultivation began. When human activities created the opportunities, such species were ready to move in. Thus large numbers of easily dispersed seeds, and rapid germination and growth, are useful characteristics for plants adapted to colonising transient habitats such as fresh soil exposed by natural erosion or periodic changes in river levels. These characters confer an advantage when bare soil is created by farmers and gardeners.

Many annual weed species are believed to have succeeded during the harsh, brief periods when glaciers left soil uncovered. Cold weather does not greatly inhibit them and they therefore continue to grow in winter. Others have sneaked into cultivated areas as passengers in food crops or as the result of various human activities. Fat hen, for example, is believed to have arrived as a weed in crop seeds. Other weeds were originally introduced for their medicinal or food qualities, or even for their ornamental value: ground elder (*Aegopodium podagraria*) was thought to alleviate gout, fennel (*Foeniculum vulgare*), horseradish and leaf beets (*Beta vulgaris*) are vegetables 'gone wild' and are now persistent weeds. Unwise introductions of garden ornamentals have given gardeners some particularly invasive and persistent weeds, most notably Japanese knotweed (*Fallopia japonica*), which was esteemed and widely planted in the nineteenth century. Waterborne weeds are solely aquatic plants, but some land weeds are carried to new homes by streams and rivers, for example giant hogweed (*Heracleum mantegazzianum*) which is a particular problem on river banks.

**Fig. 15.2** Lesser celandine (*Ranunculus ficaria*), a native woodland perennial that is a weed, spreading by bulbils and root tubers. *Photograph courtesy of the Royal Horticultural Society.*

Other ornamentals that have become weeds include slender speedwell (*Veronica filiformis*), mind-your-own-business (*Soleirola soleirolii*, Fig. 15.2) and Canadian pondweed (*Elodea canadensis*).

Better seed-cleaning technology has largely eliminated contaminated seeds, but a modern example of human-induced weed spread is the prevalence of mosses, liverworts (see Chapter 1), willow herb (*Epilobium* spp.) and hairy bittercress (*Cardamine hirsuta*) in container-grown nursery stock. Contaminated tools, vehicles and manures are also involved in spreading weeds. Not only is the manure of farmed beasts liable to contain seeds, but birds and mammals consume fruits and seeds and subsequently void viable seeds. Rabbits, for example, spread redshank (*Persicaria maculosa*) in this way and the prevalence of honeysuckles (*Lonicera* spp.) as weeds in country gardens results in part from the activities of birds. The seeds of weeds with adhesive qualities such as those of cleavers (*Galium aparine*) and wood avens (*Geum urbanum*) may be introduced to gardens from the wild on the coats of pets or the clothes of humans.

## Useful weeds

Weeds are not all bad. Some make tasty salads, purslane (*Portulaca* spp.) and dandelions (*Taraxacum officinale*) being good examples, while others such as good King Henry (*Chenopodium bonus-henricus*) may be used as vegetables, and ramsons (*Allium* spp.) is used to season food. Also, fashions change and what were once weeds, for example poppies (*Papaver* spp.) and corn cockle (*Agrostemma githago*), are now desirable plants in wild-flower gardens. Annual meadow grass makes passable lawns, as long as it is fed and watered abundantly and mown frequently and closely. Weeds sometimes act as green manures, scavenging nutrients from the soil after crops have been gathered. These are then safe from being washed deep into the soil where plants cannot use them and from where they may pollute watercourses and wells. Recycled through the compost heap or by being dug into the soil, decaying weeds may release nutrients to nourish subsequent crops.

Weeds also provide cover for useful predatory insects, especially in winter. Moreover, mixed populations of plants may be less likely to suffer devastating disease and insect attacks than populations entirely composed of one species or cultivar. The evidence for this in gardens is slim, although there is good evidence that diseases develop more slowly in cultivar mixtures of crops like wheat than in monocultures. Those involved in research on this phenomenon suggest that a few weeds in gardens may be helpful as long as they do not set seed.

## DECIDING WHETHER AN ORGANISM IS A PROBLEM

Gardeners need to consider three points to manage pests, diseases and weeds effectively and sensibly: whether control is necessary; when to take action; and what treatment to use. Science can help with many of these decisions, as described in the sections that follow and in Chapter 16.

## When is a pest a pest?

Animals only become pests when they cause damage that reduces either the yield or the quality of cultivated plants. Otherwise they may not need to be controlled. But how can gardeners tell when control is worthwhile?

The part of the plant that is affected should first be considered. If the pest attacks the 'important' part of a plant, such as the edible parts of fruit and vegetables or the attractive parts of ornamentals, the presence of relatively low numbers of individuals usually warrants control. If unimportant parts of the plant are attacked, even relatively large numbers of pests may be left uncontrolled. For example, the larvae of codling moth (*Cydia pomonella*) feed inside apples, so even relatively small infestations should be controlled, whereas caterpillars of winter moth (*Operophtera brumata* and some other species) feed on the leaves of apple trees, so serious damage does not occur until large numbers are present. Some pests may cause different amounts of damage on different crops. For example, appreciably more caterpillars of cabbage white butterflies (*Pieris* spp.) are needed to damage swedes significantly than to spoil cabbages. Conversely, small numbers of the larvae of cabbage root fly (*Delia radicum*), that feed on the roots, will disfigure swedes, but the same number does less harm to cabbages.

If the numbers of pests on crops are plotted against the amount of damage they cause, the graph nearly always takes the same basic 'S' shape (Fig. 15.3). Low numbers cause little damage and may even improve growth, comparable to the normal garden practice of pinching out the tops of plant to promote branching and flowering. As the numbers increase they reach the damage threshold level, at which significant damage starts to occur. For a time the amount of damage done is almost proportional to the numbers of the pest, although the slope of the line varies. When a small increase in the numbers of the pest results in considerably more damage the slope is steep. This happens when the pest attacks an 'important' part of the plant. Conversely, when significant damage is only caused by a large increase in the numbers of the pest the slope is shallow. Eventually, so much damage is done that the plant is severely debilitated, the numbers of the pest stabilise and the curve flattens out.

Although plants are more susceptible to infestation at some stages of growth than at others, additional factors also influence their tolerance of pests. Some or all of these should be considered when assessing the likelihood of damage or the need for control.

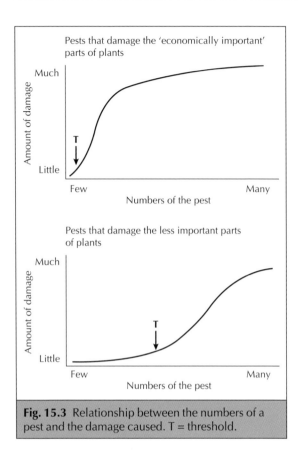

Pests that damage the 'economically important' parts of plants

Pests that damage the less important parts of plants

**Fig. 15.3** Relationship between the numbers of a pest and the damage caused. T = threshold.

For example, slugs will feed voraciously in the spring, but some plants are more susceptible to damage in the preceding autumn. This is partly because the plants are smaller then, but is also because in shorter days growth is slow, and the host cannot compensate for the damage. Flea beetles (*Phyllotreta* spp.) do more damage to the seedlings of *Brassica* spp. in hot dry summers than in cool wet ones, partly because the insects are more active, but also because the plants do not grow so well when it is hot and dry. Root-feeding nematodes similarly cause more damage in dry years than wet ones, particularly on light soils.

External factors may influence the reproduction rate and the behaviour of pests and their predators differently, and this affects the success of biological control. Two-spotted spider mites (principally *Tetranychus urticae*) prefer hot, dry conditions, but the predatory mite used to control them (*Phytoseiulus persimilis*) prefers cooler, humid conditions. The pest therefore breeds more quickly than the predator in hot dry weather, but this is compounded by the contrasting behaviour of the two mites in these conditions. When it is hot and dry the predators stay on the cooler and more humid lower leaves, but the spider mites congregate at the tops of plants where it is hot and dry, incidentally escaping their predators. Gardeners may not be able to alter the temperature or the day length, certainly out of doors, but in hot weather they can water and damp down in the greenhouse, thereby tipping the balance in favour of predators, which dramatically improves control.

An understanding of the factors affecting the balance between plants, pests and any treatments enables the gardener to anticipate the degree of damage a pest may cause, to decide whether control measures are necessary and, if so, to apply the most appropriate treatment at the optimal time. Appreciating the importance of these interacting factors comes with experience, but having this 'feel' for plants is a logical process that may be learned or taught.

### The importance of temperature

The development of most invertebrates (i.e. the rate at which they change from egg to adult), and the growth plants and the growth of plants is governed mainly by temperature. Other factors may affect the health, size, quality, fecundity, reproduction and survival of animals and plants, but it is the cumulative effect of the temperature over the season that is critical. Above a certain minimum temperature, called the base threshold, which varies with species, the amount an insect develops in a day is proportional to the average temperature that day. If the base threshold is 6°C, an insect develops three times as quickly at 9°C (3°C above base threshold) as it will at 7°C (1°C above base threshold). Information about temperature is used by farmers and growers to predict when insects will emerge, so they can time the application of control measures accurately.

The emergence of the spring cabbage root fly (*D. radicum*) and carrot fly (*Psila rosae*) may be predicted by recording the daily temperature from February until April or May. However, many carrot growers also put out sticky traps to catch the flies, to determine the size of the population, which is a function of emergence. Temperature data may also be used to calculate when turnip moths (*Agrotis segetum*), the caterpillars of which are the familiar 'cutworms', are likely to start laying eggs and when these eggs will hatch.

Temperature is also responsible for synchronising the development of some insects and their host plants, and this may be turned to the gardener's advantage. Cow parsley (*Anthriscus sylvestris*) is an important source of the pollen which female cabbage root flies need before they can lay many eggs. Both have similar temperature requirements, so the flies always emerge as the plants begin to flower. This happens in different parts of the country and in early and late seasons alike. Noting when cow parsley starts to flower is a much easier way for gardeners to predict when cabbage root fly will emerge than recording the temperature continuously from February to April, and is just as accurate.

Insect numbers increase geometrically from one generation to the next, so an insect laying 100 eggs has the potential to give rise to 100 offspring in the first generation, 10 000 in the second and 1 000 000 in the third. The more generations the insect completes in a season, the larger the number of offspring, which is why pests that breed continuously, like aphids, spider mites and thrips, are generally more numerous in hot seasons than in cool ones.

Insects such as glasshouse whitefly (*Trialeurodes vaporariorum*) sometimes seem to appear overnight. The relatively low numbers present early in the season are easily missed, but if it is hot and reproduction is rapid, swarms of adults appear when the next generation hatches. This might seem like spontaneous generation, but is in reality a simple function of mathematics; it also illustrates why it is so important to check apparently healthy crops carefully to detect the very earliest stages of infestation by a potential pest.

### When to control pests

One should never underestimate the scientific worth of local knowledge and past experience when deciding whether and when pests need to be controlled. There are generally sound reasons for the local variations in the need or timing of pest control, even though these may not be fully understood. Pests are affected by local conditions that are unlikely to change overnight, although climate change appears to be having a widespread effect. A number of insects that live only in the south of England may now be found about 20 km further north than was the case 10 years ago. Thus it is important to think about the logic behind old sayings and traditions, temper them to suit changing situations and individual seasons if necessary, but never to dismiss them.

## When is a disease a problem?

As with pests, the importance of a disease depends on many factors, notably the purpose for which the plant is being grown, the time of disease attack in the plant's life cycle and the type of damage that results from infection. Apple scab (*Venturis inaequalis*) disfigures the fruit surface and minor attacks have no effect on edibility, but even the most minor cosmetic damage may reduce the value of a commercial crop to the extent that it is not worth selling, which can bankrupt the grower: an example of minor damage having a very serious effect. Conversely, potato blight (*Phytophthora infestans*) may attack the foliage late in the season after the tubers are formed, and if these are harvested promptly before the fungus reaches them they will be healthy even though the foliage may be destroyed; in this case a serious disease has little effect, although it is devastating if the tubers do become infected, or if the attack occurs so early that it prevents them forming.

Appearances of severity may be very deceptive: cherry laurel (*Prunus laurocerasus*) is often very obviously attacked by powdery mildew (*Sphaerotheca pannosa*), which deforms the leaves but has negligible effect on the growth of such a robust tree, whereas *Phytophthora* root rot can kill it with no visible presence above ground.

### Identifying diseases

Symptoms may have more than one cause. For example, fungal infection of roots or of conducting tissue causes wilting, as does insect damage to roots, or drought. The interpretation of symptoms is therefore not always straightforward. This is particularly the case with root disease, where the symptoms in the upper parts of the plants are seen by gardeners but the actual damage to the plant occurs below ground and out of sight. Similarly, yellowing leaves may be caused by virus or phytoplasma infection in the leaves, or a deficiency in the supply of nutrients arriving at the leaf. Such a deficiency may have a number of causes: a genuine lack of the vital nutrient in the soil; an inappropriate soil chemistry that does not allow the nutrient to be taken up, such as lime-induced iron deficiency; poor root development, because soil factors such as a hard pan restrict root growth; attack by nematodes or other soil animals; root disease that damages normal root function; or stem disease that decreases nutrient flow to the leaves. Appearances can thus be very deceptive.

Because pathogens are microscopic they may not be detectable by direct observation, although some micro-fungi produce characteristic visible structures en masse: the dusty white spores of powdery mildews (Erysiphales) and the orange, brown or black pustules of rusts (Uredinales), for example, are unmistakeable. However, diagnosis often depends on indirect methods, the simplest being examination with a lens or light microscope. In the field a hand lens is indispensable, but in a diagnostic laboratory the first stop for a plant with a problem is usually the dissecting microscope. This low-power light microscope allows diagnosticians to identify some fungi by direct observation, especially if spores are being produced. If closer examination is needed, pieces of suspect material may be picked off, teased out in a drop of stain and examined under the compound light microscope at up to ×2000 magnification. At this magnification, fungi may usually be seen in enough detail to be identified if the appropriate stages of the life cycle are present. Bacteria may also be detected, but viruses cannot be seen with even the most powerful light microscope.

Difficulties of identification may arise even with fungi if they are not producing their characteristic spores, only very broad characterisation being possible from sterile hyphae. In such cases it may be necessary to keep the specimen for a few days at high humidity, in the hope that the spores will appear, or to culture the fungus on a suitable culture medium, usually a simple solution of nutrients and perhaps antibiotics to suppress bacteria, solidified with agar gel. A pure culture obtained in this way may be subject to further tests, for example to induce spore production or pigment formation by supplying or withholding light. Modern molecular methods of identification involve digesting chosen parts of the DNA with enzymes and running out the fragments in a gel matrix in an electric field to generate characteristic banding patterns. These methods can usually only be used when a pure culture of the organism is available.

For bacterial diseases a culture is essential for accurate diagnosis. The identification is then based on the character of colonies on a culture medium, the chemistry of the cell walls, the pigments secreted into the culture medium and the enzymes they contain. Tests used in diagnosis include the cell wall staining

reaction known as the Gram stain, which differentiates so-called Gram-positive bacteria, whose cell walls retain a crystal violet-iodine stain, from Gram-negative ones whose cell walls do not. The enzyme tests used involve placing some of the culture on a chosen substrate and observing changes that may occur. A species may be differentiated on its ability to reduce nitrate to nitrite, or whether it can digest the amino acid asparagine. Although the underlying chemistry may appear complex, the tests are often very simple. For example, a positive result may simply mean the test medium becomes cloudy because of the presence of bacterial cells, or changes colour because their activity alters the acidity, which changes the colour of a dye added to the substrate.

Diagnosis of virus infection is more difficult and time-consuming. Viruses cannot be grown except in the cells of their host plants. If a virus infection is suspected on the basis of the symptoms in a plant, such as mottling, chlorosis or distortion, there are various ways that virologists may try initially to confirm its presence. First, some infected material may be ground up and examined with an electron microscope. In a light microscope, the maximum magnification is determined by the wavelength of visible light. An electron microscope uses a beam of electrons with a much shorter wavelength and therefore allows much greater magnification. The specimen must be placed in a vacuum and the image projected on to a screen. If the virus particles are numerous, they may be seen using this instrument, but the electron microscope can only be used to examine very small amounts of material and if the viruses are not numerous, or are present in only a few parts of the plant, they may be missed.

A second approach to virus detection uses serology. Like the viruses that infect mammals, plant viruses cause mammal immune systems to react by the production of specific proteins called antibodies, which combine specifically with the protein components of the outer surface of the virus and inactivate it by preventing replication. It is interesting to speculate why mammalian immune systems should have evolved to do this with organisms that are non-infective and apparently pose them no threat; in fact mammals produce antibodies to very many alien proteins, including those of plant pathogenic fungi. Thus, purified plant viruses injected into rabbits will cause the production of virus-specific antibodies, which may then be extracted from a drop of the

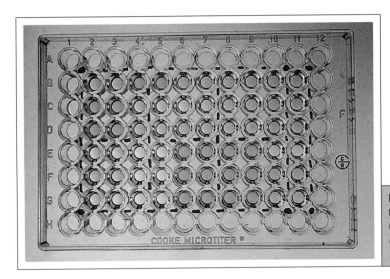

**Fig. 15.4** A DAS-ELISA diagnostic test plate showing some positive (dark yellow) reactions. *Photograph courtesy of N. Spence, Horticulture Research International.*

rabbit's blood. The blood product containing the antibodies is called an antiserum.

Antisera may be used in various ways, of which one of the most elegant is the DAS-ELISA test (double antibody sandwich enzyme-linked immunosorbent assay; see Fig. 15.4). A drop of antibody is first added to a small plastic well, where it attaches to the plastic base. A separate batch is chemically joined to an enzyme, alkaline phosphatase, that reacts with a chosen chemical substrate to produce a distinctive colour. The ground-up plant material is then added to the well and the antibody traps the virus at the bottom. The well is then washed out, but when the virus is present it attaches to the antibody and is retained. The enzyme-linked antibody is then added and attaches to the trapped virus. After another wash, the substrate is added. A colour change in the substrate indicates the presence of the enzyme, which is only present where the virus is too. Although chemically complex to prepare, such tests are simple to carry out. The reaction is assessed by a light-sensitive plate reader which compares the intensity of the colour against a known standard.

Serology can only identify viruses to which antibodies have already been raised. If no antibodies are available, virologists may carry out plant inoculation tests. Some plants, such as *Chenopodium amaranticolor*, are easy to grow and react to virus infection in characteristic ways. Inoculations on these standard test plants may reveal symptoms that identify the virus, or at least confirm that a virus is present.

The tests are particularly useful if virus levels in the infected material are low, and difficult to detect by the two methods already described, because a susceptible plant allows the virus to multiply to detectable levels.

Modern techniques of molecular biology have raised hopes of more rapid and specific tests for microorganisms. Some exist, but are not yet in widespread use. They are very valuable for precise identification, but to date they are not so useful for detection, because most rely on obtaining a DNA sample from the organism. To do this one usually needs a pure culture, which may be very time-consuming to prepare. One of the main areas of research at present is developing techniques that allow the detection of the DNA of specific pathogens among the DNAs of the diverse range of other microorganisms that invariably occur in most infected plants.

### Proving Koch's postulates

If a microorganism is consistently associated with a disease, pathologists will wish to prove that it is the cause of that disease. To do this, they must prove *Koch's postulates*, put forward in the late nineteenth century by the German human pathologist Robert Koch. To satisfy his postulates, the pathologist must show that the microorganism is consistently associated with the problem, can be extracted and grown in a pure form, will cause the disease when reinoculated into the plant and can then be re-isolated. This is a very time-consuming procedure and if the organism cannot be cultured, as is the case with viruses, may be very difficult, though viruses can be purified. Diagnosticians do not confirm Koch's postulates for routine cases, but they remain a very powerful tool to discriminate true pathogens from the many suspicious organisms that are usually present on and around diseased parts of plants.

## Weed problems

### Competitive damage

Weeds restrict the yield of garden crops and ornamentals by competing for moisture, nutrients and light. If uncontrolled they may restrict yield and prevent the development of desirable plants. Bigger and/or larger numbers of weeds clearly make greater demands upon the soil. However, the bigger and better developed the cultivated plants, the more effectively they can compete with weeds, so vigorous garden plants often tolerate more weeds.

Research has shown that most vegetables and flowers tolerate annual weeds to some extent before yield is lost. Such tolerance is often dependent on the growth stage of the cultivated plants. Initially seedlings do not suffer much from seedling weeds, but vigorous weeds may 'overtake' the cultivated plants and damage them later. Ripening plants are not at much risk from weed damage at such a late stage of development, but if left the weeds are likely to shed seeds, which may infest subsequent crops. Between the seedling stage and ripening there is a middle, critical, period when weeds have a very marked effect on the growth of cultivated plants. Removing weeds is essential at this time as the effort and expense involved make a marked difference to the final yield or appearance. Commercial growers may exploit knowledge of critical periods to save on weeding. Using pre-emergence weedkillers, they may prevent weeds from damaging crops at the vulnerable stages, or may even use selective weedkillers to eliminate weeds at the critical times; weed seeds germinating in subsequent crops are of no great concern, for they can be controlled by weedkillers. However, the use of weedkillers in cultivated plants is not usually an option for gardeners, because appropriate products are not available for home use. Periods when weeds are not damaging cannot, therefore, be exploited for control and the shedding of weed seeds may cause problems later.

The size of the population of weeds at which yield is reduced varies. Research with onions showed that 21 weeds per square metre caused no losses, even if left for the entire life of the crop. More than 21 weeds reduced the yield, and the more weeds present the more quickly they adversely affected yield. Once more than 150 weeds per square metre were present, additional weeds caused little extra loss.

Weed control is often only undertaken when weed levels reach a point at which the cost and disruption caused by removing them is less than the damage they do if left. This use of a threshold level is very common agricultural practice, but less of an option for gardeners, for two reasons. First, they have fewer chemical means of removing weeds available to them and non-chemical methods are often very damaging to growing plants. Second, the threshold depends on the type of weed and the length of time it has to develop before the crop is harvested. Even a very few weeds can cause serious loss in long-term stands of cultivated plants, if the weed has the capacity to grow to a large size. Fat hen (*Chenopodium album*), for example, may grow in an *indeterminate* way and become very large, whereas groundsel (*Senecis vulgaris*) has a limit to the size it can reach. This means that quite complex decisions would be required for each weed and cultivated plant combination. Where weedkillers can be used on a field scale this is practical, but for a garden it is excessively complicated.

Perennial plants typically suffer most from weed competition during the period following planting. At this time weeds may reduce the initial growth rates of the plants and the damage caused is often not made up in subsequent years. Weed populations need not be particularly high to inflict damage: even a few weeds may be very harmful if they are vigorous and the cultivated plant in question is unable to compete. The effects of weed competition in the early years of perennial flowers or crops may be quite subtle. By restricting the uptake of nutrients and water, overall growth is reduced, but plants often compensate for this by producing flower and fruit at the expense of foliage, so overall performance is not impaired. In fact, *cover crops* are sometimes used to produce a similar effect as, for example, with grass grown on orchard floors. It is risky, however, to use weeds to substitute for cover crops as they may persist into subsequent crops or spread to adjacent areas. In conclusion, eliminating every weed is sometimes more trouble than it is worth, although severe weed infestations should always be dealt with.

## Interference effects

Weeds interfere with cultivation. Annual meadow grass (*Poa annua*), for example, exerts little competitive effect, but removing it uproots seedlings, and annual nettles (*Urtica urens*) make weeding and harvesting unpleasant. Weeds also interfere with the harvesting and consumption of produce, as when fragments of chickweed (*Stellaria media*) contaminate lettuce (*Lactuca sativa*) at harvest.

Weeds damage buildings, walls, paths and other structures. They impair the function of garden equipment and features, blunting mowers, blocking drains and clogging ponds. They may also have physical effects on humans by stinging, pricking and even poisoning them; the sap of giant hogweed (*Heracleum mantegazzianum*), for example, sensitises the skin to sunlight, resulting in a severe rash. Some weeds poison desirable garden plants by secreting chemicals that inhibit their growth, a phenomenon known to botanists as allelopathy. Few UK garden weeds are parasitic on other plants, although worldwide parasitic weeds such as dwarf mistletoes (*Arcuthobium* spp.) and witchweeds (*Striga* spp.) are a menace to forestry and agriculture.

Some weeds impair the visual quality of gardens. Lawns suffer especially from broad-leaved weeds but grass weeds may also be a problem, as when annual meadow grass invades, its pale colour, abundant seedheads and susceptibility to drought making it most unwelcome. Gravel and paving are not enhanced by the weeds which colonise them. Borders of ornamental plants are rendered unattractive by an undergrowth of weeds. Even wild areas are sometimes colonised by weeds which out-compete the desirable wild vegetation, especially in fertile soil, a notable example being invasion by *Rhododendron ponticum*.

Not all weed damage is obvious. For example, weeds may act as hosts for harmful organisms, which may spread into adjacent susceptible plants, or weeds may act as green bridges between susceptible crops, undoing many of the helpful effects of rotation. Cucumber mosaic virus, which has a wide host range but is especially damaging to courgettes, is harboured by chickweed (*Stellaria media*), for example, while beet western yellows virus overwinters in weeds such as groundsel (*S. vulgaris*), ready to infect lettuces the

following season. Club-root disease, caused by the fungus *Plasmodiophora brassicae*, persists in weeds belonging to the Brassicaceae, such as shepherd's purse (*Capsella bursa-pastoris*). On the other hand, weeds are not usually important in the persistence or spread of pests, because ornamental plants, especially vegetatively propagated ones, provide alternative green bridges in many gardens.

The importance of weed control to reduce carry-over of diseases and pests is not likely to be great. Crop residues such as carrots and potatoes left in the ground or not thoroughly composted, present a much greater risk.

## POISONOUS AND IRRITANT PLANTS

A number of plants other than weeds have features that make them undesirable. These include plants that are toxic to mammals or are highly irritant to the skin and/or eyes. Many plants in this category are widely grown as ornamentals (e.g. *Laburnum* spp.) or may be valuable because they yield a product that is important in medicine, such as morphine from *Papaver somniferum*.

Gardeners are becoming increasingly aware of the large numbers of poisonous plants commonly grown in gardens or found in the wild. The primary role of toxins in plants is to act as feeding deterrents for herbivores (see Chapter 12), but there is growing concern about their potential danger to people, especially children who may be attracted to colourful berries (e.g. *Daphne* spp.) or seeds (e.g. *Laburnum* spp.). A great many plant families contain some poisonous species, although the Euphorbiaceae (spurge family), Ranunculaceae (buttercup family) and Solanaceae (potato family) have more than most. Toxins may be present in the whole plant, or only in a part, as for example in seeds of *Laburnum anagyroides* or the foliage (but not the edible petioles) of rhubarb (*Rheum × hybridum*).

Many poisonous plants contain alkaloids, cyanogenic glycosides or other nitrogen-containing secondary plant products. All these, if taken in sufficient quantities, are toxic to humans but a few are highly poisonous even in very small amounts. Highly toxic alkaloids include coniine from hemlock (*Conium*

**Table 15.3** Plants that may cause irritation or allergic reactions.

| Irritant sap | Allergens* | Phototoxins† | Hazardous hairs |
|---|---|---|---|
| Dieffenbachia spp. | Alstroemeria spp. | Ruta graveolens (rue) | Echium spp. |
| Schefflera spp. | Euphorbia pulcherrima | Heracleum mantegazzianum | Pulmonaria spp. |
| Spathiphyllum spp. | (poinsettia) | (giant hogweed) | Symphytum spp. |
| Monstera deliciosa | Ficus benjamina | Apium graveolens (celery) | Myosotis spp. |
| Euphorbia spp. | Primula obconica | Pastinacia sativa (parsnip) | Fremontodendron californicum |

Based on Bostock, H. (2006) Plants that may cause irritation or allergic reaction. *The Garden* **131**, 433.
\* Allergens do not affect everyone but individuals may acquire sensitivity to them, especially if they handle the plants regularly.
† Some plants have sap that renders skin sensitive to strong sunlight, resulting in severe localised blistering when contact is followed by exposure to the sun.

*maculatum*) and atropine from deadly nightshade (*Atropa belladonna*). Ricin from the castor oil plant (*Ricinus communis*) is also highly toxic, the highest concentration being in the seeds. However, in lower doses some toxic alkaloids have useful pharmacological properties, such as morphine from the opium poppy (*Papaver somniferum*). Other highly toxic compounds are the cyanogenic glycosides which, when crushed, release hydrogen cyanide (HCN) gas. In the plant the cyanogenic compounds are separated from the enzymes that act upon them and so only release HCN when crushed, as in *Lotus corniculatus*.

Other potentially harmful plants are those that cause problems on physical contact, including skin reactions ranging from mild irritation to severe blistering. In stinging plants such as nettles (*Urtica* spp.) the toxic compounds are injected by surface hairs. Plants containing allergens often cause a delayed sensitisation reaction, for example *Chrysanthemum* cultivars and poison ivy (*Toxidendron radicans*). Other well-known plant allergens are the wind-blown pollens, especially from grasses, which cause hay fever. Table 15.3 gives a list of some common plants that may cause irritant or allergic reactions.

There is now a Code of Recommended Practice for the labelling of potentially harmful plants by garden centres and nurseries, while many gardening books give lists of those that are potentially poisonous or harmful in other ways. A helpful leaflet entitled *Take Care Be Aware* is also available from the Royal Horticultural Society. It should be remembered that although most poisonous plants are easily recognisable, there are problems of potential misidentification in a few cases. For example, hemlock could be confused with wild parsley.

## CONCLUSION

Despite the great diversity of undesirable organisms that may be present in or among garden plants, these do not necessarily affect the yield or performance of cultivated plants. Correct identification of the offending pest, pathogen or weed, and careful monitoring of its development in relation to the development of the plant or crop and of the prevailing weather conditions, help the gardener to decide whether or not the application of control measures is appropriate or necessary. Gardeners should be aware of the dangers of poisonous and irritant plants.

For Further reading see the end of Chapter 16.

# 16

## Controlling the Undesirables

### SUMMARY

In this chapter the control of pests, pathogens and weeds is considered. First, the criteria used to decide when control is necessary are described for each group of undesirable organisms. Next, the various methods of control available to gardeners are reviewed, including the use of chemical pesticides, fungicides and herbicides, the deployment of biological control agents and the use of specialised cultivation methods. The importance of the correct identification of a pest, pathogen or weed, and of an understanding of its life cycle relative to that of its host, are emphasised. Special consideration is given to integrated pest management, in which all available approaches to the control of pests, pathogens and weeds are used to best advantage, thereby reducing both the range and quantities of chemical agents required.

## WHEN TO APPLY CONTROL MEASURES

### Direct observation of pests

Crops should be monitored regularly, so that pest, disease and weed problems may be identified and controlled in good time. It is important to learn to recognise the damage that pests cause, as this is often easier to see than the culprits themselves, which may be small or only present on the undersides of leaves. Distorted leaves and growing points indicate the presence of aphids (Hemiptera), leaf speckling is typical of spider mite (Acerina) and leafhopper (Hemiptera) damage, and black sooty moulds are often present on the honeydew produced by sucking insects such as whitefly, aphids and scale insects (Hemiptera).

It is a good strategy to try to get a general impression of the health of the whole batch of plants first, before examining individual ones. The experienced eye will quickly spot an atypical plant, which may then be examined more closely, ideally with a ×10 hand lens. Pests prefer certain species and varieties of plants to others, so identifying those that are particularly susceptible and concentrating on these so-called indicator plants when checking for pests is an efficient approach. It may even be worth considering applying additional control measures to such plants.

### Trapping and monitoring

Sticky traps are useful for monitoring some pests, particularly in glasshouses, because they catch insects before they reach levels that can realistically be detected by eye. Yellow traps catch aphids, whitefly and thrips (Thysanoptera), while blue ones catch mainly thrips. Unfortunately, yellow traps also catch beneficial insects, so they should be used sparingly in a glasshouse where biological control is being deployed (see below). Traps should be examined regularly, although it is not necessary to make detailed counts. Instead it should be noted whether the numbers of pests being caught change significantly. Sticky traps should be changed regularly

239

while it is still possible to assess numbers easily, generally every 2 to 4 weeks.

Traps are also useful for monitoring nocturnal pests. The numbers of slugs (Mollusca) trapped under a flat stone or a tile baited with a teaspoonful of slug pellets may be used to decide whether treatment is justified. This technique may, however, underestimate the numbers present, because if the weather is too cold or dry, slugs stay underground and are not trapped. Catches, therefore, only reflect the numbers of slugs active on the surface of the soil.

## Pheromones

Pheromones are chemical messengers that insects produce to communicate with each other in different ways. Some female insects produce volatile sex pheromones to attract males. These chemicals, although highly specific so that they only attract males of the same species, are easily synthesised artificially and are used in a number of ways to control insects. Sticky traps, baited with artificial sex pheromone to attract and catch male moths, have been used for many years by commercial growers to time the application of conventional sprays correctly to control a number of pests. Some traps are also available to gardeners to control pests such as codling moth (*Cydia pomonella*; Fig. 16.1), and red plum maggot (*Cydia funebrana*).

**Fig. 16.1** A codling moth pheromone trap.
*Photograph courtesy of Royal Horticultural Society.*

Recently, traps in commercial crops have been baited with sex pheromones and fine powders that coat visiting male moths. Not only are they confused, but other males mistake them for females and become similarly affected when they try to mate, a technique known as auto dissemination. In other trials, traps baited with pheromones and either a chemical or biological pesticide were as effective as much larger amounts of pesticide sprayed over a similar area.

Parasites of aphids detect their hosts partly by scent, and the pheromone involved can be synthesised. Research is in progress to see whether such chemicals can be applied to crops to attract the parasites from hedges and surrounding areas early in the year, so that they are already present on the plants when the aphids arrive. If successful, the technique has considerable scope for use in gardens.

## Weather

Past and current weather conditions have both direct and indirect influences on many pests, and may affect the amount of damage they cause. Average temperature and rainfall are the most important factors, but wind strength and direction, and extremes of temperature, may also be critical. Pests are more active when it is warm, but slugs prefer it warm and wet, whereas other pests such as aphids, spider mites and thrips prefer it warm and dry. Many soil-dwelling pests are largely unaffected by moisture themselves, but if they feed on roots may do more damage in dry conditions.

Previous weather conditions may have a greater impact on pests than those that prevail when the actual damage to plants occurs. Leatherjackets, the larvae of crane flies (daddy-long-legs, *Tipula* spp.) cause most damage in spring, but the extent of this damage depends largely on the amount of rain the preceding autumn, when the eggs were laid. The eggs and young larvae do not survive well in dry soil, so leatherjacket damage is normally only serious after a wet autumn.

Two-spotted spider mite (*Tetranychus urticae*) is mainly a pest of glasshouse plants, although in hot summers it damages outdoor crops such as strawberries, raspberries and runner beans. It over-winters as specially adapted adult mites, in cracks and crannies in glasshouses and in the stakes and bamboo canes used to support raspberries and runner beans. Vast numbers of mites may go into hibernation after a hot summer, and if this happens early serious attacks may be expected the following spring. A simple ploy to avoid damage to runner beans is to have two batches of support canes and to use these in alternate years, because the mites will not survive without food for two seasons.

Most native pests survive just as well in cold winters as they do in mild ones, but aphids are an exception. Some species over-winter both as eggs and as wingless adults in mild years, but only the eggs survive cold winters. Attacks of cabbage aphid (*Brevicoryne brassicae*) and the peach-potato aphid (*Myzus persicae*) are usually earlier and more severe in the spring following a mild winter than they are after a cold one.

Many small insects, such as aphids and thrips, are carried a long way by the wind. These insects will not take off from plants if the wind speed is more than 5–6 km per hour, but once airborne they have little control over where they are blown. Larger, more powerful insects may also be blown long distances. In the UK, serious attacks in mid-summer of caterpillars of the diamond-back moth (*Plutella xylostella*) and the silver-Y moth (*Autographa gamma*) generally occur only after a prolonged period of south-easterly wind earlier in the year, which blows adults over from mainland Europe.

# Diseases

The fungi and bacteria that cause disease in annual plants, or in the deciduous parts of perennial plants, must survive the winter in temperate climates, and the dry seasons in the tropics, when there is no suitable host material present. The absence of the host, therefore, imposes on the pathogens the need for a survival strategy. Just as plants have seeds, most fungi and bacteria have resting spores for this purpose (Fig. 16.2), or they form other types of resting structures such as sclerotia (see Chapter 15). This reduction of the organism to a single, tough and microscopic speck imposes on the pathogen a need to sense the right time to attack the host. The more specialised the pathogen, the more critical this is, and some wonderful examples of co-dependence exist. The camellia-petal-blight fungus (*Ciborinia camelliae*), recently arrived in the UK, infects the flowers of camellias, grows in them and forms sclerotia in the

**Fig. 16.2** A scanning electron micrograph of a pustule of rose rust (*Phragmidium* sp.) on the underside of a rose leaf, showing a mass of large resting spores (*teliospores*) which carry the fungus through the winter and germinate in the spring to reinfect roses. *Photograph courtesy of Georgina Godwin-Keene (CABI Bioscience); Royal Horticultural Society.*

dead, fallen flowers. These stay dormant on the ground until the following spring, when they germinate at the precise moment when the flowers open, release infective spores and reinfect. If the fungus misses the flowering period, of only a few weeks each year, it cannot survive. Climatic factors such as temperature, or chemicals released by the host, usually act as triggers to resting spore or sclerotium germination. For example, the sclerotia of the onion white rot fungus, *Sclerotium cepivorum*, are induced to germinate by allyl sulphides leaking from the roots of young onion plants.

The most specialised adaptations occur among those pathogens that are completely dependent on their hosts, notably the biotrophs that feed only on the living tissues of a host and have little or no ability to live apart from it. This is not as restrictive as it appears, because such sophisticated parasites, exemplified by the rusts (Uredinales) and powdery mildews (Erysiphales), have evolved not to damage the host so much that it dies, since they would die with it. The most destructive pathogens, in fact, are frequently those that are able to live independently of the host, often on its decaying tissues, but also on other substrates, so a period of dormancy is unnecessary. These include honey fungus (*Armillaria* spp.), the most destructive pathogen in gardens. Honey fungus invades the roots of almost any woody perennial, kills them and then feeds on the decaying wood. Keeping the host alive confers no advantage, and this voracious life cycle makes the fungus particularly difficult to control because there is no weak point to attack; it is never dormant, rarely reduced to a single spore survival stage and usually underground, out of sight.

Many bacteria also form resting spores to survive when the host is not present, but they have a less complex life cycle: resting spores germinate to form bacterial cells and in turn form resting spores when they run out of food. Those that do not form resting spores either survive at low levels in perennial host tissues, or can compete as saprotrophs, usually in the soil, in their active state.

Viruses cannot grow except in the cells of their host plants, so they survive outside the host only in their vectors, or passively in the environment. This appears to make them very amenable to control by attacking their vectors, usually insects or nematodes. Nonetheless, viruses are very successful organisms. In particular, they are extremely infectious and this makes control by exterminating vectors much more difficult than it might appear. Even eliminating 99% of the vectors, which is difficult even with the most effective chemical insecticides, will not eliminate a virus, which has only to survive in a tiny fraction of the vector population to break out again, relying on the unerring ability of its vector to find a host plant, which that vector must do if it is to survive.

Given that pathogens have a typical life cycle of multiplication, reproduction, rest and multiplication again, much advantage may be gained by the gardener by attacking at the most vulnerable stage. Accurate knowledge of the life cycle is crucial, yet the life cycles of some pathogens have been very hard to unravel. In particular, what happens to some temperate-climate fungal pathogens in winter is still not completely known: the peach-leaf-curl fungus *Taphrina deformans* is thought to spend the winter hidden in peach tree buds, waiting to invade the newly formed leaves in spring, but pathologists are still not completely confident that this is so. Much research is still required to elucidate the life cycles

of many common plant pathogenic fungi. In general, the ability of fungi to survive at low levels in their plant hosts is very important for the life cycle, even though such survival may be in a quiescent state that is not easily detectable. In recent years there has been renewed interest in this pattern of behaviour with the demonstration that plants host a wide range of previously unsuspected endophytic fungi. These are embedded in the tissues and have all sorts of unexpected side effects. One that has been known about for a long time is *Cryptostroma corticale*, the cause of sooty-bark disease of sycamore. Undetected while the tree is in good health, it breaks out during times of stress, especially very hot summer weather, to kill the bark and produce masses of black dispersal spores.

Correct diagnosis is obviously crucial to control, because it may then be possible to identify the weak point in a pathogen's life cycle at which it can most easily be attacked. This may be the survival stage, an attractive target because the pathogen is present only in very low numbers. On the other hand, resting spores may be very tough and difficult to kill. If the pathogen infects via wounds, as in the case of silver leaf (*Chondrostereum purpureum*), the gardener may protect wounds in the host chemically to prevent the spores from colonising the wood, or prune during summer months when spores are not produced. If damage to the host depends on high levels of infection at a particular growth stage, good protection may be achieved by spraying during this stage. For example, a timely fungicide application to outdoor tomatoes will not prevent infection by blight (*Phytophthora infestans*), but it may delay the build up of disease sufficiently to allow a crop of tomatoes to be harvested from an infected plant.

Where disease control depends on fungicides, as it often does in commercial agriculture and horticulture, the timing of applications is crucial if the control is to be effective: too early and the chemical inhibition is diluted by the growth of the plant; too late and the pathogen has too strong a hold. Disease prediction to reduce wastage of chemical controls and maximise effectiveness is therefore attractive and much effort has gone into developing predictive models, mathematical equations that combine key data on climate and pathogen availability. The concept is easy enough to understand: in a crop such as a potato field or an apple orchard, spore traps monitor airborne spores continuously and climate sensors monitor temperature, rainfall, humidity and leaf wetness. The data are analysed concurrently using a predetermined model that has shown that when an amount of fungal spores (*x*) is present during a period of temperature (*y*) and leaf wetness (*z*), a dangerous level of disease will develop. Just before the danger level is reached, farmers are warned to spray. Commercial systems do exist and some are very effective, but they are not as widely used as could be wished, because local variations in climate, and imperfections in the models, make them somewhat unreliable and expensive to implement. Gardeners may benefit from warnings to farmers of potato blight (*P. infestans*) and apple scab (*Venturia inaequalis*) attacks, if they are in the right area for an accurate forecast, but of course someone has to pay for this advice, which is expensive to produce. Also, gardeners may not have a useful range of control measures available to them. For example, they do not have access to the most effective chemical fungicides for potato blight that are available to commercial growers, and control will therefore be more difficult where it has to depend on less effective products for home use.

## When to control weeds

Most gardeners regard all weeds as unacceptable and try to keep their numbers very low. This is partly because weeds are thought to be unsightly, but also because experience has shown that even low levels of weeds can quickly result in heavy infestations once seeding occurs. This approach may have a high price because the removal of weeds is often a very lengthy and labour intensive process. The disruption to soil, by cultivating and pulling weeds, the compaction of soil resulting from treading on the ground and the side effects of weedkillers all reduce the growth of the crop. Sometimes the damage caused by weeding is greater than that caused by the weeds themselves.

Weeds are highly adaptable and even low populations may soon lead to massive plants shedding large numbers of seeds. Partial destruction, by hoeing in weather that allows the weeds to re-root, for example, may be no more effective than leaving them alone. When time and resources are limited, weed-control decisions have to be given careful consideration.

Weed-control tactics depend on forecasting the likely consequences of weed infestations. In practise it is very difficult to know what bare ground or

apparently clear plantations have in store, and research on the seed content of newly planted soil has shown that it is rarely possible to predict subsequent weed crops accurately. Weed seeds are hard to find and count in soil, and if found they may be so dormant that they do not germinate for years. Trying to germinate the weed seeds in a soil sample is no more effective, as populations of weed seeds include individuals with different levels of dormancy, so germination occurs in flushes through the year.

Producing many seeds allows even a few survivors to maintain populations and allows the possibility of dispersal to fresh habitats. Annual meadow grass (*Poa annua*), for example, may produce as many as 13 000 seeds per plant and groundsel (*Senecio vulgaris*) as many as 38 000. The disadvantage, for the weed, of producing many highly mobile seeds is that the seeds have to be small. A small seed has few reserves of nutrients to enable the seedling to become established in adverse conditions, and even if it does become established it is likely to be weak and easily out-competed by surrounding vegetation. To cope with this, weeds have evolved several mechanisms that allow seeds to 'sense' their surrounding and germinate when conditions are optimal for survival (see Chapter 9).

For example, light is required for many annual weed seeds to germinate, ensuring that they do not germinate until the soil is disturbed. Annual nettles (*Urtica urens*) and annual meadow grass are garden examples, where buried seeds do not germinate until they reach the surface. On the other hand, weed seeds germinating in the light, but in the presence of competitors, would be unlikely to survive had not mechanisms evolved to cope with such a situation. Thus, where sunlight passes through vegetation, red and blue light is absorbed and used in photosynthesis, while green light is reflected; the light that is transmitted is mainly in the far-red part of the spectrum and inhibits germination of the seeds of weeds such as chickweed (*Stellaria media*) that cannot cope with shade.

Temperature fluctuations are greatest near the soil surface and some weed seeds are able to sense these, again ensuring that germination is only triggered when the soil is disturbed. For further information on the importance of light and temperature in triggering seed germination, see Chapter 9.

Persistence of dormant seeds in the soil, as a *seed bank*, is a common attribute of weeds. Weed seeds may remain viable in the soil for very long periods:

field speedwell (*Veronica agrestis*), for example, may survive for at least 30 years and common poppy (*Papaver rhoeas*) may last 26 years or more. On the other hand, some weed seeds are short lived: seeds of groundsel, for instance, persist for only three years. Some weeds germinate soon after being shed while others may germinate at a low frequency over several years, when conditions favouring rapid growth and development occur. Such conditions are often required in complex combinations. For example, disturbed soil and warmth trigger annual nettle seeds, while alternating temperatures, a preceding period of chilling at low temperatures, nitrates in the soil, moisture and light are required for the seeds of annual meadow grass to break their dormancy. Such fine-tuning greatly enhances the persistence of weeds by ensuring that dormant seeds only germinate when conditions are optimal for seedling growth and development.

Other mechanisms to ensure survival include the production of different types of seed. Fat hen (*Chenopodium album*), for example, produces many black and fewer brown seeds. The brown seeds are thin-walled and germinate immediately after separating from the parent plant, late in the season. The small number of seedlings so produced that survive the winter form large plants, shedding five times the normal number of seeds. The black seeds germinate over a long period, beginning the following spring, when the seedlings are less likely to be killed by adverse weather conditions.

Weed seeds may also exhibit a variety of other dormancy mechanisms (see Chapter 9), examples being hard seed coats, which require rotting or abrasion before germination will occur, the presence of inhibitory substances that need to be leached out by water, and slowly developing embryos.

Seed-bank control is an essential gardening practice. Leaving weed seed on the surface promotes early germination, rendering seedlings vulnerable to weedkillers and cultivation.

Perennial weeds have bud banks: these are the buds on perennating structures, such as rhizomes, that perform a similar function to seeds and also need to be controlled. Buds are often less resistant to elimination measures than seeds, but exceptions are those of the bulbous weeds, such as ramsons (*Allium ursinum*) and lesser celandine (*Ranunculus ficaria*), which form robust small bulbs. Forecasting problems

with perennial weeds is often easier than for annual weeds, for they persist from year to year in refuges where they are safe from attack, for example beneath or among buildings, paving slabs, permanent plantings or deep in the soil.

## HOW TO CONTROL

### Integrated pest management

Pest and disease control is often seen as a choice between natural, 'organic' methods and the use of manufactured pesticides. Both approaches have advantages and disadvantages, and the best of both worlds may be achieved in gardens by using an *integrated pest and disease management* (IPM) approach, developed over several decades in agricultural systems. Originally devised as a reaction against excessive and unnecessary routine use of chemical pesticides, IPM emphasises that all available techniques should be used to best advantage, and that chemical control agents should only be used when needed, when monitoring techniques indicate genuinely threatening levels of pest or disease and when no effective alternative exists. More recently, the emphasis in agriculture has shifted to integrated crop or farm management, to take account of the fact that pests do not exist in isolation from the crops and the farm environment.

Application of IPM techniques has produced some startling reductions in, or even eliminations of, pesticide and fungicide use, but many farmers still spray routinely for many problems. Home gardeners have probably been more ingenious practitioners of IPM than farmers, being forced to rely more on non-chemical means of control because only a limited range of pesticides and fungicides is available to them. IPM will always make best use of plant resistance to attack or infection, cultural techniques and biological controls before deploying chemical control agents, but it is emphatically not a synonym for organic gardening; growers who adopt IPM will use chemical agents, but only as one weapon among many. The real key to successful IPM for farmers and home gardeners is accurate and complete information about the target pest, disease or weed.

# Controlling pests

## Chemical controls

Chemical pesticides may give rapid control of pest problems, providing that they are appropriate for the problem and are thoroughly applied at the right time. Most garden insecticides have a broad spectrum of activity, dealing with a variety of pests but also killing non-target insects, including bees and predators and parasites of pests. Those pests that survive treatment breed more prolifically if their natural enemies have been eliminated. Moreover, frequent use of the same type of chemical may result in the selection of pesticide-resistant pests, particularly glasshouse pests such as whitefly (*Trialeurodes vaporariorum*), red spider mite (*Tetranychus urticae*) and some aphids.

Beneficial insects and mites are susceptible to most insecticides. Fatty acids, plant oils, rotenone (derris) and pyrethrum are easily degraded and therefore persist for only short periods. This limits the harm they do to non-target insects, so their use, where appropriate, helps to conserve predators and parasites of pests.

## Biological control: predators and parasites

Predators and parasites sometimes keep pests at a low level and therefore make the use of insecticides unnecessary. Unfortunately, in gardens they are often not present sufficiently early in spring and early summer to prevent damaging infestations from developing. Moreover, although most pests have at least one natural enemy, many major pests lack predators or parasites that give effective control. The artificial application of natural enemies or biological controls may be valuable, and is most successful in greenhouses. Glasshouse whitefly, two-spotted spider mite and mealybugs (*Pseudococcus* spp.) are not native to Britain, and good control has been achieved by using predators and parasites from the countries of origin of these pests. Other predators and parasites are commercially available for controlling aphids, thrips (*Thrips* spp.), fungus gnats (*Bradysia* spp.) and some glasshouse scale insects (Coccoidea). Pathogenic nematodes, which transmit bacterial diseases fatal to some invertebrates, are available for controlling pests such as slugs (*Arion hortensis*), leatherjackets, chafer grubs (e.g. *Phyllopertha* and *Hoplia* spp.) and vine weevil grubs (*Otiorhynchus sulcatus*). A bacterium,

*Bacillus thuringiensis*, produces a highly specific toxin and when applied like a conventional spray causes a fatal disease in moth or butterfly caterpillars that eat treated plants. Unfortunately, this is not currently available for garden use.

## Resistant plants

Most popular garden plants are available in a range of cultivars and these may show variation in susceptibility to pests. Growing cultivars that are less attractive to pests or are better able to tolerate damage is part of integrated pest control. Some plants have been bred for resistance to particular pests or diseases; these have a high degree of resistance, often making them immune to the problem organism. New techniques in plant breeding, such as genetic modification, open up the possibility of artificially inserting genes for resistance, sometimes from unrelated organisms. For example, the gene controlling production of the insecticidal toxin in *Bacillus thuringiensis* spores has been introduced into some plants, such as tobacco, maize and potato, to produce cultivars resistant to caterpillars. No genetically modified cultivars are currently available in the UK, but this method of plant breeding will develop and may ultimately provide pest- and/or disease-resistant plants for home garden use, providing that appropriate legislation is enacted.

## Barrier methods

Rabbits may be excluded from gardens or flower beds with wire netting. Insects can also be kept away by covering plants with a finely woven material such as horticultural fleece (Fig. 16.3), which allows light and rain to reach the plants but denies access to all but the smallest of insects. This technique is mainly used in vegetable gardens to protect brassicas from cabbage root fly (*Delia radicum*), cabbage butterflies (*Pieris* spp.) and moth (*Mamestra brassicae*), or on carrots to exclude carrot fly (*Psila rosae*). Crop rotation must be practised as well, since many pests over-winter as pupae in the soil; emerging adults will be trapped under the fleece with their host plants if the growing positions are not changed.

## Companion planting

Certain plant combinations are said to be beneficial partnerships because volatile chemicals produced by one plant make the other less susceptible to attack

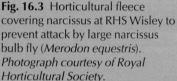

**Fig. 16.3** Horticultural fleece covering narcissus at RHS Wisley to prevent attack by large narcissus bulb fly (*Merodon equestris*). *Photograph courtesy of Royal Horticultural Society.*

by a pest or disease. Frequently quoted examples are onions and carrots to repel onion fly (*Delia antiqua*) and carrot fly respectively; African marigold (*Tagetes erecta*) to repel glasshouse whitefly from tomatoes; and garlic (*Allium sativum*) to protect plants against diseases such as peach-leaf curl (*Taphrina deformans*) and rose blackspot (*Diplocarpon rosae*). These claims are largely based on anecdotal evidence and scientific trials have failed to show worthwhile protection. There is, however, some evidence of how plants may interact. Some plants produce a chemical called *methyl jasmonate*, which mediates the biochemical mechanisms that lead to the synthesis of compounds conferring resistance to pests and diseases; for example, proteinase inhibitors which reduce feeding by pests, such as caterpillars and leaf beetles, and phenolic compounds which suppress fungal diseases such as rusts. Some plants, such as *Artemisia* spp., produce so much methyl jasmonate, which is volatile, that they may release it into the air and stimulate increased resistance in neighbouring plants. The level of control shown so far is inadequate to prevent damage, but there may be some truth in the notion of companion planting. The observations of gardeners are an important means of providing evidence for and against this method of control.

## Using biological controls

Using natural enemies to control pests is not new, although most examples of successful application were developed in the twentieth century. Worldwide there have been some spectacular successes in the control of pests and weeds, mostly species introduced accidentally into countries where they have not previously occurred. In such countries there is often an absence of natural enemies capable of providing control, so the introduced species reproduce unchecked and become a major problem unless a predator is also introduced. A well-established industry has developed since the 1960s for supplying predators, parasites and pathogenic organisms for controlling pests; these are mainly for glasshouse pests, but some may also be used in outdoor situations (Fig. 16.4). Glasshouse whitefly and two-spotted spider mite had by the 1970s become resistant to a wide range of formerly effective pesticides, so the use of biological control organisms has provided an important alternative. When used correctly, biological methods give control that is as good as, or better than, chemical methods, at a competitive cost.

The biological controls listed in Table 16.1 are all available by mail order to gardeners. The addresses of suppliers may be found in advertisements in gardening magazines or on the Royal Horticultural Society website (www.rhs.org.uk/advice/biocontrol.asp.)

To be effective, biological control agents must be used in an appropriate manner. They are pest-specific, so it is essential to identify the pest correctly. The pest must also be present when the biological control is introduced, otherwise there is nothing for the predator to feed on or the parasite to breed in. They are not miracle workers and, unlike pesticides, cannot give an immediate reduction in pest numbers. Predators and parasites should be introduced before

**Fig. 16.4** The predatory mite, *Phytoseiulus persimilis*, feeding on red spider mite. *Photograph courtesy of Royal Horticultural Society.*

**Table 16.1** Biological controls available by mail order to amateur gardeners.

| Pest | Control organism |
|---|---|
| Aphids | A fly larva predator, *Aphidoletes aphidimyza* |
| | Parasitic wasps, *Aphidius* or *Praon* spp. |
| | Lacewing larvae, *Chrysoperla carnea* |
| Two-spotted spider mite | A predatory mite, *Phytoseiulus persimilis** |
| Glasshouse whitefly | A parasitic wasp, *Encarsia formosa* |
| | A ladybird predator, *Delphastus catalinae* |
| Mealybugs | A ladybird predator, *Cryptolaemus montrouzieri* |
| Soft scale | A parasitic wasp, *Metaphycus helvolae* |
| Fungus gnats/sciarid flies | A predatory mite, *Hypoaspis miles* |
| Western flower thrips | A predatory mite, *Amblyseius* spp. |
| Slugs | A pathogenic nematode, *Phasmarhabditis hermaphrodita** |
| Vine weevil grubs | A pathogenic nematode, *Steinernema kraussei** |
| Leatherjacket | A pathogenic nematode, *Steinernema feltiae** |
| Chafer grubs | A pathogenic nematode, *Heterorhabditis megidis** |

All may be used in greenhouses, conservatories or on houseplants but those marked with an asterisk (*) can also be used out of doors.

heavy infestations develop, to prevent damage occurring before the biological agents achieve control. All predators and parasites need warm sunny conditions if they are to be active and outbreed the pests. Their period of use is generally April to October, when daytime temperatures are at or above 21°C. If pesticides need to be used before introducing biological controls, short-persistence fatty acids, plant oils, rotenone (derris) or pyrethrum are the preferred choices; most other insecticides kill predators and parasites as well as target insects and some, such as

bifenthrin, remain active for 8 to 10 weeks after their last use.

Pathogenic nematodes are watered into the soil to control slugs, leatherjackets, chafer grubs or vine weevil grubs. The nematodes enter the bodies of the host and release bacteria which cause septicaemia. The slug and vine weevil nematodes can be used when soil temperatures are above 5°C; the slug nematode is best used in spring, when seedlings and soft emergent shoots need protection; the best time for vine weevil control is late summer. Leatherjacket and chafer grub nematodes need higher soil temperatures (14–21°C); the leatherjacket nematode is best applied in late summer; the chafer grub nematode is best applied in early summer, before the grubs are large enough to cause serious root damage. All biological controls are harmless to humans and animals other than the intended pest targets.

## Safe use of pesticides

Before any insecticide, fungicide or herbicide is approved for garden use it undergoes extensive testing and trialling to assess its effectiveness, toxicity to humans and wildlife, persistence and environmental effects. Careful attention to the manufacturers' instructions, especially the dilution rate and types of plants on which they should be used, will enable safe use to be made of these substances. Regulations made under the Food and Environment Protection Act 1985 give legal force to the manufacturers' instructions, making it illegal to use the wrong dilution rate or to use a product for purposes other than those stated. Prosecution could follow, for example, if careless spraying of plants in flower results in a neighbouring beekeeper losing bees. Only products approved by the government may be used as pesticides; home-made pesticides brewed from rhubarb, cigarette butts or even washing-up liquid, once popular with gardeners, are now outside the law. Moreover, gardeners may not purchase, store or use chemicals marketed for commercial growers. Finally, chemicals must be stored in their original containers, so it is illegal to split a batch of pesticide among several gardeners.

# Controlling diseases

Disease can seldom be eliminated and gardeners, like commercial growers, must learn to manage disease to keep levels within acceptable limits. Commercial growers manage disease to maximise profits and, because public tolerance of defects in their fruit and vegetables is very low, aim for very low disease levels. As a result, their reliance on chemicals is often very heavy. Gardeners have different objectives and fewer chemical control agents available to them, but more time at their disposal.

## Quarantine

A gardener's first line of defence against disease is quarantine. Much damage to plants is prevented on a world scale by national and international legislation to prevent the spread of plant disease: in Europe, the European Plant Protection Organisation (EPPO) ensures that national authorities are kept aware of the exotic pathogens that pose a threat, and publishes lists of those that are currently completely absent from a country (A1 organisms), or of limited distribution (A2 organisms). Despite these international efforts, the spread of pests and diseases is inexorable and is increasing all the time with the massive increase in air travel and the abolition of trade restrictions.

Relatively recent arrivals in the UK have included: chrysanthemum white rust *Puccinia horiana* in 1963, which is now established, attempts to eradicate it having been abandoned; the new and virulent pathogen *Ophiostoma novo-ulmi* causing Dutch elm disease in the 1970s; and camellia-petal-blight (*Ciborinia camelliae*) in the 1990s. More recent introductions have been box blight (*Cylindrocladium buxicola*) and ramorum die-back, also known as sudden oak death (*Phytophthora ramorum*), which infects *Viburnum* spp. and *Rhododendron* spp. More recently, plant pathologists have discovered another invasive *Phytophthora* in the UK for the first time, *Phytophthora kernoviae*, which infects common beech (*Fagus sylvatica*). Still absent, but ever-threatening, are sweet chestnut blight (*Cryphonectria parasitica*), which is present in France, and oak wilt (*Ceratocystis fagacearum*) and its beetle vector, which have not yet crossed the Atlantic from North America. *Phytophthora ramorum* appeared almost simultaneously in northern Europe and western California in the 1990s from an unknown origin. It infects many woody hosts in these two outbreak areas. Since it was previously unrecognised as a pathogen, its movement could not have been prevented by international plant quarantine regulations (see Brasier, 2005). Invasive pathogens represent a very real threat to the horticultural industry and private gardens, and also to the native British flora; this latter threat is well exemplified by the widespread loss of mature elm trees in the countryside since the 1970s to Dutch elm disease. They are, therefore, of great concern to conservationists, an issue that will be dealt with more fully in Chapter 18.

Within countries, legislation may exist to limit the spread of pathogens. For example, fire blight (*Erwinia amylovora*) must be absent from commercial sources of susceptible plants in the UK. Some diseases must be compulsorily notified if found, such as wart disease of potatoes (*Synchytrium endobi-*

*oticum*), a serious disease which is well controlled by a combination of legislation and resistant varieties.

## Hygiene

Sometimes diseases may be avoided completely by excluding them. The use of clean seed – and all reputable commercial seed sources are produced to very high standards of purity – will ensure almost complete freedom from such diseases as bacterial blight of beans (*Pseudomonas syringae* pv. *phaseolicola*). Attacking seed-borne diseases by treating seed with chemical control agents is also an extremely economical and efficient use of a chemical, with minimal hazard to the environment. Sometimes it is impractical to eliminate disease completely from seed, but good hygiene can still reduce levels effectively, as in the case of powdery scab on potatoes (*Spongospora subterranea*), which is controlled by ensuring that seed potatoes are infested only at the very low levels which are known not to result in significant disease in the following crop.

Most common garden diseases must survive, usually over winter, when the host is not present and good hygiene, such as the destruction of diseased plants and the careful washing of pots and other containers, is therefore very important. By removing the sources of inoculum and ensuring that there is a clean start in the spring, onset of the disease can be delayed. This may be true even for wind-blown pathogens which will inevitably reinvade eventually from over the neighbour's fence, because delaying the arrival of the pathogen by even a few weeks can make all the difference to the health of the crop. For example, blight on potatoes and tomatoes (*Phytophthora infestans*) can survive over winter in infected tubers or fruit left on the ground; careful removal will ensure a disease-free start. Spores will inevitably blow in later, but rapid application of a protectant fungicide will suppress and delay the infection, and a delay of one or two weeks may allow a crop to be harvested before the disease reaches tubers or fruit. Loss of leaves at this late stage is unimportant, but if the disease develops unchecked the entire crop may be destroyed.

## Resistant plants

Most plants are immune to most diseases and susceptibility to a particular pathogen is the exception. When plants are attacked, wild relatives are usually less affected than garden cultivars; the inherent resistance of the wild ancestors has often been sacrificed, or simply allowed to dissipate unnoticed, during the relentless selection by breeders for yield, flower size, sweetness or flavour. Breeders are nevertheless very aware of the value of resistance to disease, not least because when it is effective it offers disease control to the grower for only the price of the seed. Resistance is conferred by biochemical mechanisms and is under the control of many genes, some of which confer high levels of protection and some less. Good for gardeners, but hard for breeders to manipulate, is multigene resistance, where many genes each contribute a small amount to the overall resistance of the plant. This may not confer a very high overall resistance, but is more stable. Spectacular breakdowns of resistance can occur when breeders place too much emphasis on too few genes, or even only one, the so-called major-gene resistance, for example in the case of rust of wheat (*Puccinia graminis*) and blight of potatoes (*Phytophthora infestans*). In these cases resistance is very good, because the breeders have selected one gene which has a major effect, but the pathogens in turn rapidly develop mutant genes which overcome this host gene, and the host is then susceptible.

Resistant cultivars should always be considered by gardeners and are vital for those who do not wish to use chemical control. However, best-loved varieties of many plants are often disease-susceptible (Fig. 16.5); for example the popular apple 'Cox's Orange Pippin' is very susceptible to apple scab (*Venturia inaequalis*). In some cases good levels of resistance do not exist at all in a species: for example, all cultivars of strawberry are more or less prone to grey mould (*Botrytis cinerea*). Research efforts are now under way to improve resistance by using genetic modification techniques to introduce the appropriate genes into strawberries from other sources.

## Biological control and soil management

The science of biological control is much less advanced for diseases than for insect pests. Some products have come on to the commercial market which rely on antagonistic bacteria and fungi to control disease, but most are for commercial crops and they have yet to make as much impact as the biological control systems for insect pests in glasshouses. Product efficacy and stability have been difficult problems to overcome.

More promising for gardeners are systems of soil management that promote high levels of naturally occurring antagonists of pathogens. Soil contains non-pathogenic bacteria, fungi and protozoa which live on

**Fig. 16.5** Cultivars of *Iris germanica* hybrids in a demonstration at RHS Wisley, showing large differences in resistance to iris rust (*Puccinia iridis*). The one on the left was severely affected and would require a level of fungicide protection that most gardeners would find unacceptable; the other two were hardly affected. *Photograph courtesy of Royal Horticultural Society.*

soil organic matter, which includes dormant pathogen spores. As a general principle, the higher the levels of microbial activity in soil, the more rapid the removal of pathogen spores. Organic manure and compost provide a better food source for such antagonists than inorganic fertilisers, and thus may help to suppress diseases. However, gardeners who place heavy reliance on organic soil amendments should be aware that attempts at disease control using these materials have not always been successful, and successes have not always been repeatable. Composted organic waste that has been heated properly will be free of most disease spores, but poorly prepared compost may simply add large amounts of disease from the previous season. Gardeners should therefore avoid composting diseased plant material. The heating process sterilises compost and microorganisms recolonise it as it cools. This is an uncontrolled process and may explain why the effects of compost on disease suppression are unpredictable. Gardeners should also be aware that products sold for soil improvement may not be subject to such rigorous regulatory requirements as products sold as pesticides or fungicides.

**Chemical control**

The range of chemical fungicides available to gardeners is very limited compared with the total range of products available to commercial growers. Indeed, the number of fungicides available to gardeners is steadily decreasing as public pressure for ever higher, and more costly, standards of safety renders the market increasingly uneconomic. In addition to

a reduction in active ingredients, most products are now subject to precise, statutory labelling to identify the disease targets and must be used only for the purposes stated on the label, as with pesticides. Nonetheless, where fungicides are available they can be very effective. Systemic compounds such as myclobutanil and penconazole penetrate the plant tissues and give protection from within, safe from being washed off the plant by rain. They may also have some curative effects on disease already present. Protectants, such as copper formulations and mancozeb, do not penetrate the tissues and are not curative. They are prone to being washed off in rain, but may still be very effective, particularly in the case of Bordeaux mixture, which actually sticks very well to leaves if allowed to dry before rain arrives. Fungi do not normally become resistant to the metallic protectants, whereas they have frequently adapted their biochemistry to detoxify some of the organic, systemic compounds.

Where chemical control apparently fails, there are three likely reasons. The pathogen may have developed resistance, in which case another chemical should be tried if a suitable one is available. The application may have been made too late, and gardeners should remember that disease often develops invisibly within a plant well before symptoms appear. Also, many garden sprayers are relatively crude, and gardeners need to exert considerable skill in ensuring an even coverage of chemical, particularly underneath the leaves. Most manufacturers advise users to spray to 'run-off', but it is easy to see that this results in less material being retained on the leaf than spraying to just *before* run-off. At the moment of run-off, much of the accumulated fungicide runs off onto the ground; the ideal application is, therefore, a dense, even distribution of discrete droplets over all the parts that need protection.

## How to control weeds

Most research on the control of weeds has been carried out with commercial growers in mind. Although the results must therefore be treated with caution by gardeners, the basic principles that have been reported may be useful in devising controls for garden weeds.

### Weed control in cultivated soil

Cultivating the ground has several uses; it uproots and buries weeds, gets rid of residues, mixes in lime, fertiliser and manure and breaks the soil into structures that favour crops. Weed vegetation is completely destroyed by cultivation, and must therefore regenerate from survival organs, mainly seeds. The drawback is that cultivation buries some weed seeds, sending them into dormant states so that they survive longer, whilst at the same time bringing to the surface other seeds that were previously buried and exposing them to light and other dormancy-breaking factors. Also, where weeds are merely uprooted rather than buried, a proportion may survive long enough to re-root or set seed.

The majority of weed problems are already present in most gardens, concealed in seed or bud banks; imported seeds usually represent only a fraction of those already in the ground. Repeated cultivation ought, in theory, to expose all the weed seeds in the soil to dormancy-breaking factors, leading to a reduction and perhaps elimination of soil weed seeds. Research has shown this to be broadly the case, but the process would take 7 years to reduce the seed bank to less than 1% of the original population. This suggests that digging and other cultivation methods are poor ways to control weeds, as well as being time-consuming and hard work. Shallow cultivation has some value, being easily accomplished and effective; hoeing is the usual method, ideally with a tool such as a Dutch hoe that slices weeds off just below the surface, rather than uprooting them as cultivators and ordinary hoes do. Hoeing works well between rows or widely spaced plants, but less well within rows.

Here weeds can thrive and seed unless laborious hand weeding is undertaken.

No-digging methods of gardening avoid bringing weed seeds to the surface and thereby replenishing the seed bank. Moreover, weed seeds that fall onto the surface are either exposed to seed-eating wildlife or, not being pushed into dormancy, germinate quickly and thus become vulnerable to destruction. Some caution is needed in adopting this approach, however, because under agricultural conditions no-tillage has sometimes resulted in a weed flora that thrives in the absence of ploughing. On a garden scale, annual meadow grass (*Poa annua*) and other weeds that seed when young are hard to spot, resist hoeing and may therefore become prevalent.

Finally, research has shown that the seed bank in untilled ground declines at half the rate it would in tilled ground. It would therefore probably take up to 4 years to erode the seed bank significantly. The conclusion has to be that with a no-digging regime it will take several years for the full weed-control benefits to emerge.

When devising a strategy to control the weeds present in a garden, prevention of seeding is the first priority. Imported weeds are unlikely to add significantly to the weed problem, unless a new weed species is inadvertently introduced. Depletion of the seed and bud bank by optimising rotation, cultivation and cultural techniques are important longer-term aims. To hard-pressed gardeners this might sound unrealistic, but researchers have found that a high standard of cultivation in vegetable cropping resulted in a fall from 40 000 viable seeds per square metre to around 3800 seeds per square metre over 4 years. However, there were complications, for in a subsequent wet year the seed bank temporarily tripled.

Heating weed seeds is an effective way of controlling the seed bank, but is only applicable to the ingredients of sterilised potting media, especially loam. A procedure called *solarisation*, in which soil is covered with a plastic sheet, effectively traps sunlight and in warm conditions may kill weed seeds. Solarisation has been successfully used in warm climates, but in cooler regions such as the UK is unlikely to be useful. Chemical sterilisation of the soil is also possible, but the most effective chemicals are only available to commercial growers.

Heat and chemical sterilisation have important pest and disease control actions, and the expense of using them is probably more easily justified on these grounds than for weed control. Chemical control using selective weedkillers is not an option in cultivated garden soil, since the need to match the correct herbicide with the crop species and appropriate growth stage is not practical on a garden scale. Total weed control by destroying weeds with a contact weedkiller before planting or sowing can be effective in reducing later weed infestations, so long as few seeds are brought to the surface in subsequent hoeing or other cultivation activities. An alternative to a chemical weedkiller is a flamegun.

Birds, ants and other seed eaters may also eliminate useful numbers of weed seeds, especially in no-digging gardens, where weed seeds are left on the surface. Cultural weed-control methods include rotations, where it may be possible to leave ground fallow for a period to eliminate weeds more easily, close spacing to suppress weed growth and seed germination, and the use of smother crops where the thick canopy of leaves keeps weeds down. Rest periods of perennial plants, soft fruit for example, where mulches and herbicides can eliminate annual weed problems, provide another cultural control strategy.

## Weed control in perennial plantings

In fruit plantations and perennial borders, weed control by soil disturbance disrupts the root activities of the crop. Managing the weed population without cultivation therefore improves growth and yields. Selective weedkillers have potential in such situations; for example, those based on dichlobenil are available to eliminate weeds from established woody plantings. It is often possible to use directed sprays of weedkillers (based on glyphosate, paraquat, diquat or glufosinate) to control weeds in perennial plantings, but the weedkiller should not come into contact with young bark. The label recommendations should always be followed assiduously. Herbicides that are translocated from leaves to roots, such as dichlobenil or glyphosate, are especially useful where perennial weeds are abundant, for they kill the entire plant.

Weedkillers have many different modes of action: some interfere with cell division, whereas others, such as glyphosate, prevent synthesis of certain amino acids, and hence proteins. Many gardeners are uneasy about using such agents and prefer alternatives, such as flame weeders, which sear off vegetation. Research suggests that burning the straw mulch from between strawberries will destroy many, but not all weed seeds.

Mulching is ideal for gardens where annual weeds are the main problem. Taking advantage of the inability of weed seeds, especially small ones, to survive if they germinate below the surface, a covering of at least 50 mm of weed-free organic material may prevent annual weeds from growing. Where perennial weeds occur, opaque plastic sheets make effective mulches. Plastic membranes that are permeable to water may be used to suppress weeds in new plantings of shrubs, especially in gravel gardens.

Cover crops, usually mown grass, are often used to suppress weeds in permanent plantations. They are also visually pleasing, help to control erosion, boost soil structure and improve access. Experiments suggest, however, that they may also act as weeds themselves, depriving crops of nutrients and, especially of water. A compromise is to grow strips of cover in the alleys between crops, leaving the soil beneath the crops bare or mulched. In this case there is a reduced loss of yield, but most of the benefits of the cover crop are retained.

## Weed control in special cases

Lawns, almost uniquely in gardens, may be weeded efficiently with selective weedkillers. These eliminate broad-leaved weeds, leaving the grass unharmed. Where weeds are resistant to such weedkillers, selective cultural measures sometimes work well as alternatives. Examples include liming to reduce woodrushes (*Luzula* spp.) and improving drainage to suppress mosses. In extreme cases, as with highly resistant weeds such as mind-your-own-business (*Soleirolia soleirolii*), removal of all vegetation using herbicides or cultivation, followed by re-sowing, may be the only remedy.

Chemical control of water weeds is not very satisfactory; algicides are only partially effective, and environmental considerations limit the use of other chemicals. In such situations cultural controls are often used, such as removing rampant weeds mechanically and endeavouring to suppress unwanted species by shading them with highly desirable ones like water-lily (*Nymphaea* spp.). Special pumps containing ultraviolet light and/or biocides are becoming increasingly available for controlling aquatic weeds in garden water features.

Unplanted paved areas are readily kept clear of weeds by killing established species with a translocated herbicide and preventing re-establishment by using a residual herbicide. Some residual herbicides travel in the soil, reaching the roots of adjacent plants, although most modern products are only slightly soluble and are therefore safe in this respect. An alternative to the use of weedkillers is to lay paving on weedproof membranes, while flame weeders offer an alternative means of killing annual weeds.

## Biological control of weeds

The destructive capacity of some plant pests and diseases may be turned to advantage to control weeds biologically. Many plants which are harmless in their native homes have become invasive weeds when grown in other countries. This is primarily because, at home, such plants have many natural enemies, particularly insects and fungi that reduce their vigour. When the plants arrive at a new location, often as seed, they escape these natural enemies and, released from the constraints the enemies impose, grow more vigorously and become weeds. A well-established principle of what has come to be called 'classical' biological control is to collect natural enemies in the weed's centre of origin and then release them in the weed's new habitat. Early attempts concentrated on plant-feeding insects, perhaps the most spectacular success being the control of prickly pear cacti (*Opuntia* spp.) in Australia by a combination of host-specific insects, the most important being the scale insect *Dactylopius ceylonicus* and caterpillars of the moth *Cactoblastis cactorum*. The latter was responsible for the effective control of the most invasive species, *Opuntia bentonii*, in over 20 million ha in the early part of the twentieth century.

More recently pathogenic fungi have been used as biocontrol agents. Rusts are particularly favoured because of their high degree of host specificity. In modern biological control programmes, the specificity of potential agents is determined by a rigorous programme of centrifugal testing. The plant species taxonomically closest to the target weed are tested first, then progressively more distant relatives until no infection is detected. Important crop plants may be included even if taxonomically very distant, and they may also be tested under different environmental regimes. Rusts may be so specific that more than one rust genotype will be required. For example, the first introductions from Europe into Australia of the rust *Puccinia chondrillina* to control the weed *Chondrilla juncea* were so specific that they did not infect all the biotypes of the weed. More recently there has been spectacular success

**Fig. 16.6** Japanese knotweed (*Fallopia japonica*) invading a garden boundary. *Photograph courtesy of R. Shaw (CABI Bioscience).*

with another rust, *Maravalia cryptostegiae*, to control the weed *Cryptostegia grandiflora* in northern Australia. This attractive plant was introduced from Madagascar into Australia during the gold rush to cover mining spoil heaps. It became very invasive, over-running riverine vegetation and blocking access to rivers for cattle to drink. Collections of the rust from Madagascar were screened under quarantine in the UK for host specificity before release in Queensland in 1995. By 2004 the intensive defoliation caused by the rust had killed up to 75% of the weed plants at some sites. Such control is permanent, and compared to alternative mechanical or chemical control methods, relatively cheap and a one-off cost. It is the only economically viable method for weeds of low-value cropping systems such as rangelands.

In the UK most weeds are native species and biological control has had very limited application so far. There have been extensive studies on the possibility of controlling bracken (*Pteridium aquilinum*) using leaf-feeding insects collected from southern African populations of this fern, but no releases have yet been made. More recently, there has been increasing interest in the biological control of two other introduced plants which have become weeds, giant hogweed (*Heracleum mantegazzianum*) and Japanese knotweed (*Fallopia japonica*; Fig. 16.6; see www. cabi.org./BIOSCIENCE/japanese_knotweed_allianc e.htm for more information on current proposals). It is possible that in the future more biological control agents effective against weeds in UK gardens will be developed, for research on this control strategy is still in its infancy.

## CONCLUSION

The same scientific principles underlie pest, disease and weed control for professional growers and gardeners alike. However, gardeners have different needs and opportunities, and this is reflected in their approaches to managing their problems. They work on a smaller scale, sometimes have more time, and are often driven by enthusiasm rather than economic necessity. Also, they have fewer chemical control agents available to them, less-effective application methods, and in many cases no enthusiasm to use them. All gardeners should make the best use they can of the scientific knowledge that is available, if for no other reasons than that a simple dependence on control chemicals is often not their best option, and that the integration of all available methods requires a subtle and complete knowledge of the plants and the multitude of organisms that interact with them.

## FURTHER READING FOR CHAPTERS 15 AND 16

Agrios, G.N. (2005) *Plant Pathology*, 5th edn. Academic Press, San Diego, CA.

Aldrich R.J. & Kremer, R.J. (1997) *Principles in Weed Management*, 2nd edn. Iowa State University Press, Ames, IO.

Alford, D.V. (1999) *A Textbook of Agricultural Entomology*. Blackwell Science, Oxford.

Alford, D.V. (2003) *A Colour Atlas of Pests of Ornamental Trees, Shrubs and Flowers*. Timber Press, Portland, OR.

Alford, D.V. (2006) *Fruit Pests*. Manson Publishing, London.

Brasier, C.M. (2005) The threat to the UK environment and horticultural heritage from invasive plant diseases. *The Plantsman* **4**, 54–7.

Buczacki, S. (2000) *Plant Problems*. David and Charles, Newton Abbot.

Buczacki, S. & Harris, K.M. (2005) *Collins Photoguide: Pests, Diseases and Disorders of Garden Plants*. HarperCollins, London.

Gratwick, M. (ed.) (1992) *Crop Pests in the UK*. Chapman and Hall, London.

Greenwood, P. & Halstead, A. (2007) *The RHS Pests and Diseases*. Dorling Kindersley, London.

Holm, L., Doll, J., Holm, E., Pancho, J. & Herberger, J. (1997) *Weeds: Natural History and Distribution*. John Wiley & Sons, New York, NY.

Huxley, A. (ed.) (1997) Poisonous plants. In *The New Royal Horticultural Society Dictionary of Gardening*, 4 volumes, pp. 669–70. Macmillan Reference, London.

Ingram, D.S. & Robertson, N.F. (1999) *Plant Disease*. The New Naturalist. HarperCollins, London.

McGiffen, M.E. (1997) Weed management in horticultural crops. American Society for Horticultural Science & Weed Science Society of America Joint Workshop, 6–7 February 1997. American Society for Horticultural Science, Alexandria, VA.

Salisbury, E. (1961) *Weeds and Aliens*. Collins, London.

Stephens, R.J. (1982) *Theory and Practice of Weed Control*. Macmillan, London.

Thompson, P. (1997) *The Self Sustaining Garden*. B.T. Batsford, London.

Williams, J.B. & Morrison, J.R. (1987) *ADAS Colour Atlas of Weed Seedlings*. Wolfe Publishing, London.

# 17

# Maturation, Ripening and Storage

## SUMMARY

In this chapter the scientific rationale for and the application of procedures for the commercial storage of vegetables, fruits, flowers and seeds are described, and future trends outlined. The ways in which aspects of these procedures may be adapted for use by gardeners for the non-commercial storage of produce are given, together with descriptions of more traditional storage methods.

## INTRODUCTION

Gardeners generally harvest their produce for immediate consumption when it has reached the right stage of maturity or ripeness. By choosing varieties that mature at different times, or by delayed planting or repeat sowing, the seasonal nature of production from the garden can be maintained. There are, however, circumstances when harvested produce needs to be stored. If seeds are to be saved for sowing the next year they must be kept in a viable condition over the winter. Late-ripening cultivars of apples can be maintained in good condition for some time if stored in the right way. Even cut flowers may be kept for longer than normal if treated appropriately.

In commercial horticulture, post-harvest handling and storage of produce is required to provide consumers with fresh products irrespective of the distance from the growing area, and time of year. The need to develop better techniques for maintaining product quality and extending storage life has led to continuing advances in understanding the changes that take place after harvest.

This chapter is written to provide the gardener with some understanding of the physiological changes that occur in storage structures, seeds, fruit and flowers during maturation and following harvest, especially those that affect their longevity in store. Although gardeners do not normally have access to commercial storage facilities, some knowledge of the theoretical basis for the most important aspects of commercial practice contributes to our understanding of the processes that occur during ripening and after harvest, and is therefore included in this chapter.

## CHANGES IN METABOLISM AND ETHYLENE SYNTHESIS

### Respiration and water loss

When a plant tissue has reached maturity, its respiration rate will either remain roughly constant or decrease slowly with age. An exception to this behaviour is the marked rise in respiration, known as the climacteric, that accompanies the onset of ripening in many fruits and senescence in flowers. Both ripening and the climacteric are triggered by the endogenous production of ethylene, and both processes may be stimulated to occur prematurely by the application of this gaseous hormone (see below).

Although synthetic (anabolic) processes occur during maturation, the balance of metabolism is destructive, or catabolic, and ultimately produce becomes senescent. The *respiration rate* of a product is an indicator of the rate of metabolism, and in general higher respiratory activity is associated with a shorter storage life. Moreover, fruit and vegetables with a higher content of stored materials that can be utilised in respiration will, on the whole, remain in good condition for longer than those with only small amounts.

Once harvested, fruits and vegetables can no longer take up water and gradually become dehydrated and shrivel. Thus, preventing water loss as much as possible can contribute significantly to their post-harvest life. Similarly, although cut flowers may still absorb water through their stems, its uptake can be severely reduced if the water-conducting vessels become blocked. Seeds are different in that they lose water during ripening, and here storage life depends on keeping them in a dry, dormant condition (see Chapter 9).

The objective of commercial storage is to delay senescence, preserving the desired characteristics of fresh horticultural produce, and providing the market with a continuous supply of products with high nutritional value and acceptable sensory quality. To do this, technologies are used to reduce respiration and conserve moisture within the product. Many of these techniques are not available to gardeners, but an understanding of the underlying physiology, especially the importance of preventing water loss, the role of ethylene and the need to reduce the rate of respiration, can be used by them to increase the life of harvested produce.

## Ethylene

During the nineteenth century, when coal gas was used for street lighting, it was observed that trees near the lamps shed their leaves earlier than other trees. Later, ethylene was identified as the active compound. Natural gas does not contain ethylene, so gas-induced leaf loss would not be observed today. Ethylene ($CH_2{=}CH_2$) is a colourless gas that has particular significance in the ripening of climacteric fruits, and in the senescence of leaves and flowers. The dramatic increase in ethylene concentration in climacteric fruits such as apple (*Malus domestica*),

pear (*Pyrus communis*), plum (*Prunus domestica*) and tomato (*Lycopersicon esculentum*) is required to trigger and synchronise the ripening processes, and does not occur until they approach full development. From an evolutionary viewpoint, this could safeguard against premature ripening at a stage when the seeds are immature and the attractiveness of the fruit for consumption by animals to facilitate seed dispersal is less than optimal. When unripe climacteric fruits are treated with ethylene the onset of ripening is hastened. In contrast, fruits such as citrus and grapes (*Vitis vinifera*) do not exhibit a dramatic rise in ethylene production and are called non-climacteric. In these cases the application of ethylene does not trigger ripening.

Clearly, ethylene has important commercial implications in the storage and marketing of fruit, vegetables and cut flowers. The ability of ethylene to initiate ripening is exploited in the marketing of several types of fruit such as avocado (*Persea americana*), banana (*Musa* spp.) and tomato. The commercial ripening of bananas is a routine operation carried out in importing countries to provide fruit to consumers at a specified colour stage. Harvesting bananas in an unripe (green) condition and maintaining cool temperatures (13–14°C) during transport facilitates their supply to distant markets; subsequent ripening with ethylene ensures uniformity in ripening and product quality. There are some instances where ethylene treatment is useful for non-climacteric fruit, for example in the de-greening of citrus fruits where certain cultivars become edible before the green colour of the peel has disappeared. In such cases ethylene gas is used to encourage the degradation of the chlorophyll in the peel.

The gardener may also make use of the knowledge that the ripening of climacteric fruits harvested in an immature condition is hastened by treatment with ethylene. For example, it may be useful to stimulate ripening of immature tomatoes at the end of the growing season by enclosing the green fruits with red, ripening fruits or other ripe climacteric fruits such as apples or bananas, all of which will be emitting ethylene gas as part of their own ripening process. The greatest stimulation of ripening is achieved in warm temperatures.

Although ethylene gas may be used to improve the quality of certain types of fruit, its presence in the atmosphere may also have detrimental effects. The build-up of ethylene generally reduces the quality of

**Fig. 17.1** The effect of ethylene gas on Asiatic lilies has caused them to lose all of their flowers (right). This bud drop has been effectively prevented by treatment with the anti-ethylene agent silver thiosulphate (left). *Photograph courtesy of Sarah Snow, School of Biological Sciences, University of Reading.*

vegetables, although there is a range in sensitivities to the hormone. Vegetables that display high sensitivity include broccoli (*Brassica oleracea*), Brussels sprouts (*Brassica oleracea*), Chinese cabbage (*Brassica rapa*), cauliflower (*Brassica oleracea*), cabbage (*Brassica oleracea*), cucumber (*Cucumis sativus*), endive (*Cichorium endivia*), sweet corn (*Zea mays*), lettuce (*Lactuca sativa*) and spinach (*Spinacea oleracea*). Adverse effects include loss of visual quality due to accelerated de-greening, senescence and impaired eating quality, such as toughness in asparagus (*Asparagus officinalis*) or bitterness in carrots (*Daucus carota*).

Vegetables may be exposed to high concentrations of ethylene during distribution and retailing, chiefly because of the presence of ripening fruit. Care is therefore taken to avoid unnecessary mixing of vegetables with fruit. Similarly, gardeners are advised not to store pome and stone fruits, which produce large amounts of ethylene, in the same refrigerator as ethylene-sensitive vegetables. However, the cool temperatures in a domestic refrigerator do reduce ethylene production and slow the overall rates of deterioration of sensitive vegetables.

Another effect of exposure to ethylene is the accelerated senescence and fading of cut flowers. An example of this that older gardeners may remember is the 'sleepiness', or failure of flowers to open fully, that used to occur in carnations (*Dianthus caryophyllus*) that had inadvertently been exposed to ethylene from coal gas. Nowadays one of the biggest risks for many species is pollution by ethylene in the

emissions from motor vehicles during transport. Gardeners should ensure that ethylene-sensitive flowers are not stored with climacteric fruits such as apples and pears, since the presence of ethylene in greenhouses, storage rooms or anywhere in the distribution chain can have serious detrimental effects on the keeping quality of flowers. These include the downward bending of leaves (*epinasty*), withering, ageing, yellowing and the abscission of flower parts and leaves (Fig. 17.1).

Various types of technology exist to reduce ethylene concentrations around stored fruit and vegetables: these include its removal by ventilation with fresh air, oxidation using chemical agents such as potassium permanganate and ozone or physical/chemical methods such as with heated catalyst systems. Although important benefits of ethylene removal have been established for apples, it is not at present commercially practical in the UK. Gardeners who store apples should ensure that the fruit is as well ventilated as possible.

Both ethylene production and ethylene action may be blocked by specific inhibitors. Silver ions applied as silver thiosulphate are potent inhibitors of ethylene action, an effect that cannot be achieved by using any other metal ion. In cut flowers, silver ions delay senescence and may be given as a post-harvest pulse treatment of silver thiosulphate in the nursery prior to dry packing and transport (Fig. 17.1). Pulse treatments with silver thiosulphate were mandatory for many cut-flower species sold in auctions in the Netherlands, although recently there have been environmental concerns about the use of heavy metals such as silver. There is now increasing interest in more environmentally friendly alternatives such as 1-methylcyclopropene (1-MCP), a gaseous compound that inhibits a range of plant responses to ethylene, including senescence of carnation flowers.

Although less efficient than silver, carbon dioxide at high concentrations, in the range of 5–10%, inhibits many responses to ethylene, such as the induction of fruit ripening. This is discussed below in relation to the controlled-atmosphere storage of fruits.

## MATURATION AND RIPENING

### Seeds

The embryos of all seeds develop to a particular stage that is characteristic of the species. When the embryo reaches full size, the deposition of food reserves is completed (see Chapter 9), physiological maturity is attained and ripening or dehydration begins. In hot summers the moisture content of the seed may be reduced to 10%, although a water content of 15% is usual in an average summer. Once this stage of desiccation is reached, the seed is usually shed from the parent plant. By this time it is physiologically inert and will remain in a dormant state (see Chapter 9) until the conditions are right for germination.

### Fruits

The botanical definition of a fruit is a 'seed receptacle developed from an ovary' (see also Chapters 2 and 9). A consumer definition is more likely to concentrate on sensory qualities, the aromatic flavours and sweetness that distinguish a fruit from a vegetable. In nature, the aromas and palatable nature of most fruits are important to ensure that seeds are consumed and dispersed by animals. The tissues that form various fruits are derived from different parts of the flower. Some examples from temperate fruits that are common in gardens include the receptacle (strawberries, *Fragaria × ananassa*), accessory tissues (apples) and the placental-tissue septum (tomato). Ripening is a term reserved for fruits and relates to changes in colour, texture, flavour and aroma. The state of

ripeness at the point of harvest has particular implications for producers and consumers alike. Generally, fruits that are allowed to ripen on the plant achieve a higher sensory quality, but are more susceptible to damage during post-harvest handling and may have a reduced storage life. The biochemical changes associated with fruit ripening and the controlling mechanisms continue to attract much research attention. A better understanding of the ripening processes may eventually provide strategies for the more effective control of ripening.

Although fruit is important in a healthy diet, most people also eat fruit for enjoyment. The perception of flavour is an important part of sensory quality (see Chapter 12), and it is vital that procedures to extend the period of availability of fruits do not result in an unacceptable loss of taste and flavour. The taste and flavour of fruits is derived from a complex mix of sugars, acids, phenolic compounds and a wide range of volatile chemicals. The sugars and acids derive from photosynthesis, and in some fruits such as apples and pears the sugars are converted to starch during fruit development. Fruits that accumulate carbohydrates as starch are often picked at an immature stage and only achieve an acceptable taste and flavour during subsequent ripening when the starch is broken down again to form the sugars glucose, fructose or sucrose. Fruits continue to accumulate sugars from the plant during maturation and ripening and for some, such as strawberry, this contributes significantly to flavour at harvest. Strawberries that are harvested too early will never develop sufficient sweetness and flavour for commercial acceptability.

Malic and citric acids are the main organic acids of fruits, and these are significant components of taste. It is therefore important to achieve the correct balance between the acid and sugar concentrations in the fruit. Excessive acidity is often associated with harvesting fruits in an immature or unripe state. The concentrations of acids decline during ripening and are complemented by corresponding increases in sugar concentration. Phenolic substances such as tannins are responsible for astringency and generally result in an adverse reaction from consumers. Although most fruits lose their astringency during ripening, this remains a major quality problem in fruits such as persimmon (*Diospyros kaki*) because of the presence of water-soluble tannins.

Specialised volatile flavour compounds are produced in ripening fruit and these provide the unique sensory character associated with the different types of fruit. In apples and pears, the major volatiles are aliphatic esters, although terpenoids and aldehydes may also contribute to flavour. In strawberries esters, alcohols, carbonyls and sulphur-containing compounds are important. In stone fruit, mostly of the genus *Prunus*, lactones have a role in flavour development, particularly in peaches (*Prunus persica*) and nectarines (*P. persica* var. *nectarina*).

The visual, structural and chemical changes that occur during ripening in soft fruits such as strawberry and raspberry (*Rubus idaeus*), stone fruits such as cherry (*Prunus avium*) and plum, and pome fruits such as apple and pear are associated with an orchestrated sequence of biochemical changes that are under genetic control. The extent to which these underlying changes progress determines the visual and eating quality of the product at the point of sale and consumption.

## Vegetables

Vegetables are derived from virtually all types of plant tissue, and include underground organs such as swollen roots (e.g. carrots) and swollen stem tubers (e.g. potatoes, *Solanum tuberosum*), and above ground parts such as leaf blades (e.g. spinach), axillary buds (e.g. Brussels sprouts), swollen inflorescences (e.g. broccoli) and seeds (e.g. sweet corn). From the consumer's point of view, vegetables also include some fleshy fruits such as tomatoes and cucumbers, and immature fruits such as peas (*Pisum sativum*) and bean pulses (*Phaseolus* spp.). The part of the plant that the vegetable develops from has important implications for its behaviour during storage. Young tissues that are respiring and transpiring actively, such as the leaves of lettuce and spinach, are likely to lose quality particularly rapidly, and in commercial practice immediate steps are taken to reduce their rate of metabolism and transpiration by reducing the temperature.

The stage of development is also critical in the maintenance of quality, particularly in biennial vegetables where some kind of storage organ is formed in the first season, followed by flowering and seed formation in the second year. Although biennial vegetables are extremely varied morphologically, they all cease growth in the autumn. The rate of metabolism of their storage organs declines

naturally, which makes them well adapted for long-term storage.

## Cut flowers

There is a large variation in the structure of flowers grown commercially for cutting. Examples include corymbs (spray carnation), umbels (belladonna lily, *Hippeastrum* spp.) and spadix plus spathe (anthurium, *Anthurium andraeanum*). The functional life span of the flowers of different species varies from a few hours to several months, although for most commercial species the approximate storage period is a few days to a few weeks. Generally inflorescences have a low carbohydrate reserve and in this regard are similar to many leafy vegetables. They also have a large surface area in relation to their mass and this makes them very vulnerable to water stress. Flowers that have lost 10–15% of their original fresh weight are normally wilted.

Ethylene has a major role in the senescence of some cut flowers such as carnation, but some other species are insensitive to this hormone. These include *Chrysanthemum* cultivars, aster (Michaelmas daisy, *Aster novi-belgii*), sunflower (*Helianthus annuus*), nerine (*Nerine bowdenii*), liatris (*Liatris spicata*), zinnia (*Zinnia elegans*) and rudbeckia (*Rudbeckia hirta*).

Senescence in cut carnation flowers is characterised by a rise in respiration rate and ethylene production, and in this respect the carnation behaves much like a climacteric fruit. In carnation and some other species of flowers, pollination stimulates ethylene production and causes senescence of the petals. The functional significance of pollination-induced senescence may be to save energy and sugar reserves that would otherwise be used in maintaining elaborate flower structures, and to prevent further visits by pollinators such as bees.

The physiological processes that cause the deterioration of cut flowers have many similarities with those that operate in fruits and vegetables. This is not surprising, for some vegetables are derived from flower buds (e.g. artichokes, *Cynara scolymus*) or swollen inflorescences (e.g. broccoli). Decorative foliage plants used by florists (florist greens) are likely to behave in a similar way to leafy vegetables.

There are many causes of deterioration in cut flowers. Depletion of food reserves reduces the amount of energy available to maintain cell structure and function and, as with all fresh horticultural crops, there is a risk of attack by bacteria and fungi. Excessive loss of moisture by transpiration causes wilting of both flowers and foliage. Bruising and crushing during handling may stimulate respiration and hasten the senescence process. Exposure to warm temperatures is likely to increase respiration and accelerate senescence, while in some flowers low temperatures may result in chilling injury (see below). Colour changes or fading may affect flower quality and acceptability, and in sensitive varieties the accumulation of ethylene in the storage environment may have serious consequences by accelerating the ageing process (Fig. 17.1).

## PRE-HARVEST INFLUENCES ON STORAGE QUALITY

### Diseases and disorders

Fresh fruits and vegetables are highly perishable and spoilage during storage may occur as a result of attack by fungi, bacteria and viruses or by the development of functional or physiological disorders (see Chapters 15 and 16). Soft rots caused by bacteria such as *Erwinia* and *Pseudomonas* spp. may affect practically all vegetables and especially carrot, celery (*Apium graveolens* var. *dulce*) and potato, but fungi cause most of the spoilage that occurs in fruit and vegetables during storage. Genera such as *Penicillium*, *Botrytis* and *Sclerotinia* commonly affect most fruit and vegetables, and are likely to affect produce stored by the gardener. Some of the physiological disorders of apple such as superficial scald, a skin-browning disorder, and bitter pit (Fig. 17.2b), an internal calcium-deficiency disorder characterised by brown 'corky' lesions, are also likely to occur.

Stems and leaves of cut flowers and florist greens are usually contaminated with bacteria and quickly contaminate clean vase water. Bacteria isolated from cut-flower containers include species of *Achromobacter*, *Bacillus*, *Micrococcus* and *Pseudomonas*. There is little information on the loss of quality or level of wastage in cut flowers that can be attributed to pathogens, but good hygiene combined with the use of biocides is usually sufficient to prevent the significant decay of most cut flowers.

## Climatic factors

The potential of fruits and vegetables to develop particular diseases or disorders is influenced by climatic and field or orchard factors. The extent to which disease can develop in stored crops is determined by the number of viable pathogenic microorganisms present and their ability to infect the plants in question. Since the propagules of many of the important fungal diseases of apple are dispersed by rain, it is not surprising that wet growing seasons are associated with a higher incidence of storage rots in this crop. High rainfall in August and September is associated with higher rotting in stored 'Cox's Orange Pippin' apples due to *Gloeosporium* spp. Infection of apples by *Phytophthora syringae* occurs during periods of heavy rain just prior to or during harvest, spores being spread from the soil to the fruit by either direct contact or rain-splash.

Weather conditions during their development and harvesting also affect the susceptibility of vegetable crops to disease after harvest. For example, *Phytophthora* rot of winter white cabbage (*Phytophthora porri*) is most likely to develop when cabbages are harvested in very wet field conditions. Similarly, harvesting root crops in wet weather may result in considerable amounts of soil and disease inoculum being taken into store with the produce.

Climatic conditions during the development of the fruit affect the susceptibility of stored apples to physiological disorders. Cool wet summers are known to reduce their tolerance of low storage temperatures, and in some years it may be necessary to raise the storage temperature slightly to avoid a deterioration of the fruit flesh, referred to as low-temperature breakdown. Dull sunless summers may increase the risk of core flush, which as the name implies is a physiological disorder that results in a pink discoloration in the core of the fruit. Warm dry summers increase the risk of superficial scald, bitter pit and water core. In such years it is particularly important to apply chemical antioxidants to the harvested fruit to prevent scald development later in the storage period. To reduce the susceptibility of apples to bitter pit and water core, calcium sprays, usually in the form of calcium chloride, should be applied in the orchard to supplement the natural uptake of calcium into the developing fruit. Mathematical models based on temperature and rainfall measurements made in

**Fig. 17.2** (a) Blossom-end rot of tomatoes (*Lycopersicon esculentum*) and (b) bitter pit of apples (*Malus* spp.). *Photographs courtesy of Royal Horticultural Society.*

the orchard during fruit development have now been developed to predict the risk of physiological disorders in different consignments of 'Cox's Orange Pippin' apples.

In many cases, control of post-harvest diseases is only possible by eradicating the causal organisms, or protecting the crop in the field or orchard before harvest. Examples of this include such fungal dis-eases as grey mould of strawberry (*Botrytis cinerea*), so-called bull's-eye rot of apple (*Gloeosporium* spp.) and neck rot of dry bulb onions (*Botrytis allii*).

As most cut flowers are produced as protected crops under glass, it may be less relevant to consider pre-harvest climatic influences because these are largely under the control of the grower. If flowers are exposed to high light conditions before cutting, sugars accumulate and storage life is improved. Any subsequent exposure to light may also be beneficial, since cut flowers retain their capacity to photosynthesise and produce carbohydrates.

Seed vigour (see Chapter 9) and seed longevity are closely related, and other things being equal high-vigour seeds will retain their viability for longer than low-vigour seeds. Weather conditions during ripening may sometimes adversely affect seed vigour and storage potential. Excessive rainfall may lead to an increase in fungal infection, whereas cycles of wet and dry periods may damage the seeds as they swell and contract under the changing moisture conditions.

## Field factors

A great deal of research has been carried out on the effects of orchard factors on the storage quality of apples and pears. There have been numerous investigations of the effects of tree factors such as rootstock, age, cropping level and pruning, and of soil and nutritional factors and the use of orchard sprays such as fungicides, insecticides and growth regulators. For some cultivars of apples nutritional factors have a major influence on the keeping quality of the fruit. The importance of achieving adequate calcium nutrition in fruit and vegetables is universally recognised. Calcium deficiency impairs membrane permeability in plant cells, and membrane leakage is followed by a major disintegration of membrane structure and reduction in growth of meristematic tissues. An under-supply of calcium results in, for example, bitter pit of apple, blossom-end rot of tomato (Fig. 17.2a) and blackheart of celery. The most effective remedy is the repeated application of orchard sprays containing calcium chloride or calcium nitrate. The importance of nutrition during the development of apples is so great in relation to the storage quality that mineral analysis standards are available for growers, to help them select the most suitable storage and marketing strategy for any particular consignment of fruit.

Calcium deficiency is a major cause of losses in vegetable crops. In the USA it is estimated that in the period 1951–60 annual losses caused by calcium-dependent disorders such as blossom-end rot in tomato, blackheart in celery and tipburn in lettuce amounted to US$4.5 million. Although these disorders develop before harvest, internal disorders may

not be apparent at that stage and other problems may develop during storage, such as tipburn in white winter cabbage.

There is relatively little information on how cultural practices affect the post-harvest performance of cut flowers, although excessive applications of fertiliser, especially nitrogen, are thought to encourage more rapid deterioration of flowers.

For gardeners who rely on self-saved seed for the next crop, the best way to produce high-vigour seeds that store well is to ensure that the mother plant has adequate nutrition, so that the seeds mature fully on the parent plant.

## Varietal factors

The strong genetic basis for the storage potential of different apple cultivars is well recognised, and despite the application of modern storage techniques maximum storage varies from a few weeks for cultivars such as 'Discovery' to up to 10 months for cultivars such as 'Bramley's Seedling'. There are also marked differences in the storage potential of pear cultivars and these are taken into account in storage recommendations provided to the UK fruit industry. Regardless of variety, however, the storage of pears is difficult for gardeners, as the fruits bruise easily and are hard to ripen to a satisfactory quality.

## HARVESTING, HANDLING AND PREPARATION FOR STORAGE

### Time of harvest

The physiological maturity of a vegetable does not always coincide with its commercial maturity. With the exception of biennial types, most vegetables are picked at a stage of maturity determined by the demands of consumers or processors rather than at a stage that might provide a more sustained storage or shelf life. Thus there is often a distinction between *physiological maturity* and *commercial maturity*. Physiological maturity refers to progress through a series of developmental stages that typically include the growth, maturation, ripening and senescence of an organ or organism. Commercial maturity reflects the stage at which the market requires the plant organ. For example artichoke, broccoli and cauliflower are harvested when the inflorescence is sufficiently well developed. Cucumbers, green beans (*Phaseolus vulgaris*), okra (*Abelmoschus esculentus*) and sweet corn are marketed as partially developed fruits. The time of harvesting of biennial types is determined by their progress to the dormant state and harvesting is undertaken when growth ceases in the autumn. At this time the diminished rate of metabolism (respiration) makes them suited for long-term storage.

There may be other factors that determine the time of harvest, including the development of pest and disease problems in vegetables left in the ground and the likelihood of adverse weather conditions that may affect the ability to harvest. Time of harvest may be critical to prevent loss of quality in the harvested product. For example, late harvesting of onions (*Allium cepa*) leads to a greater proportion of split and shed skins and to the staining of bulbs left in the field. In commercial horticulture onions are therefore harvested in the green. This is made possible, however, by facilities for fast and efficient drying, and unfortunately such facilities are not usually available to gardeners.

Most of the ripening changes in non-climacteric fruits such as strawberries and cherries take place while the fruits are still attached to the plant, so these are normally harvested when they are fully ripe, at optimum consumer quality. Although climacteric fruits such as plums, apples and pears will also ripen on the tree, they are commonly harvested in an under-ripe condition but at a stage where optimum consumer quality will be achieved after storage. Avocados are an exception and only ripen when detached from the plant.

For the long-term storage of apples, the fruit must be picked before the onset of the climacteric rise in respiration rate. This applies to storage by both gardeners and commercial growers. Commercial judgements of the correct time to pick apples for long-term storage are based primarily on starch content, firmness and sugar concentration. Where there is no requirement for storage, apples are left on the tree to ripen, leading to superior eating quality.

Although the main objective with cut flowers is the longest possible vase life, the flowers must reach their full floral development and be fully open. Some types such as carnations and daffodils (*Narcissus pseudonarcissus*) may be picked at the bud stage, but others such as orchids (Orchidaceae) should be fully developed before they are cut.

Seeds collected too early may not be viable, whereas mature seeds left too long on the plant may lose quality. Seeds for saving should therefore be collected as soon as they are fully developed. There are, however, a few exceptions: immature seeds of Ranunculaceae produce seedlings more readily than mature seeds, whereas some capsules, such as those of hellebores (*Helleborus* spp.), release their seed before they are completely dry, so must be collected early.

## Handling

Fruits and vegetables are susceptible to decay caused by a wide range of bacteria and fungi. Consequently great care is needed in both harvesting and post-harvest handling to avoid damage that might allow infection by pathogens. Commercial fruit crops destined for the fresh market are picked by hand to avoid damage but those for processing, such as black-currants (*Ribes nigrum*), may be harvested mechanically. Harvesting field vegetables by hand may not be feasible, particularly with root crops, and particular care is required with mechanical harvesting as damage may result in direct loss of visual quality, internal damage such as bruising in potato, and access to pathogens.

Cut flowers are highly perishable and should be handled promptly and carefully. Any physical impacts will damage and bruise the blooms, thereby affecting their visual quality and reducing vase life by stimulating ethylene production. Hygiene is particularly important: dead or decaying plant material remaining in containers used for fresh cut flowers is likely to be a source of ethylene and of microorganisms that may cause decay. Some of these, particularly bacteria, are implicated as a cause of stem blockage in freshly harvested flowers, reducing water uptake and hastening the onset of wilting.

Seeds need to be handled carefully as they do not store well if damaged. After collection they should be dried quickly at a relatively cool temperature before being transferred to storage conditions.

## Post-harvest treatments

Various treatments may be applied to harvested horticultural crops before they go into storage; all are aimed at maintaining quality or minimising decay. Some crops such as dry-bulb onions require fast and efficient drying until the necks of the bulbs are tight and dry. This may be achieved by blowing heated air through the store until the onions have lost about 3–5% of their original weight. In countries where fine weather after harvest is normally experienced, adequate curing of onions may be achieved in the field over a period of 2–4 weeks. In subsequent storage a comparatively low relative humidity (65–70%) is required to prevent re-rooting and growth of shoots, and to minimise the development of neck rot.

Potato tubers require 'curing' after harvesting, to stimulate the production of suberised corky periderm tissue and thereby reduce moisture loss during storage, and to increase resistance to infection by *Fusarium* spp. and other rot-forming organisms. The periderm, comprised of corky cells, that forms around wounds inflicted during harvesting effectively heals cuts and bruises and thereby forms a physical barrier to infection. Potatoes are cured at temperatures of 10–15.5°C and 95% relative humidity for up to 2 weeks. Washing may be important to remove surface deposits of debris, dirt and sap from certain crops such as carrots, but in others contact with water may increase the spread of disease. Where washing is required, clean water is essential and disinfectant may be added to the rinse water to kill bacteria and fungal spores.

Other post-harvest treatments include waxing, to enhance appearance and reduce water loss in commodities such as apples, citrus and sweet potatoes (*Ipomoea batatas*), and the application of chemicals to suppress sprouting. The sprout inhibitor tecnazene is normally applied to potatoes after loading into store, while maleic hydrazide is applied to onions as a pre-harvest spray when 50–80% of the foliage is down.

Insect disinfestations using compounds such as methyl bromide are important in the international trade of a wide range of fruit and vegetable crops. Other post-harvest chemical treatments include the use of specific fungicides, antioxidants and calcium. Fungicides such as carbendazim are applied to apples and pears as a pre-storage drench when a significant risk of rotting in store is predicted for particular consignments.

Chemical antioxidants such as diphenylamine and ethoxyquin are applied to cultivars of apples and pears that are susceptible to the form of skin browning known as superficial scald. This physiological disorder affects many apple cultivars grown in hot dry conditions but in the UK is potentially a serious problem only on the culinary cultivar 'Bramley's Seedling'. It occurs following the death of cells in the epidermal and hypodermal layers of the fruit as a result of the oxidation of α-farnesene, which occurs naturally in the cuticle, into harmful conjugated triene compounds. Antioxidants prevent this occurring.

Post-harvest treatment with calcium is restricted mainly to apples, and in particular to cultivars such as 'Cox's Orange Pippin' and 'Egremont Russet', which are susceptible to a range of physiological disorders caused by insufficient accumulation of calcium in the fruit during development, notably bitter pit (Fig. 17.2b). Although affected fruits are edible, the pitted areas are bitter in taste and severely affected fruits may be unpleasant to eat.

The use of preservative solutions throughout the marketing chain is particularly important to maintain the quality and vase life of cut flowers. Several commercial preservative materials are available. These usually contain sucrose to provide the energy that the flowers require to maintain normal cell functions and to complete development, if this is not already completed at the time of harvest. *Biocides* such as chlorine and quaternary ammonium compounds are included, to kill microorganisms. An acidifying agent such as citric acid is also included to lower the pH. This has the effect of improving water uptake, although the mechanism is unknown. The use of ethylene inhibitors such as silver salts as a pulse treatment after harvest has already been described (see above).

## THE STORAGE ENVIRONMENT

## Fresh produce

The deterioration of fruits, vegetables and cut flowers after harvest may be expected to occur under conditions that accelerate the rate of respiration and promote water loss, notably where temperatures are high and humidity low. The consequences of such conditions will vary according to the extent to which a particular type of produce is adapted to withstand stressful conditions.

Temperature is one of the most important factors affecting the keeping quality of horticultural produce. Lowering the temperature reduces the rate of respiration and other metabolic processes, and thereby reduces the rate of senescence and of ripening in fruits. The highest freezing point for fruits varies between −3.0°C (pomegranates, *Punica granatum*) and −0.3°C (avocados), and for vegetables between −2.2°C (Jerusalem artichokes, *Helianthus tuberosus*) and −0.1°C (endive and escarole, *Cichorium endivia*). The highest recorded freezing point for blooms varies from −0.5°C for Easter lilies (*Lilium longiflorum*) and roses (*Rosa* spp.) to −0.7°C for carnations.

It might be assumed that the longest storage or shelf life of fruits, vegetables and cut flowers is likely to be achieved by maintaining temperatures at slightly above their freezing points, and the recommended temperature for the storage of most types of vegetables of temperate origin is 0°C. This applies to many of the field vegetables grown in the UK such as roots and onions, brassicas and legumes. Even at 0°C the storage life of vegetables varies from 5 to 8 days (sweet corn), to up to 8 months (onion).

There are a number of plants, however, particularly those originating in tropical or sub-tropical regions, that are injured when stored at temperatures well above freezing. For such plants the full potential benefit of refrigeration cannot be realised. Chilling stress leads to an altered metabolism and the development of a range of injury symptoms. Among vegetables affected by chilling injury, probably the most economically significant are squash (*Cucurbita* spp.), cucumber, aubergine (*Solanum melongena*), tomato, snap beans (*Phaseolus vulgaris*), sweet pepper (*Capsicum annuum*) and potato. The storage of potatoes at temperatures below about 6°C results in the hydrolysis of starch to form sugars. This 'low-temperature sweetening' imparts an unacceptable dark colour to potato crisps and chips.

Chilling-sensitive fruits that are most familiar to UK consumers include avocados, bananas (*Musa* spp.), grapefruits (*Citrus* × *paradisi*), lemons (*Citrus limon*), mangoes (*Mangifera indica*), melons (*Cucumis melo*) and pineapples (*Ananas comosus*). Of the fruit types important in UK production (apples, pears, plums, cherries, strawberries, raspberries and blackcurrants) only apple is regarded as susceptible to chilling injury and this is highly dependent on the cultivar. Cut flowers are commonly kept at about 4°C

at the wholesale level and during transport, but there are some types that require much higher temperatures to avoid chilling injury. These include some orchids such as *Cattleya* (7–10°C) and *Anthurium* (13°C). Chilling injury may result in failure of flowers to open properly or discoloration of sepals and petals.

## Low-temperature storage methods

Various types of mechanical refrigeration systems are used for the commercial storage of fruits and vegetables. The construction and refrigeration systems required for any particular application are dictated by the characteristics of the produce, and in particular by the rate of cooling required to maintain quality after harvest. Suggested cooling times vary from 3 hours for highly perishable products such as soft fruit, sweet corn, asparagus, calabrese (*Brassica oleracea*) and spinach to up to 6 weeks for potatoes and onions. Quick methods of cooling such as forced-air cooling, vacuum cooling and hydro-cooling are required to remove the field heat rapidly from perishable crops. Forced-air cooling has been used increasingly to pre-cool cut flowers prior to shipment under continuous refrigeration. Forced-air systems are available that maintain a high humidity (95–98% relative humidity) while cooling to the required temperature within 1 hour.

Precise control of the temperature is a major factor limiting the ability of the gardener to store fruits and vegetables effectively. However, commercially, refrigeration is of prime importance in extending the storage and shelf life of fresh horticultural crops. It is particularly important in the distribution of highly perishable vegetables such as asparagus and lettuce and soft fruits such as strawberries and raspberries, although the use of refrigeration for these types of crops cannot extend the season significantly. This can only be achieved by the choice of cultivar, repeated sowing and cultural techniques. The use of refrigeration as a means of extending the marketing period is most appropriate for perennial fruit trees that produce one crop a year.

Root crops such as carrots and parsnips (*Pastinaca sativa*) are generally kept in the ground; they are naturally dormant during the winter, when lower soil temperatures reduce respiration rate and developmental changes, and high soil moisture prevents desiccation. Ground storage of carrots has become quite sophisticated. Polythene sheeting is used to cover the crop, followed by a layer (30–60 cm) of straw. This provides protection against rain and frost and delays the warming up of the soil in the spring with the result that storage is possible through to May.

Refrigeration is the major method of slowing deterioration and extending the life of cut flowers, and is used extensively during transport and distribution. It is perhaps more appropriate to talk about refrigerated transport than storage of cut flowers. For most species maximum storage life may be achieved without water in moisture-retentive containers at −0.5 to 0.5°C. This dry-pack method of storage has extended the storage life of many cut flowers such as carnations and roses.

## Controlled-atmosphere storage

The effectiveness of refrigeration in extending the storage life of produce is limited by the susceptibility to chilling injury of each type of fruit or vegetable. An important advance was the discovery that modifying the storage atmosphere, by increasing the concentration of carbon dioxide and decreasing the concentration of oxygen, could extend the life of fruit and vegetables beyond that obtainable by cold temperatures alone. Franklin Kidd and Cyril West, who worked initially in the Botany School at Cambridge, UK, and subsequently at the Ditton Laboratory in Kent, UK, are recognised as the pioneers of *controlled atmosphere storage* (CA storage). However, the concept of controlling ripening by modification of the atmosphere was conceived prior to their systematic investigations.

Lowering the oxygen concentration depresses the rate of aerobic respiration by decreasing the amount of oxygen available for oxidative reactions. The oxygen concentration recommended for CA storage of apples is typically 1–2% (air is normally 21% oxygen). However, if the oxygen concentration is too low, anaerobic respiration may occur, leading to the production of fermentation products such as ethyl alcohol and ethyl aldehyde that would impart off-flavours to the fruit. Consequently, precise control of the atmosphere in the storage chamber is extremely important.

The carbon dioxide concentration for stored apples may be as high as 10% (air is approximately 0.03% carbon dioxide). Since carbon dioxide is the end product of aerobic respiration, increasing the ambient concentration may lead to a reduction in respiration rate due to feedback inhibition. However, even at a concentration of 5% there is little effect on respiration rate and it is probable that carbon dioxide also increases storage life by inhibiting the action of ethylene, as discussed above.

CA storage has major advantages over cold storage in air for the preservation of many types of fruits and vegetables, and the technique is widely practised in many of the world's production areas for horticultural crops. Recommended concentrations of oxygen and carbon dioxide are available for apples, pears, Nashi (Asian pear, *Pyrus pyrifolia*), fruits other than pome fruits and vegetables. Several thousand tonnes of UK-grown onions are stored in CA to extend the period of availability. White winter cabbage is also kept for long periods in CA storage to provide continuity of supply to the fresh market and for processing into coleslaw. Although CA was practised traditionally in purpose-built stores the technology has progressed to include shipping containers and the packaging of produce in semi-permeable films to generate CA conditions within the package.

CA technologies are generally unavailable to the gardener, although storing apples and pears in polythene bags has been recommended as a means of reducing water loss and preventing shrivelled or leathery fruit. The bags should not be tightly sealed, otherwise carbon dioxide will accumulate and oxygen will deplete to damaging concentrations.

CA storage is not generally recommended for cut flowers because of the small margin of safety between effectiveness and phytotoxicity, the small volumes of any one cultivar and the high cost of treatment.

## Seeds

Seeds are stored for a variety of reasons: to maintain seed stocks for growing from one season to the next, to keep stocks for sale, to maintain breeding lines and to conserve genetic material. Food and grain seeds are stored before processing and consumption. The basic difference between storing fresh produce and seeds is that the latter are already dry, and storage methods aim to restrict water uptake rather than to reduce water loss.

Seed vigour and seed longevity are closely related, and other things being equal, high-vigour seeds will retain their viability for longer than low-vigour seeds.

**Table 17.1** Estimates of the probable regeneration intervals* for seeds stored at −20°C and 5% moisture content.

| Plant | Cultivar | Probable regeneration interval (years) |
|---|---|---|
| Barley | 'Proctor' | 70 |
| Rice | 'Norin' | 300 |
| Wheat | 'Atle' | 78 |
| Broad bean | 'Claudia Superaquadulce' | 270 |
| Pea | 'Meteor' | 1090 |
| Onion | 'White Portugal' | 28 |
| Lettuce | 'Grand Rapids' | 11 |

From Roberts, E.H. and Ellis, R.H. (1977) Prediction of seed longevity at sub-zero temperatures and genetic resources conservation. *Nature* **368**, 431–3.
* The regeneration interval is the predicted time for viability to fall to 95% of its initial value. Thus, it will take the barley cultivar 'Proctor' 70 years to fall from 99% viability to 94% viability, if stored under optimum conditions.

Thus the factors which increase vigour (see Chapter 8) will also normally increase longevity. Similarly, the genetic constitution affects both vigour and storage potential. Seeds of onions and lettuce have low vigour and store less well than, for example, the high-vigour, starchy seeds of round-seeded peas. However, the conditions under which the seeds are dried and stored are also extremely important in determining longevity.

Since all seeds lose vigour and viability during storage, it is important to establish the optimum storage conditions for different types of seeds. In 1973, Roberts divided seeds into two broad categories, depending on their storage behaviour. Seeds which can be dried to low moisture contents and stored at −18°C for long periods are termed orthodox seeds, whereas those that are killed if their moisture content is reduced below some relatively high level (12–31%) are termed recalcitrant seeds. The latter include many tropical species, especially tropical fruits such as mangoes (*Mangifera indica*), coconuts (*Cocos nucifera*) and jackfruit (*Artocarpus heterophyllus*), as well as several large-seeded temperate species such as oak (*Quercus robur*), horsechestnut (*Aesculus hippocastanum*) and sweet chestnut (*Castanea sativa*). More recently an intermediate group has been identified, where drying below 10–12% moisture results in early loss of viability; coffee (*Coffea arabica*) belongs in this group.

Damaged seeds do not store well and it is important to minimise damage during and after harvesting. A common cause of post-harvest damage with consequent loss of viability is drying the seeds at too high a temperature. There is no real 'safe' temperature at which to dry seeds, but seed banks minimise deterioration and thereby safeguard longevity by drying at 15°C and 15% moisture content of the air.

Once the seeds have been dried, their longevity is strongly influenced by their moisture content and by the storage temperature. Seed longevity is related directly to the interaction of a number of environmental factors, notably temperature, moisture content of the seed and oxygen levels. The relationship between temperature and seed moisture content is an interesting one. Generally, the lower the moisture content and the lower the temperature the greater the seed longevity, as under these conditions the seeds are maintained in a state of suspended animation. Thus the wetter and warmer the seeds are in storage, the faster they lose viability and vigour. Therefore orthodox seeds should be stored under dry and cool conditions. To maintain the viability of orthodox seeds for as long as possible, the ideal conditions are hermetic storage at a temperature below −18°C (the lowest limit of a domestic freezer) and a moisture content below 5%. In these conditions metabolic processes are reduced to a minimum level. Predictions of storage longevity under these conditions are given in Table 17.1.

# FUTURE TRENDS IN COMMERCIAL STORAGE

Over the past decades there have been major changes in the storage requirements for horticultural crops.

Storage initially provided consumers with fresh produce during the winter months, and the major objective was to provide producers with the means to store various crops without undue wastage. In due course, advances in storage technology for permanent structures and for transportation contributed to international trade and eventually the seasonal aspect of fruit and vegetable production was eroded. In a highly competitive global market it was increasingly important to provide fresh produce of the highest visual and eating quality. The development of storage recommendations reflected the changing need for high-quality standards, in addition to the control of diseases and disorders.

In the future it is likely that consumer demands for quality will change. In the more health-conscious society of today, storage techniques will need to be developed that maintain or improve components of food that contribute to a healthy diet. Research will be directed towards maximising sensory attributes such as taste, aroma and texture, essential nutritive compounds such as carbohydrates, proteins, vitamins and minerals, and bioactive substances such as polyphenols, carotenoids, phyto-oestrogens and dietary fibre. There is also likely to be increased consumer awareness of undesirable attributes such as *mycotoxins* and pesticide residues.

Changes in consumer awareness of the health, nutritional and ecological aspects of food quality will lead to continual changes in the market place; future developments in pre- and post-harvest management practices will have to reflect these changing needs. To provide fruit and vegetables in the freshest condition possible there will be increased use of cool-chain marketing, chilled display cabinets in retail shops and perhaps an increased use of polymeric film packaging. The demand for ready-to-use fruit and vegetables is likely to grow, with consumers willing to pay for both quality and convenience.

## NON-COMMERCIAL STORAGE

Although this chapter was not intended to provide practical guidance on the storage of garden products, it may be useful to summarise key aspects of post-harvest biology and storage that may help gardeners to manage their produce effectively after harvest.

## Seeds

These should be collected as soon as they are ripe and laid out to dry in seed trays lined with paper. Where seeds are thrown from the ripening pods as with *Erodium* spp., or shaken out as with *Papaver* spp., the seed heads should be collected as they approach maturity and ripened in paper bags.

The best way to store seeds is to keep them cool and dry. A popular and effective method is to keep them in an air-tight container, preferably with a desiccant, in a refrigerator. However, it is important that the container is hermetically sealed, otherwise the seeds will equilibrate with the moisture content of the refrigerator. This will hasten their deterioration, even though they are being stored at a low temperature.

When the seed is removed from the refrigerator, it should be allowed to equilibrate with the air temperature before the container is opened. This ensures that the warm ambient air, which has a higher humidity than the air in the container, does not rapidly enter the seeds, possibly causing membrane or cell damage.

For long-term storage the best practice is to dry seeds thoroughly, place them in an air-tight box over a desiccant such as silica gel and keep them in a freezer at a temperature of −13 to −15°C. If it is necessary to after-ripen the seeds (see Chapter 9) they should be kept for 2–3 months before being frozen. For short-term storage, when seeds are to be sown the following year, it is sufficient to put completely dry seeds into sealed paper bags and store them in a cool dry place.

## Soft fruits

Soft fruits such as strawberries and raspberries are highly perishable and should be picked mature but not over mature. Eating quality will not improve after harvest. Fruits should be picked during the coolest part of the day and stored in a refrigerator until required. It is important not to overload the refrigerator, because the heat produced by the metabolising fruit may raise the temperature to a level that is not only inappropriate for fruit storage but also exceeds that recommended for other stored foods.

## Stone fruits

Stone fruits also benefit from refrigerated storage, but it is unlikely that a domestic refrigerator has sufficient capacity to store significant quantities. Cherries should be picked when they have achieved the desired eating quality but plums should be harvested slightly under-ripe, when the background colour is green/yellow as opposed to yellow, if they are to be stored. Only perfect fruits should be kept, as any physical damage or lesions caused by insects or disease are potential sites for infection by rot-forming organisms.

## Pome fruits

These, especially apples, are most amenable to storage, particularly cultivars that mature late in the season, from the end of September onwards. Apple cultivars differ markedly in their rate of ripening and senescence and consequently in their storage potential; they also differ in the ways that they interact with the storage environment. Gardeners should not attempt to store early varieties of apple and it is not worth taking too much trouble with mid-season varieties. Varieties picked from the middle of September onwards are likely to keep for periods varying from several weeks to several months. Of the major commercial cultivars currently in production in the UK, the most suitable for storage by gardeners would include 'Cox's Orange Pippin', 'Red Pippin' ('Fiesta'), 'Bramley's Seedling', 'Gala' and 'Jonagold'.

Gardeners do not have access to the sophisticated methods of determining harvest maturity that are used commercially. The general recommendation is to pick apples for storage when the fruits can easily be detached from the tree by gentle twisting, and before the good eating quality in terms of sugar content and aromatic properties has been achieved.

The need to select only perfect apples for storage cannot be over-emphasised. They should be kept in a well-ventilated, cool (4°C) frost-free cellar or out-building. Storage in perforated polythene bags is preferred to prevent the shrivelling that some cultivars are especially prone to. The bags must not be sealed completely, otherwise the oxygen in the bag will be depleted and alcoholic off-flavours will develop. Since the apples are in contact, there is the potential

for the spread of disease within the bags. Regular inspection is necessary to remove any rotted fruits. A more traditional method is to wrap the fruits and store them in single layers in wooden, slatted trays.

Gardeners may improve the storage potential of apples by spraying their trees with 0.8% (w/v) flake calcium chloride at intervals of 10–14 days from late June to harvest. Damage to the foliage may be avoided by spraying at temperatures below 21°C, preferably in the evening. Post-harvest application of calcium compounds such as calcium chloride provides an additional means of supplementing calcium in the fruit.

The storage of pears presents difficulties for the gardener, particularly in handling and ripening to a satisfactory quality, regardless of cultivar. Pears do not ripen properly when left on the tree and are difficult to store because they ripen unless kept below 0°C. Sound pears should be kept as cool as possible and inspected regularly. The extent of softening may be gauged by gentle squeezing. Ripening should be completed by transferring the fruits to room temperature. Although polythene bags may be used for pears, it is particularly important to provide sufficient ventilation, otherwise the build-up of carbon dioxide and the depletion of oxygen may damage the fruit.

## Vegetables

The traditional method of storing vegetable root and tuber crops is to make a *clamp* or 'pie'. This was the original method of storing potatoes following their introduction into the UK during the sixteenth century. Garden clamps are used mostly for potatoes, but also for other vegetables such as beets, carrots, swedes (*Brassica napus*) and turnips (*Brassica campestris*). Clamps are piles of roots or tubers on a straw base, often triangular in vertical cross-section, that are covered with a layer of straw and a layer of soil to provide protection from frost and rain. Clamps should be constructed on well-drained land and have a trench dug out around the structure to provide drainage for rain running off the sides. Ventilation holes plugged with loose straw are required in the ridge of the clamp to prevent a build-up of carbon dioxide and the depletion of oxygen through respiration. Only healthy roots or tubers should be placed in the clamp. When storing potatoes, it is particularly important to exclude light to prevent greening and the concomitant build up of toxic alkaloids in the tubers. Exposure to below-freezing temperatures leads to the breakdown of starch and the development of 'sweetening'.

Onions are particularly suitable for storage by the gardener. They should be harvested when the foliage is brown and brittle, pulled or dug from the ground on a fine sunny day and left on the surface of the soil to dry. When they are dry, remove any soil, dead roots and loose, dry skins. Store the onions in trays in a cool dry place, or rope them if preferred. Check the condition of the bulbs at intervals and use before sprouting signals the break of dormancy.

## Cut flowers

The gardener is not likely to need to store cut flowers, but more general methods of extending the vase life of flowers from the garden or from retail stores may be relevant. Cutting flowers at the bud stage may be appropriate for flowers such as roses and gladioli. Flowers should be handled carefully and placed in fresh clean water, with leaves removed from stem sections below the water level. Microorganisms sometimes block the water-transporting tissues and lead to premature wilting, and the practice of cutting the base of stems every few days helps to ensure that stems continue to take up water.

There is no advantage in a depth of water greater than 10–15 cm. Proprietary preservative products should be used where available, and excessively warm dry atmospheres avoided.

## CONCLUSION

Wherever possible, practical advice has been provided on the storage of fruits and vegetables that might be grown by gardeners. In addition, the many other procedures used by commercial growers and also described in this chapter may provide the more

adventurous gardener with a basis for experimentation. One of the joys of gardening is to keep experimenting, learning from one's failures and rejoicing in one's successes.

## FURTHER READING

Blanpied, G.D., Bartsch, J.A. & Hicks, J.R. (eds) (1993) *Proceedings from the Sixth International Controlled Atmosphere Research Conference.* Cornell University, Ithaca, NY.

Ellis, R.H., Hong, T.D. & Roberts, E.H. (1985) *Handbook of Seed Technology for Genebanks*, vol. 1, *Principles and Methods.* PBPGR, Rome.

Fidler, J.C., Wilkinson, B.G., Edney, K.L. & Sharples, R.O. (1973) *The Biology of Apple and Pear Storage.* Research review no. 3. Commonwealth Agricultural Bureau, Slough.

Hardenburg, R.E., Watada, A.E. & Wang, C.Y. (1986) *The Commercial Storage of Fruits, Vegetables, and Florist and Nursery Stocks.* Agriculture Handbook No. 66. US Department of Agriculture, Washington DC.

Johnson, D.S. (1999) Controlled atmosphere storage of apples in the UK. Proceedings of the International Symposium on Effect of Preharvest and Postharvest Factors on Storage of Fruit. *Acta Horticulturae* **485**, 187–93.

Kays, S.J. (1991) *Postharvest Physiology of Perishable Plant Products.* Chapman & Hall, London.

Ministry for Agriculture, Fisheries and Food (1979) *Refrigerated Storage of Fruit and Vegetables.* Reference book 324. The Stationery Office, London.

Roberts, J.A. & Tucker, G.A. (eds) (1985) *Ethylene and Plant Development.* Butterworths, London.

Seymour, G.B., Taylor, J.E. & Tucker, G.A. (eds) (1993) *Biochemistry of Fruit Ripening.* Chapman & Hall, London.

Shewfelt, R.L. & Bruckner, B. (eds) (2000) *Fruit and Vegetable Quality: an Integrated View.* Technomic Publishing Company, Lancaster, PA.

Thompson, P. (2005) *Creative Propagation*, 2nd edn. Timber Press, Cambridge.

Wills, R., McGlasson, B., Graham, D. & Joyce, D. (1998) *Postharvest: an Introduction to the Physiology & Handling of Fruit, Vegetables & Ornamentals*, 4th edn. CAB International, Oxford.

# 18

## Conservation and Sustainable Gardening

## SUMMARY

First, some of the major global environmental changes relating to conservation and sustainability are outlined: climate change, changes in the ozone concentrations of the atmosphere, disruption of nutrient cycles and the erosion of biodiversity. Information is provided on how gardeners, acting locally, may contribute to international efforts to minimise these changes. Next, the conservation of water and energy in the garden, and of garden plant diversity, are considered. The role of horticulturists, in collaboration with other specialists, in the restoration and creation of natural and semi-natural habitats, is outlined. Finally, the threats posed by the release into the wild of invasive non-native plants, animals, pathogens and pests are summarised, and the role that gardeners may play in preventing such releases is emphasised.

## INTRODUCTION

When considering conservation and sustainability in the garden, it is necessary to think globally and act locally. Two issues of overriding importance in this regard are climate change and the erosion of biodiversity. Linked to these and also of great significance are changes in the concentrations of ozone in the stratosphere and troposphere and disruption of the functioning of nutrient cycles. These all form part of a much larger matrix of causes and consequences of environmental change on a global scale, sometimes referred to collectively as global environmental change.

### Climate change

This term refers to any change in the climate, whether natural or induced, local or worldwide. Examples include the ice ages of the geological past, the so-called Little Ice Age, which in Europe lasted from 1450 to 1850, leaving a legacy of snow scenes on Christmas cards, and the *global warming* now accelerating as a result of rising concentrations of *greenhouse gases* in the troposphere, the part of the atmosphere closest to the surface of the Earth.

Global warming is nothing new, having occurred in the past as a result of natural processes. The current acceleration, however, results from massively increased emissions of greenhouse gases from human activities. The most important greenhouse gases are water vapour ($H_2O$), carbon dioxide ($CO_2$), methane ($CH_4$), nitrous oxide ($N_2O$) and chlorofluorocarbons (CFCs). Those that have been increasing most rapidly are $CO_2$ from the burning of fossil fuels and $CH_4$ from livestock farming, although the latter may have stabilised somewhat in the last few years. Having accumulated in the troposphere, the greenhouse gases contribute to the so-called greenhouse effect.

The Earth receives radiation of a wide range of wavelengths from the sun, a relatively hot body, and as a relatively cool body radiates energy of relatively long wavelengths (heat). Before passing into space, the long-wavelength energy is trapped for a period in the atmosphere by naturally occurring greenhouse gases. This trapped energy heats the lower atmosphere, making it sufficiently warm for living organisms to survive on the Earth's surface. Without the trapped energy the surface temperature of the Earth would be about $-15°C$. The process is said to be analogous to that which traps heat in a greenhouse (see Chapter 14), hence the term greenhouse effect, although the analogy is a poor one. It is this normal process that has recently been distorted by anthropogenic greenhouse gases. It has been estimated, for example, that since the start of the Industrial Revolution, $CO_2$ has risen by approximately 30%, $CH_4$ has more than doubled and $N_2O$ has risen by approximately 15%.

Climate results from complex interactions between the atmosphere, the oceans and the land, notably the soils, vegetation and human settlements. The extent and effects of climate change cannot be predicted accurately, for various reasons. First, there are scientific uncertainties about, for example, the working of climate systems, especially the role of water vapour and clouds in the exchange of heat. Second, there are social and economic uncertainties about the rate of greenhouse gas emissions over the next hundred years.

It has been predicted that, on a world scale, warm seasons are likely to become longer and cold seasons shorter; northern latitudes will probably have wetter autumns and winters and drier springs and summers; and rainfall will increase in the tropics and decrease in the sub-tropics. There will thus be a shift in the main climatic zones of the world and this is likely to be associated with a rise in the frequency of climatic perturbations such as floods, droughts, typhoons, tornadoes, hurricanes and gales. Also, sea levels are likely to rise as a result of temperature-induced expansion of oceanic waters and the melting of the ice caps and glaciers.

The Intergovernmental Panel on Climate Change (IPCC) has predicted that if the emission of greenhouse gases continues to rise at present rates, the average temperature of the Earth overall will rise $2.5°C$ by 2050, a rate of change considerably in excess of the natural rate of temperature changes experienced over the last 8000 years. The rate of change for the future is predicted to be about $0.3°C$ every 10 years, which some have suggested is about three times faster than the rate that many ecosystems are capable of adjusting to. This means that, worldwide, ecosystem functioning could be disrupted, with significant concomitant damage to biodiversity.

In the UK, in the recent past, increases in temperature have been higher than the world average. In central England and eastern Scotland temperatures appear to have increased by about $1°C$ in the twentieth century. The 1990s were the warmest decade since records began, and since 2000 the number of frost-free winter days and unusually hot summer days has increased, and the growing season has become longer. Overall, the amount of rainfall has not changed, but winters seem to have been getting wetter, with more of the precipitation occurring as heavy downpours. Summers were slightly drier during the twentieth century.

It is especially difficult to model the future effects of climate change in individual countries. However, the UK Climate Impacts Programme (UKCIP) has developed four potential UK climate-change scenarios, based on low, medium low, medium high and high greenhouse gas emissions. According to these scenarios, temperatures will rise between 2 and $3.5°C$ by the 2080s, with greater warming in the South-east than in the North-west. There will probably be more warming in the summer and autumn than in spring and winter. In the south east, summer could become $5°C$ warmer under the high scenario. Spring temperatures could occur from 1 to 3 weeks earlier by 2050, and the onset of winter could be delayed by similar amounts. Rainfall may decrease slightly overall, perhaps by up to 15% by the 2080s. The trend for winters to be wetter and summers drier, especially in the south east, is likely to continue. Under the high scenario, summer rainfall may decrease by up to 50% in the South-east. In winter, periods of heavy rainfall may become more frequent, giving a higher proportion of rain in winter. Finally, sea levels could rise significantly. In the short term there could be some benefits such as the better growth of some plants resulting from increased $CO_2$ levels in the air (see Chapters 8 and 14) and warmer temperatures. In the longer term, droughts in summer, leaching of soils by heavy rain in winter, and damage resulting from storms and flooding will probably far outweigh these benefits. In all the scenarios there could ultimately be serious damage to UK ecosystems as a result of changes in temperature and patterns of precipitation, while sea-level rises would lead to extensive loss of coastal habitats.

Tackling climate change requires concerted effort on a world scale. The Kyoto Protocol is an international agreement that sets targets for individual governments on the level of emission reduction of greenhouse gases. Some flexibility is built in to allow countries to emit more greenhouse gases if they plant more trees, which take up $CO_2$ from the atmosphere during photosynthesis and therefore constitute so-called carbon sinks, or if they can demonstrate that they have reduced emissions in other countries. Richer, developed countries have agreed to reduce their emissions by 5.2% by 2010, but developing countries have not been set formal limits, in part so as not to inhibit economic growth. Discussion of these matters is continuing and targets may change again. Under the Convention on Biological Diversity (see below) countries have an obligation to identify and address cases of, or threats to, biodiversity, including climate change. Possible approaches to reducing emissions include reducing demand for energy by changing social habits such as the excessive use of private cars, developing more energy-efficient public transport systems, meeting more of a country's energy needs from renewable sources such as biofuels and solar, wind and wave power, changing livestock farming activities so as to reduce methane emissions, conserving existing natural sinks for carbon, especially forests, and planting more forests.

At a local level, gardeners may contribute to the international efforts to slow global warming by:

1. Adapting planting strategies to cope with the expected changes in environmental conditions, especially reductions in the availability of water during the summer in the South-east (see below and Chapters 8 and 13), rather than relying on the excessive use of water for irrigation.
2. Developing ways of conserving water, especially in the South-east (see below).
3. Reducing energy use, especially from non-renewable sources, for glasshouse heating and power tools.
4. Reducing the use of inorganic fertilisers, pesticides, fungicides and plastics, all of which require large amounts of energy for their manufacture (Chapters 6, 7, 15 and 16 include advice on effective strategies for managing the deployment of garden chemicals).
5. Recycling through composting of green waste, taking care that compost heaps do not become anaerobic, resulting in the emission of methane, a greenhouse gas (see Chapter 7).
6. Avoiding the use of peat, since peatlands are major sinks for carbon. Using peat in horticulture therefore adds to the $CO_2$ load.

## Ozone concentrations in the stratosphere and troposphere

Ozone ($O_3$) is a highly reactive gas which accounts for 0.00006% of the Earth's atmosphere. There are two very different environmental problems concerning ozone: depletion in the stratosphere and increasing concentrations in the troposphere.

In 1985, a 40% seasonal thinning of the ozone layer of the stratosphere over the Antarctic continent was first discovered. This thinning, usually referred to as a 'hole in the ozone layer', was later found to be growing bigger year on year. Moreover, thinning over the Arctic and parts of Western Europe was also discovered. Although some of the thinning was caused by natural aerosols from the eruption of the Mount Pinatubo, a volcano, in 1991, it is now generally agreed that most is caused by pollution of the air with the greenhouse gases chlorofluorocarbons (CFCs), used until the 1990s as propellants in aerosols, as coolants in refrigerators and air conditioners and in the manufacture of boxes for take-away food.

CFCs are broken down by ultraviolet (UV) light in the stratosphere, thereby releasing chlorine which destroys ozone molecules. Since the ozone layer is important in sheltering the Earth from UV radiation, there was great international concern about its thinning. As a result, most industrial nations agreed, under the Montreal Protocol of 1987, to phase out the production of CFCs by the year 2000. Unfortunately, however, some CFCs are very long-lived, remaining active for up to 100 years. Moreover, although good progress has been made in phasing out the manufacture and release of CFCs, they are still being produced by some countries.

The effects of the thinning of the ozone layer will include scorching of plant life, especially crops, with concomitant falls in yield, and damage to the phytoplankton of the oceans, which at present constitute a major sink for carbon dioxide. Moreover, the effect on human health could be considerable, with increasing numbers of radiation-induced cataracts, skin cancers and damaged immune systems.

Gardeners may contribute locally to the increasingly effective global efforts to tackle thinning of the ozone layer by ensuring that any remaining sprays containing CFC propellants, and any old cooling devices, are disposed of through local-authority disposal schemes designed to prevent the release of CFCs into the atmosphere. They should also take the health threat very seriously and protect themselves from strong sunlight with barrier creams and/or UV-proof hats, clothes and sunglasses when working out of doors.

A second environmental issue concerning ozone, and one that is of great significance to plants, is the increasing concentration of low level ozone in the troposphere. Some of this results from natural factors, but the major part results from pollution, especially as a result of photochemical reactions involving exhaust fumes from vehicles. Excess ozone in the troposphere is a constituent of smog, may inhibit plant metabolism and is partly responsible for some of the damage caused by *acid rain* (see below). Gardeners acting locally may contribute to efforts to reduce this problem by reducing direct and indirect energy use in the garden (see below).

## Disruption of nutrient cycles

Nutrient cycles represent the natural circulation of chemical elements, such as carbon and nitrogen and compounds containing them, from the abiotic components of an ecosystem (see Chapter 19) into the organic molecules of the biotic components, and back again to the abiotic components. The disruption of these cycles may have far-reaching consequences. For example, the carbon cycle is at present being disrupted by changes in land use, especially the clearance of natural vegetation, resulting in a decrease in the natural sinks for $CO_2$. Levels of $CO_2$ are therefore building up in the atmosphere and this is contributing to the greenhouse effect described above.

The nitrogen cycle is also being disrupted, by the release of nitrogen oxides ($NO_x$) as components of air pollution. These are gaseous compounds of nitrogen and oxygen, produced directly or indirectly by the combustion of fossil fuels and from certain synthetic chemical processes. They are greenhouse gases, and after conversion to nitrogen dioxide react with water vapour to form nitric acid, a component of acid rain. The other principle components of acid rain are derived from products of sulphur dioxide, mainly from coal-fired power stations.

Acid rain has a pH of less than 5.6 and may be deposited many thousands of kilometres from its site of formation, either in a dry form or combined with rain and snow. It causes acidification of soils, water bodies and vegetation such as grasslands and forests.

This leads to, among other things, tree die-back and death, inhibition of plant growth and damage to phytoplankton. After many delays, the problem of reducing acid rain by reducing and controlling emissions is now being tackled internationally. Gardeners may make their contribution locally by reducing direct and indirect energy use in the garden (see below).

## Erosion of biodiversity

Biodiversity is the summation of all life on Earth. The term embraces the diversity of ecosystems (see Chapter 19), of species and of the genetic variation within species, including cultivars (see Chapters 4 and 5). These different levels of diversity cannot be separated, as each interacts with, and influences, the others. The survival of species diversity depends on healthy ecosystems, and the evolution of new species depends on the richness of the genetic diversity at the subspecies level.

The diversity of living organisms – plants, animals and microorganisms – provides for the most basic of human needs: food, fuel, fibres, medicines and building materials. The diversity of ecosystems – oceans, lakes, rivers, wetlands, grasslands, forests and deserts – acts as a sink for $CO_2$ (see above), provides pollinators for crops and plays a major role in stabilising the climate and protecting human habitations from floods and drought. The genetic diversity among species and within species provides the new material for the breeding or development of new livestock, crops and ornamental plants, and the development of new drugs and products for the control of pests and pathogens of humans, animals and plants. And last, but not least, biodiversity at all levels is of inestimable spiritual, cultural and recreational value to societies worldwide.

This biodiversity is currently under considerable threat from human activities. Estimates for the rate of loss of habitats or/and extinction of species vary greatly, and may ignore completely the as-yet relatively unquantified biodiversity of soils and the oceans. One reliable and conservative estimate of the 'normal' or background rate of extinction of species in the absence of human interference is one species per million per annum. This would be balanced by the evolution of new species. The same author suggests that current human activity in felling rainforests alone has increased the rate of extinction by between

1000 and 10 000 times. Figures such as this suggest that the Earth is undergoing one of the most significant extinction periods of biological history. The reasons for the erosion of biodiversity are complex, but include the following:

1. *Habitat loss, fragmentation and degradation* as a result of exploitation for timber and other products, providing land for commercial agriculture and forestry, industry, and expanding cities and other settlements.
2. *The spread of invasive alien species* of plants, animals and microorganisms (see below) which compete with native species for nutrients and habitat space, or cause other types of damage such as epidemics of pests and diseases.
3. *Over-exploitation of species* for food, medicines and luxury goods.
4. *Pollution* from agriculture, forestry and especially city living and other human activities, which severely damages ecosystems.
5. *Climate change and other environmental changes* (see above), which are already disrupting ecosystem functioning.

Recognising these threats, the first *Earth Summit*, at Rio de Janeiro, led to the introduction in 2002 of the international *Convention on Biological Diversity*, usually abbreviated to the *CBD*. The CBD includes the following elements:

1. An assertion that the conservation of biodiversity is a 'common concern of humankind'; whether diversity exists within a country's borders or those of another country, an obligation is placed on the signatories to work together to conserve it.
2. A reassertion that countries have 'sovereign rights' over their biological resources; this does not mean 'ownership': it is left to individual countries to assign ownership, how access will be regulated and how the CBD's objectives will best be achieved.
3. An assertion that individual countries are responsible for conserving their biodiversity, and for using it in a sustainable way.
4. A call for a 'preventative approach', in which measures are put in place to anticipate and prevent the causes of loss of biodiversity; and a 'precautionary approach', whereby measures to avoid or minimise threats to biodiversity are not postponed because of scientific uncertainty.

In implementing these elements, the CBD promotes an ecosystem approach: strategies for the integrated management of land, water and living organisms to promote conservation and sustainable use in an equitable manner. Therefore, the conservation of genetic diversity, species and ecosystems in their natural surroundings (*in situ conservation*; see Chapter 20) is seen as the first priority. This approach necessitates the establishment, for example, of protected areas and the enactment of legislation to protect habitats. The conservation of biodiversity outside natural habitats in collections such as zoos, botanic gardens, seed banks and gene banks (*ex situ conservation*; see Chapter 20) is seen as the second or supporting priority. Finally, the CBD calls for initiatives to promote the sustainable use of biodiversity; in other words, to enable it to be used while being maintained for the future without adverse effects. Initiatives might include locally based projects for the management and use of forest products and resources, sustainable tourism, sustainable fishing and harvesting, encouraging sustainable trade and investment in pharmaceutical product development. Thus the need to maintain a balance between the human use of biodiversity and the need to conserve it is recognised.

The implementation of the CBD is integrated with the work resulting from other international treaties, notably the *Convention on International Trade in Endangered Species of Wild Flora and Fauna (CITES)*. This provides an international legal framework for the regulation of trade in those endangered plant and animal species that are exploited commercially. It operates through the issue and control of export and import permits for a number of clearly defined species listed in three appendices. The plants listed are mainly orchids, cacti, cycads, succulent *Euphorbia* spp. and certain genera of carnivorous plants.

Appendix I lists plants banned from international trade (meaning movement across an international frontier). These are species that are threatened with extinction through commercial exploitation. Rare exemptions are granted under licence, for scientific research. Appendix II, the most extensive list, includes species not yet threatened with extinction, but which may be if international trade grows to an unsustainable level. Trade of both wild and artificially propagated material of these is permitted, providing that an appropriate licence is obtained. Material for scientific research may be exported without a licence between CITES-registered institutions. Appendix III backs up national legislation in individual countries by providing a mechanism to monitor trade in certain species protected by those countries. The species on this list, which are relatively few in number, are threatened with local extinction as a result of commercial exploitation and are subject to trade control within the countries concerned.

Legislation concerning the conservation of species and habitats arising from a range of global and European Union wildlife obligations is enacted by the UK and devolved governments. These are advised by Natural England (NE), Scottish Natural Heritage (SNH), the Countryside Council for Wales (CCW) and the (Northern Ireland) Council for Nature Conservation and the Countryside (NCC) in their respective countries, and by the Joint Nature Conservation Committee (JNCC) in relation to the UK as a whole. Also, non-governmental organisations (NGOs) such as local wildlife trusts, Plantlife International, the Royal Society for the Protection of Birds, the National Trust, the National Trust for Scotland and the National Biodiversity Network Trust, play significant roles in advising government.

The UK strategy for nature conservation in recent decades has been based on two principles:

1. Legal protection for exploited, persecuted or endangered species.
2. Legal protection for important nature conservation sites, and the establishment of nature reserves by many of the statutory bodies and NGOs listed above.

The trend for the future is likely to be more holistic, whereby efforts to mobilise conservation action in the wider countryside are married to efforts to maintain the beauty of the countryside and also the ecosystem services which the countryside provides, for example as a recreational amenity and as a source of income and employment.

By acting locally, gardeners may collectively make a major contribution to furthering the aims of the CBD and of the UK conservation bodies and NGOs in the following ways, by:

1. Limiting the use of inorganic fertilisers, herbicides, fungicides and pesticides, which if used to excess may pollute soil and groundwater, thereby damaging ecosystems, and which require large amounts of energy for their production,

thereby contributing to global warming (see Chapters 6, 7, 15 and 16 for advice on the best strategies for managing the use of garden chemicals and inorganic fertilisers). The excessive or inappropriate use of organic fertilisers and manures may also be damaging (see Chapters 6 and 7 for advice on how best to manage the use of these in the garden).

2. Limiting the use of power tools and heat in greenhouses which consume energy derived from fossil fuels, resulting in greenhouse gas emissions and other forms of atmospheric pollution (see below).

3. Recycling materials where possible, especially plastics, all of which require energy for their production and most of which break down only very slowly and must, if not recycled, be incinerated or disposed of in landfill sites.

4. Using locally recycled hard landscaping materials such as rocks, gravel, stone mulches, building materials and timber, to reduce the damage caused by extraction and the consumption of energy in the transport of such materials.

5. Composting all green waste for use in soil improvement (see Chapter 7).

6. Limiting the use of peat, since its extraction may involve the destruction of wetland habitats; also, peatlands are important sinks for $CO_2$. Many alternatives to peat are available commercially (see Chapter 14).

7. Conserving energy in greenhouses (see below), again reducing reliance on the burning of fossil fuels.

8. Ensuring that any rare plants purchased carry appropriate CITES certificates and have been obtained legally (see above).

9. Not removing plants, plant parts or seeds from the wild unless licensed to do so in a sustainable way by the appropriate authorities of the country in question. If in doubt, it is always best to adopt the precautionary principle and not collect. The impact of removal of plants from the wild by an individual may seem small, but when practised by many becomes significant.

10. Ensuring that invasive alien plants, pests and pathogens are not released into the wild as these are major causes of damage to natural ecosystems through competition with native species or as the cause of epidemics of pests and diseases (see below).

11. Supporting bodies and organisations dedicated to the conservation of cultivated plants, including national and international collections (see below) and botanic gardens (see Chapter 20).
12. Managing the garden environment to provide food, refuges and corridors for wildlife (see Chapter 19).
13. Collaborating with professional ecologists and conservationists in restoring damaged habitats such as footpaths, recreating habitats damaged by commercial or industrial activities such as farming, forestry or mining, and creating new habitats, especially in cities (see below).

## CONSERVATION OF WATER AND ENERGY

## Water

With the most conservative of climate-change models, it is clear that water is likely to become increasingly scarce in the south-eastern region of the UK. Here, water conservation will be a very high priority, for this is where temperatures are likely to rise most significantly, where the greatest decline in annual precipitation may be expected and where, being the most densely populated area of the UK, there will continue to be the greatest consumer demand.

Even in areas of the UK that appear assured of a plentiful supply of water for the foreseeable future there will be pressure to conserve water, partly because the expected increase in abnormal weather events may lead to disruption of supplies, but also because there are likely to be calls to redistribute water supplies as is already done, for example, in providing water for some large cities.

It might be argued that since garden watering in the UK is only responsible for 6% of annual water use or less, depending on the region, it is 'unfair' that gardeners should have conservation forced upon them by such measures as hosepipe bans. During a long dry spell, however, garden use can amount to 50% of consumer demand, potentially leading to a disruption of supplies. Moreover, a reduction in the water demands of gardeners must be seen as part of an overall approach to water conservation, in every walk of life.

Water conservation by home gardeners requires an integrated strategy, embracing the following elements.

1. *The use of grey water*. Much of the water used in the home for personal washing and preparing and cooking vegetables, termed light-grey water, can be collected after use for garden watering, for it is relatively free of pollutants. Laundry and dishwashing water usually contains pollutants such as bleaching or whitening agents, and boron from detergents, washing powders and liquids. If such dark-grey water is to be used in the garden it is essential that eco-friendly washing agents are used. Light-grey water that has been used for personal washing, and dark-grey water from eco-friendly washing and dishwashing activities, should only be used on ornamentals and woody fruit trees and bushes, not on leafy or root vegetables that are soon to be eaten. Grey water containing organic matter such as soap or food particles may smell if stored, because of the activity of anaerobic bacteria. It is thus best used relatively quickly.
2. *Rainwater harvesting*. Rainwater may be harvested from the roofs of houses and outbuildings, and other hard surfaces. It may be stored in containers above or below ground. Organic matter such as leaves should be excluded, as this may lead to anaerobic bacterial respiration, resulting in the emission of methane, a greenhouse gas, and other unpleasant smelling gases.
3. *Avoiding evaporation and run-off*. The evaporation of water from containerised plants may be reduced by positioning them in the shade whenever the plants they contain makes this possible, lining porous containers with plastic (ideally recycled), and adding organic matter (but not peat) to the soil. In the open garden, water retention of the soil may be enhanced by the incorporation of, and/or mulching with, large quantities of non-peat organic matter such as compost. Pierced plastic soil covers may also be valuable, providing they are biodegradable or can be recycled. Water should be applied at the beginning or end of the day when maximum penetration and a minimum of evaporation is likely to be the norm. Sloping gardens should be terraced to avoid run-off.
4. *Use of irrigation devices*. Wasting water may be minimised by employing devices that deliver water accurately to the plants that require it. The simplest such device is a watering can, but more

sophisticated irrigation systems may be used to deliver accurately measured amounts of water at predetermined times or, if a suitable sensor is fitted, according to need. Sprinklers should not be used, as these distribute water, on foliage and between plants, where it is not required and is lost through evaporation. (See also Chapters 6 and 7 for information on the requirements of plants for water, the rates at which they use water and the appropriate watering frequencies; and Chapter 8 for comments on best watering practice.)

5. *Management or elimination of lawns.* Lawns require less water if drought-tolerant grasses are used, if they are not cut short and if the clippings are not collected, but left to form a mulch (although excessive mulch could increase run-off). The complete removal of lawns and their replacement with drought-tolerant plantings (see below and Chapters 8 and 13) is even better. It is important that lawns are not replaced with non-porous surfaces such as patios and hard-standing, since rainfall simply evaporates or runs off these and is wasted rather than being absorbed into the garden soil. If such surfaces already exist, however, they may be modified to allow run-off water to be collected for use elsewhere (see point 2 above).

6. *Ponds.* These may be constructed to perform a dual function by combining their role as garden features with an additional role as water-storage reservoirs, swimming ponds or reed beds for filtering grey water. Swimming ponds attract wildlife, whereas traditional swimming pools do not. If the latter are built, they should be covered when not in use to reduce evaporation.

7. *Planting appropriate plants.* The most effective way of conserving water, and the least labour-intensive, is to grow plants adapted to Mediterranean-type climates, where summers are hot and dry and any rain falls in the winter months. Plants adapted to such conditions have their origins in Europe, South Africa, the Chapparal of California and the temperate parts of Australia, New Zealand, Chile and Mexico; also, new drought-tolerant plants are being developed by plant breeders (see Chapters 8 and 13 and Further reading in this chapter for details). However, care should be taken to ensure that invasive, non-native species are not introduced in this way (see below).

For commercial growers, economic and practical considerations may preclude the use of the simple strategies for water conservation available to the home gardener. Nevertheless, many of the basic principles underlying the suggestions outlined above may be applied to commercial operations, although they would need to be costed carefully. Limiting water use may actually reduce costs in the long term, since the price of mains water is likely to increase significantly in the future, while the use of private boreholes and the extraction of water from natural watercourses is likely to come under ever more stringent controls. Different watering practices are now being examined experimentally and these may lead to novel strategies for limiting the use of water in commercial horticulture (see Chapter 8).

## Energy

The release of greenhouse gases and other pollutants from the burning of fossil fuels is the main cause of global warming and other environmental changes. Reducing the demand for energy is therefore a central plank of all strategies to combat these problems. Acting locally, home gardeners may contribute to the global effort by reducing the use of power tools to a minimum. Power lawn mowers are the major consumers of energy in the garden, so replacing lawns with other plantings as suggested above is a straightforward way of alleviating the problem while at the same time reducing dependence on water. If lawns are retained, drought-tolerant grasses should be used (see Chapter 8) and they should, whenever possible, be cut using hand-propelled mowers or, if this is impracticable, cut less frequently. It is also important to note that other power tools, such as hedge trimmers, consume considerable amounts of energy, and that rechargers for all power tools consume energy when left plugged into the mains for longer than is required to recharge the device attached to them.

Energy may also be conserved by home gardeners by not growing plants requiring heated greenhouses or, if heat is used, delivering it to the point where it is most needed by using soil-warming cables, fitting an effective thermostat and/or fitting double glazing, although there will be some loss of light with the latter. Supplementary lighting (see Chapter 14), which also consumes energy, should, however, be avoided if possible.

Before embarking on a strategy to reduce the use of energy from fossil fuels in the greenhouse, the reader is advised to consult Chapter 14.

As with water conservation, economic and practical constraints may preclude the use by commercial growers of the simple strategies for energy conservation available to home gardeners. However, the principle still applies, and all commercial operations should be carefully audited to reduce reliance on energy from fossil fuels.

One strategy that may eventually be more readily available to commercial growers than to home gardeners is the use of energy from renewable sources such as wind or water turbines, or biofuels. The technology involved is still in its infancy, however, and the environmental and social impacts of such energy production have still to be fully quantified. Moreover, costs are at present high. There is currently heavy investment in research on energy from renewable sources and rapid progress can therefore be expected.

## CONSERVATION OF GARDEN PLANT DIVERSITY

During the past 50 years there has been a steady erosion of the diversity of cultivated plants available to gardeners. The reasons for this decline are quite different from those leading to loss of species in the wild, the main factor being economic: this is the increasing cost to commercial growers of maintaining large numbers of cultivars, linked with changing patterns of marketing, in which the focus has been on 'novelties' and 'improved' cultivars and the retailing of large numbers of plants of a limited range of cultivars in garden centres. The rising cost of maintaining large private gardens, gardens associated with public institutions such as universities and collections associated with research institutes, has been an associated factor.

As a consequence, the charity the National Council for the Conservation of Plants and Gardens (NCCPG) was established in 1978, its main objectives being as follows.

1. Encouraging the propagation and conservation of endangered garden plants, both species and cultivars, in the British Isles. (The climate of the British Isles, and the long tradition of collecting and hybridising plants for cultivation in gardens, means that the diversity of cultivated plants in this relatively small area is especially rich.)
2. Encouraging and conducting research into cultivated plants, their origins, their historical and cultural importance and their environments.
3. Encouraging the education of the public in garden plant conservation.

'Through its membership and National Collection holders [see below], the NCCPG seeks to rediscover and reintroduce endangered garden plants by encouraging their propagation and distribution so that they are grown as widely as possible'. The most significant achievement of the NCCPG has been the establishment of the *National Plant Collections Scheme*, a unique collaboration between horticultural institutions, such as botanic gardens, university departments, the Royal Horticultural Society, the National Trust, the National Trust for Scotland, professional horticulturists and taxonomists, and amateur gardeners. There are 650 National Collections, located throughout the British Isles (see Further reading, below). Each one has a clear focus and comprises 'as complete a representation of a genus or section of a genus as possible', although the amount of in-cultivar genetic variation varies considerably among collections. Prospective collections and collection holders are carefully assessed by the NCCPG, and existing collections are properly monitored to maintain high standards of custodianship.

National Collection holders collect plants, grow them, propagate them, make them available to the public, research their details and make their knowledge widely available. Their work is currently coordinated by 47 collection coordinators. Initially, plants which were 'commonly available' or 'rare and endangered', and therefore in urgent need of conservation, were identified through nursery surveys conducted by local groups of the NCCPG. More recently the *RHS Plant Finder* has performed this function (see Chapter 20). Most of the plants held in National Collections are ornamentals, including trees. Fruit-producing genera such as *Malus*, *Pyrus* and *Ribes* are also included, although collections of endangered cultivars of vegetables are not. In 2006, Heritage Collections were introduced by the NCCPG to recognise non-genus-specific collections. The first so designated was the John Bartram Heritage

Collection, held by Painshill Park, Cobham, Surrey, UK. This is based on the seeds sent to the UK approximately 200 years ago by the American seed collector John Bartram. The NCCPG and the National Plant Collection Holders provide a unique and remarkable example of what can be achieved in plant conservation through the collaboration of professionals and enthusiastic amateurs.

The International Board for Plant Genetic Resources (IBPGR) was founded in 1974 to address the problems arising from the erosion of crop-plant genetic diversity worldwide. The development of the IBPGR Base Collection Network was a major achievement of this body. In the UK, the Genetic Resources Unit of Warwick Horticultural Research International, University of Warwick, holds IBPGR Base Collections of the following groups of horticultural importance: *Allium* (onions, garlic and their relatives), *Brassica oleracea* (cabbage and its relatives), *Brassica rapa* (turnip and its relatives), *Brassica napus* (swedes and their relatives), *Brassica juncea* (spicy brassicas), *Daucus* (carrot), *Raphanus* (radishes) and *Lactuca* (lettuces). The collections of *Brassical oleracea*, *B. rapa*, *B. napus* and *B. juncea* include only cultivated forms, but the collections of *Allium*, *Daucus*, *Raphanus* and *Lactuca* include both wild and cultivated taxa. The Base Collection of *Lactuca* is held in collaboration with the Centre for Genetic Resources, the Netherlands. The Scottish Crop Research Institute in Dundee (SCRI) holds the Commonwealth potato collection.

Some commercial seed suppliers list what are usually called heritage vegetables in their catalogues, although European Union (EU) legislation limits the seed that can be sold. The Heritage Seed Library of the Henry Doubleday Research Association (HDRA/Garden Organic), however, is an important alternative source of old varieties of vegetables for gardeners. They may be obtained by joining the library, and this avoids the limitations on commercial sales arising from EU legislation.

## HABITAT RESTORATION AND CREATION

General guidelines, based on broad ecological principles, for the establishment of gardens planted

mainly with native species are discussed in Chapter 19. Guidelines for the creation of four specific semi-natural habitat types, suitable for private gardens, that are representative of similar habitats to be found in the wild are also set out. Such habitat types provide a means of bringing the beauty and interest of the natural world into the garden, and therefore have a very significant aesthetic value. Moreover, they provide resources for private study and research. Their value in the conservation of biodiversity is, however, limited. Generally, they are likely to be too limited in extent and too low in genetic biodiversity, especially at the below-species level, to be equivalent to long-established natural or semi-natural habitats. Nevertheless, they undoubtedly have a conservation function, when taken together with other green spaces, in providing refuges for native species, especially animals, and corridors for their movement, notably in cities and areas of intensively farmed countryside.

Professional and suitably trained amateur horticulturists do have a significant role to play in the conservation of biodiversity, however, through the restoration of damaged areas of existing nature reserves, in the creation of new large-scale reserves, especially on derelict land within cities, and in the reintroduction of plant species that have become extinct at specific sites. The skills that horticulturists bring to bear in such projects relate especially to the propagation and bulking up of individual species, the design of planting schemes and post-planting maintenance strategies to ensure maximum survival and establishment, landscape design and hard-landscape construction. Skills provided by other professionals include knowledge of hydrology, soil science, population genetics (genecology), the ecology of individual organisms (autecology) and of complex communities (synecology), phytosociology and conservation science.

It is critically important that in restoring and creating habitats, every detail is matched to adjacent habitats and takes full account of local soils, and hydrological and climatic conditions. Moreover, seed should be sourced locally and sustainably, to ensure that it is representative at every level of the local flora. And careful consideration should be given to ensuring adequate levels of genetic diversity below the level of the species.

Essential sources of reference are the JNCC National Vegetation Classification (NVC), Rodwell's series of books on British Plant Communities, based on the

NVC, and Rodwell's *National Vegetation Classification Users' Handbook*. Other sources are given in the Further reading section, below.

## INVASIVE NON-NATIVE SPECIES

The term non-native species refers to any species or race of a plant, animal or microorganism that does not occur naturally in an area. An alternative term used in some countries is alien species. Some non-native species are invasive, being especially aggressive competitors for space and nutrients, and may cause significant ecological damage by replacing and/or hybridising with native species, altering successions or, in the case of plants, changing water flow in catchments and rivers. Some alien pests and pathogens are particularly virulent, causing epidemics among native species.

The reasons why some non-native species become invasive are complex. Some may have developed particular attributes relating to growth rate, reproductive efficiency or food-acquisition strategies and requirements that are appropriate in their natural habitats, enabling them to compete successfully with the species with which they co-evolved, but that make them over-aggressive in new habitats in competition with native species. Another factor may be the absence in a new habitat of the pest and pathogen load that would limit the vigour of the species in its natural habitat. Finally, hybridisation with native species may result in novel combinations of attributes associated with competition.

In the case of pests and pathogens, potential hosts in a new habitat may lack entirely genes for resistance to the alien species. Alternatively, the alien species may carry genes for pathogenicity or aggressiveness that are able to overcome the genes for resistance in hosts in the new habitat. And again, hybridisation with native species may result in novel combinations of genes for pathogenicity or aggressiveness that cannot be countered by genes for resistance in native hosts.

Globally, non-native invasive species are considered to be a major threat to biodiversity, especially for so-called island ecosystems, isolated by water or other physical barriers. The scale of the threat from non-native invasive species is often difficult to calculate because it may take a hundred years or more for them to become sufficiently established to have an impact, and because climate change may result in novel environmental conditions that result in a non-native species having a significantly increased competitive advantage over native species.

## Non-native invasive plant species

Since the Middle Ages, but especially since 1600, large numbers of non-native plant species have been introduced into the UK, principally as garden plants. Our gardens and our diet are the richer for them. Many of these have escaped into the wild, or have been deliberately planted or dumped there, and have become established as a significant component of the wild flora. Thus, of the higher plants recorded in the 2002 *New Atlas of the British and Irish Flora* (see Further reading), about 40% are of non-native origin. The majority of these are not considered invasive and a threat to the native flora. Some, however, have become invasive and are now causing major ecological damage. The best known in the UK are *Rhododendron ponticum*, Japanese knotweed (*Fallopia japonica*; Fig. 16.6) and Himalayan balsam (*Impatiens glandulifera*). There are, however, many more examples, and Table 18.1 lists the species named in Schedule 9 of the 1981 Wildlife and Countryside Act, which makes it illegal for them to be introduced into the wild by any means, and the much larger number of species that in 2006 the JNCC considered should be added to Schedule 9.

Gardeners may play an important role in the conservation of the native flora of the country or region in which they are living by minimising the threat from non-native invasive species. Valuable guidelines on how this may be achieved are issued by the Department for the Environment and Rural Affairs (DEFRA) on their website: www.defra.gov.uk. These were drawn up by the Working Group on the Horticultural Code of Practice, which is comprised of members drawn from DEFRA, the devolved administrations for Scotland and Wales, the horticultural trade, NGOs, the Royal Horticultural Society and botanic gardens. The resulting Horticultural Code of Practice is intended for use by horticultural practitioners of all types, from professional landscape designers, growers and gardeners to amateur gardeners. It recognises that not all non-native plants are

**Table 18.1** Alien vascular plant species named in Schedule 9 of the 1981 Wildlife and Countryside Act for the UK, together with the species that in 2000 the Joint Nature Conservation Committee recommended should be added to the list (see Note).

---

**Plants named in Schedule 9**

*Fallopia japonica* (*Polygonum cuspidatum*, Japanese knotweed), *Heracleum mantegazzianum* (giant hogweed)

**Plants recommended by the JNCC in 2006 for addition to Schedule 9**

*Allium paradoxum* (few-flowered leek), *Allium triquetrum* (three-cornered garlic), *Azolla filiculoides* (water fern), *Cabomba caroliniana* (Carolina water-shield or fanwort), *Carpobrotus edulis* (hottentot fig), *Cotoneaster* spp., and hybrids, except the wild *Cotoneaster integerrimus*, also known as *Cotoneaster cambrica* (cotoneasters), *\*Crassula helmsii* (New Zealand gymweed or Australian swamp stonecrop), *Crocosmia* species and hybrids (Montbretias), *Disphyma crassifolium* (purple dewplant), *\*Eichhornia crassipes* (water hyacinth), *Elodia* species (waterweeds or pondweeds); *Fallopia japonica* × *Fallopia sachalinensis* hybrid (hybrid knotweed), *Fallopia sachalinensis* (giant knotweed), *\*Gaultheria salon* (shallon), *Gunnera tinctoria* (giant rhubarb), *Hippophae rhamnoides* (sea buckthorn, in areas where it is non-native), *\*Hydrocotyle ranunculoides* (floating pennywort), *Impatiens glandulifera* (Himalayan balsam), *\*Lagarosiphon major* (curly waterweed), *Lamiastrum* (*Lamium*) *galeobdolon* subsp. *argentatum* (variegated yellow archangel), *Ludwigia peploides* (floating primrose willow), *\*Myriophyllum aquaticum* (parrot's feather), *\*Pistia stratiotes* (water lettuce), *Parthenocissus inserta* (false Virginia creeper), *Parthenocissus quinquefolia* (Virginia creeper), *Quercus cerris* (Turkey oak), *Quercus ilex* (evergreen oak), *Rhododendron luteum* (yellow azalea), *Rhododendron ponticum* (rhododendron), *\*Rhododendron ponticum* × *Rhododendron maximum* hybrid (rhododendron), *\*Robinia pseudoacacia* (false acacia), *Rosa rugosa* (Japanese rose), *Sagittaria latifolia* (duck potato), *\*Salvinia molesta* (giant salvinia), *Smyrnium perfoliatum* (perfoliate alexanders)

---

Species marked with an asterisk (*) were added to Schedule 9 by the Nature Conservation (Scotland) Act, 2004. Information derived from McLean, I.F.G. (2006) *Third Review of Schedule 9, Wildlife and Countryside Act, 1981*. Joint Nature Conservation Committee, Peterborough.

Note: Criteria for recommending the addition of a species to the list include: evidence that human activity results in or is likely to result in the intentional or accidental introduction of the species into the wild, and a belief that the species poses an actual or potential threat to natural biodiversity when introduced into the wild.

invasive, and acknowledges the great benefit that the trade in exotic plants brings to gardens and public places, and to the agricultural, horticultural and forestry industries. The aim is not to prohibit the use of non-native species, but to provide guidelines on how they may be used wisely, thereby preventing their spread to cause damage to native biodiversity. A summary of the principal elements of the Code are as follows (note: if the Code is to be put into practice, it is essential that the reader next visits the DEFRA website, where the details may be found).

1. *All gardeners should know what they are growing.* In other words, they should be aware of whether the plants they are growing or disposing of have invasive properties.

2. *All gardeners should be aware of non-native pests, pathogens and weeds/weed seeds* which may be present on the plants or in the soil surrounding the roots of the plants they are importing or buying (see below).

3. *Suppliers and retailers should know what they are supplying/selling.* Invasive non-native plants are often not good garden or pond plants. Alternatives should be sold if they are available, but if invasive non-native plants must be sold, they should be labelled accordingly.

4. *Suppliers should label all plants clearly and accurately.* Purchasers should avoid plants that are not properly and clearly labelled.

5. *Landscape architects, garden designers, design engineers, tutors, authors and publishers should know what they are specifying.* They should refer to plants by both the Latin and the common names, and should avoid where possible recommending or specifying invasive non-native plants.

6. *All plant users should dispose of plant waste responsibly.* Plant waste should never be disposed

of in the countryside. It should be composted or taken to local authority recycling centres, if appropriate. If the material is known to be from an invasive non-native species, the DEFRA website should be consulted as to the best method of disposal.

7. *Consumers and end users should know what they are buying.* The use of invasive non-native plants should be avoided.

8. *All users should take advice on the best methods for controlling invasive non-native plants that may be growing on their property.* Invasive plants are difficult to control, but timely action will reduce the size of the problem. The DEFRA website should be the starting point for advice.

9. *All users should be aware of the relevant legislation concerning the use of non-native plants and their safe control and disposal.* The DEFRA website is the best starting point for advice on the relevant UK, EU and international legislation.

10. *All users should control non-native plants safely.* The Code provides advice on health and safety and environmental safety for those wishing to control problem plants.

The Code is under active review and is subject to change at any time. Such change is facilitated by the fact that it is only issued in electronic form. The Working Group would welcome suggestions for improving the Code and information on its implementation. This may be sent by e-mail to species@defra.gsi.gov.uk.

## Non-native pathogens and pests

A great threat for the future comes from the reintroduction of non-native invasive pathogens. Most of these are microscopic or, if visible to the naked eye, are extremely small and therefore difficult to see clearly.

Natural plant communities consist of many species and populations, including potential pathogens, all occupying different niches in space and time. During evolution, some organisms have acquired the ability to cause disease in particular hosts. In South America, for example, where relatives of the potato (*Solanum tuberosum*) evolved, the pathogen *Phytophthora infestans*, cause of blight disease, can be found on wild *Solanum* populations. It does not, however, normally cause catastrophic epidemics, the

success and survival of both host and fungus populations being ensured by the fact that each is genetically diverse: in the former for resistance to infection, and in the latter for the ability to cause disease. Through co-evolution, a balance has been achieved such that each may grow and reproduce to give rise to the next generation.

Epidemics occur when circumstances intervene to upset the balance. In horticulture, for example, genetically uniform populations of plants are grown to facilitate display and good husbandry. If a pathogen mutates to overcome the resistance in an individual in such a population, it will then spread unchecked, for all individuals will be equally susceptible. An analogous situation occurs when an alien pathogen to which indigenous plants have no resistance is introduced from elsewhere, as in the case of a new strain of Dutch elm disease (*Ceratocystis novo-ulmae*) introduced to the UK, the USA and Europe, in the latter part of the last century, causing widespread destruction of native elms (*Ulmus* spp.; Fig. 18.1). Destruction also occurred when sudden oak death, or ramorum die-back (*Phytophthora ramorum*), was introduced to California and Europe, probably on nursery stock. The damage in this case was to both native species and cultivated *Rhododendron* and *Viburnum* spp. Box blight (Fig. 18.2), currently causing severe damage to plantings of *Buxus* spp., is caused by another non-native pathogen, *Cylindrocladium buxicola*. And in 2004 plant pathologists discovered a further alien *Phytophthora* species, *Phytophthora kernoviae*, which infects common beech (*Fagus sylvatica*). The consequences of this introduction are as yet unknown.

The threat from a non-native pathogen is greatly increased if hybridisation occurs between an aggressive non-native pathogen and a relatively benign but widespread native one. This appears to have happened with a new *Phytophthora* species currently afflicting alders (*Alnus* spp.). The problem is exacerbated because most pathogens have high mutation rates and are capable of dispersing large numbers of spores rapidly over wide areas. Moreover, with global warming, non-natives and non-native hybrids may remain undetected for long periods, only becoming a problem with a change of climatic conditions favouring their growth and dispersal.

With the rapidly increasing international trade in plants, usually rooted in soil, the risks from introduced non-native pathogens is now greater than ever

**Fig. 18.1** Dutch elm disease (*Ophiostoma novo-ulmae*; an invasive non-native pathogen) in Cambridge in the early 1970s. *Photograph by David S. Ingram.*

**Fig. 18.2** Box blight, caused by *Cylindrocladium buxicola*, a non-native pathogen. *Photograph courtesy of Royal Horticultural Society.*

before. The DEFRA Horticultural Code of Practice (see above) refers to non-native pathogens, and gardeners should be aware of this. For the future, however, new methods of detecting non-native pathogens that constitute a risk will be needed, for many are too small to be identified accurately with the naked eye or remain latent within the tissues of the host for long periods. Such tests are likely to be based on the use of molecular markers.

Testing will need to be complemented by better scientific intelligence of potential non-native invaders, and forecasting of the likely threats and consequences arising from the inadvertent introductions of such species. The only certain way of minimising the risks from the introduction of non-native pathogens,

however, would be a drastic curtailment of the international trade in living plants. This would have serious economic consequences, but may have to be considered in the future if damage to the native flora and cultivated plants increases significantly.

The risks associated with the introduction of non-native pests, mainly invertebrates, are similar to those outlined above for non-native pathogens. It has been estimated, for example, that in the last three decades there may have been more than 300 introductions of non-native insects into the UK; it was reported in 2004 that a shipment of tree ferns (*Dicksonia* spp.) into the UK was found to be carrying non-native invertebrates belonging to 18 distinct taxonomic groups.

Not all such non-native pest species are invasive, but some undoubtedly are. The New Zealand flatworm (*Artiposthia triangulata*; Fig. 18.3), for example, is now a major predator of earthworms in the UK, resulting in significant ecological damage to soils. This pest was almost certainly introduced in soil around the roots of imported horticultural plants. *Artiposthia triangulata* is already included

**Fig. 18.3** The New Zealand flatworm (*Artiposthia triangulate*), *left*, is much darker than the Australian flatworm (*Australoplana sanguinea*), *right*. Both are non-native pests. *Photograph courtesy of Tim Sandall/Royal Horticultural Society.*

in Schedule 9 of the Wildlife and Countryside Act, 1981; in 2006 the JNCC recommended adding the Australian flatworm (*Australoplana sanguinea*). Recently, the introduced *Cameraria ohridella*, the horsechestnut leaf miner (Fig. 18.4), has caused serious damage to the horsechestnuts (*Aesculus hippocastanum*). The symptoms of infestation by this pest were exacerbated in trees suffering water stress, highlighting the significance of climate to the level of damage caused by non-native species. These

examples emphasise the need for constant vigilance on the part of gardeners to prevent the introduction of non-native pests.

## Non-native vertebrates

Gardeners are less likely to introduce non-native vertebrates than plants and microorganisms. However, all who keep exotic fish in ponds, birds in garden aviaries and mammals, amphibians and reptiles in outdoor enclosures should be aware of the dangers associated with the release of any non-native species into the wild, for the risks are great. Moreover, the introduction of non-native vertebrates from one part of the UK to another can cause serious problems. For example, hedgehogs introduced on to the Hebridean island of South Uist have led to the massive depletion of wader bird populations by feeding on their eggs.

## Reintroduction of extinct native species

From time to time it is suggested that native species that have become extinct are reintroduced into the wild. For example, breeding populations of sea eagles have been reintroduced into the UK, with considerable success. Discussion of such reintroductions,

**Fig. 18.4** Damage caused by *Cameraria ohridella*, the horsechestnut leaf miner, a non-native pest. *Photograph courtesy of Andrew Halstead/Royal Horticultural Society.*

which are usually the subject of intense debate, is beyond the scope of this book, but an appropriate reference has been included in the Further reading list.

## CONCLUSION

Despite the massive impact of the combined effects of climate change, the erosion of biodiversity and other environmental changes, gardeners have the capacity to contribute significantly to national and international efforts to combat these trends. It is, therefore, the social responsibility of all gardeners to be aware of the issues and to modify their practices accordingly, to the best of their ability, within the limits imposed on them by financial, physical and other constraints.

## FURTHER READING

Akeroyd, J., McGough, N. & Wyse-Jackson, P. (1994) *A CITES Manual for Botanic Gardens*. Botanic Gardens Conservation International, Richmond.

Anon (2007 and subsequent years) *The National Plant Collections Directory*. National Council for the Conservation of Plants and Gardens, Wisley.

Astley, D. (2007) Banking on genes. *The Horticulturist* **16**, 2–5.

Bisgrove, R. (2002) Weathering climate change. *The Horticulturist* **11**, 2–5.

Brasier, C. (2005) Preventing invasive pathogens: deficiencies in the system. *The Plantsman* **4**, 54–7.

Clevely, A. (2007) Low on $H_2O$. *The Garden* **132**, 98–103.

Colborn, N. (2007) Perfectly adapted. *The Garden* **132**, 30–7.

English Nature (1995) *Habitat Creation – A Critical Guide*. English Nature Science No. 21. English Nature, Sheffield.

Gates, P. & Ardle, J. (2002) Climate change . . . coming soon to a garden near you. *The Garden* **127**, 912–17.

Goode, D.A. (2006) *Green Infrastructure*. Report commissioned by the Royal Commission on Environmental Pollution, London (available on www.rcep.org.uk).

Gore, A. (2006) *An Inconvenient Truth*. Bloomsbury, London.

Henson, R. (2006) *A Rough Guide to Climate Change*. Rough Guides, London.

Hickmott, S. (2003) *Growing Unusual Vegetables*. Ecologic Books, Bristol.

Hodgson, I. (2008) Gardening in a changing climate. *The Garden* (Special Issue), **133**, 1–74.

Hopkins, J.J., Allison, H.M., Walmsley, C.A., Gaywood, M. & Thurgate, G. (2007) *Conserving Biodiversity in a Changing Climate: Guidance on Building Capacity to Adapt*. DEFRA on behalf of the UK Biodiversity Partnership, London.

Houghton, J. (2004) *Global Warming: the Complete Briefing*, 3rd edn. Cambridge University Press, Cambridge.

Hulme, M., Jenkins, G.J., Turnpenny, J.R., Mitchell, T.D., Jones, R.G., Lowe, J., Murphy, J.M., Hassell, D., Boorman, P., McDonald, R. & Hill, S. (2002) *Climate Change Scenarios for the United Kingdom: the UKCIP02 Scientific Report*. Tyndall Centre for Climate Change Research, University of East Anglia, Norwich.

Ingram, D. (2005) Time to avert disaster. *The Plantsman* **4**, 58–9.

Intergovernmental Panel on Climate Change (2001) *Climate Change 2001: the Scientific Basis*. A Report of Working Group 1 of the Intergovernmental Panel on Climate Change. Cambridge University Press, Cambridge.

IUCN – The World Conservation Union (1995) *Guidelines for Re-introductions*. IUCN/SSC Re-introduction Specialist Group, Gland.

Jones, H.D. (2005) Identification: British land flatworms. *British Wildlife* **16**, 189–94.

Lynas, M. (2007) *Six Degrees: Our Future on a Hotter Planet*. Fourth Estate, London.

McLean, I.F.G. (ed.) (2003) *A Policy for Conservation Translocations of Species in Britain*. Joint Nature Conservation Committee, Peterborough.

Napier, E. (2007 and subsequent years) *Plant Heritage* (Journal of the National Council for the Conservation of Plants and Gardens). National Council for the Conservation of Plants and Gardens, Wisley.

Park, C. (2007) *Dictionary of Environment and Conservation*. Oxford University Press, Oxford.

Preston, C.D., Pearman, D.A. & Dines, T.D. (2002) *New Atlas of the British and Irish Flora*. Oxford University Press, Oxford.

Rodwell, J.S. (1991–2000) *British Plant Communities*, vols 1–5. Cambridge University Press, Cambridge.

Rodwell, J.S. (2006) *National Vegetation Classification Users' Handbook*. Joint Nature Conservation Committee, Peterborough.

Stuart, D. & Sutherland, J. (1989) *Plants from the Past*. Penguin, London.

Stuart, D. (1997) *Gardening with Antique Plants*. Conran Octopus, London.

Thornton-Wood, S. (2004) Gardening with a conscience. *The Garden* **129**, 350–1.

Vincent, M.A. (2006) Ideas for a UK Nature Conservation Framework. *Conservation Challenge No. 1*. Joint Nature Conservation Committee, Peterborough.

Williams, C., Davis, K. & Cheyne, P. (2003) *The CBD for Botanists*. The Royal Botanic Gardens, Kew (with the Darwin Initiative for the Survival of Species, DEFRA, London).

Wilson, E.O. (1993) *The Diversity of Life*. Allen Lane, London.

Wilson, E.O. (2003) *The Future of Life*. Abacus, London.

# 19

# Gardens and the Natural World

## SUMMARY

Some basic ecological terms and principles are outlined. Next, guidelines based on ecological principles are given for the creation of garden habitats for native species. Specific guidelines and suggested native plant species lists are then given for the creation of the following named habitats: woodland margins, hedgerows, wild-flower meadows and cornfield wild-flower plantings, and ponds and marshes. Finally, guidelines and species lists for supplementing existing gardens to make them attractive to wild flora and fauna are provided.

## INTRODUCTION

The gardens of the ancient civilisations of the Middle East were enclosed to exclude the natural world. The enclosed plot as a refuge from wild nature is also seen in medieval gardens, in the parterres of seventeenth-century Europe, in the fenced public parks of the nineteenth century and in the walled and hedged suburban gardens of the late twentieth century. Even the patio garden of the twenty-first century is a miniature version of the walled gardens of the ancients. Nevertheless, the natural world has always been present in these pleasure grounds, as also in gardens for the production of food, in the form of pests, pathogens and weeds. In all cases, however, the emphasis has been on the exclusion or destruction of such undesirables, with little or no attempt being made to accommodate them. Technological innovation during the nineteenth and twentieth centuries, which led to the development of more decorative or productive cultivars of plants (see Chapter 5), more efficient methods of cultivation and more effective methods for the control of undesirables (see Chapters 15 and 16), served to strengthen the distinction between wild nature and the garden world.

Naturalistic 'landscape gardens' were, of course, appreciated and available, at least to the wealthy owners of large country houses, from the eighteenth century onwards, but even in these nature was presented in a sanitised and unthreatening form.

Wild plants and animals were not, however, ignored by all gardeners. Native plants in particular were studied and prized for their medicinal properties from medieval times, being described and illustrated in herbals. Later, these wild plants of medicinal importance were cultivated in the first botanic gardens (see Chapter 20). Moreover, from the late eighteenth century onwards, exotic wild plants, notably from the Americas, Australia, New Zealand, South Africa and China, were brought back for incorporation into botanic and private gardens in the UK by plant collectors such as David Douglas (1799–1834) and George Forrest (1873–1932). Wild native flowers, however, although studied for their own sake by the early naturalists like John Ray (1627–1705) and Gilbert White (1720–93), were not perceived as having a role to play in the garden until the late nineteenth century. William Robinson (1838–1935), author of *The Wild Garden* of 1870, was a seminal figure in this new development. He despised formal gardens, seeing 'natural' gardens and the plants they contained as more important than structures and hard landscaping. This battle of styles is played out every year to the present day at the Chelsea Flower Show, for with the current accelerated loss of natural habitats such as woodlands, hedgerows, meadows and wetlands, the desire of gardeners to include wild plants and other wildlife into their own gardens has gathered considerable momentum.

Robinson knew nothing of the science of ecology, however, for although the term was invented in the nineteenth century, the science itself did not come into prominence until the twentieth century. It is concerned with the relationships of plants, animals and microorganisms to each other and to the environment. It thus cuts across the traditional academic disciplines of botany, zoology and geology. Also, in contrast to the essentially descriptive, anatomical and nomenclatural approaches to biology of earlier centuries, it is an analytical science, based on careful field observation, recording and experimentation. In its earliest days it represented such a fundamental change in the approach to the natural world that in the UK a new scientific society, the British Ecological Society, was founded for its practitioners, in 1910. The ever-deepening understanding of the functioning of the natural assemblages of organisms called ecosystems (see Chapter 1) has provided the scientific knowledge required to create gardens that are attractive to wildlife, and may even be good representations of natural habitats. Whether such gardens have a direct role in conservation is debatable, but the development of conservation science in the second half of the twentieth century further informed their design and management.

In gardens designed to accommodate the natural world, the terms weeds, pests and pathogens have no meaning, except when they refer to alien species that threaten native, natural ecosystems (see Chapter 18).

## SOME BASIC TERMS AND ECOLOGICAL PRINCIPLES

For the most part, plants will be used as examples throughout.

The place where an organism or group of organisms, a community, actually lives, is called a *habitat*. It is usually characterised by its physical properties, for example a pond, or the seashore, or by the principal species determining its properties, for example a woodland or grassland. The principal species of the latter habitats, trees and grasses respectively, are usually said to be dominant, meaning, within the habitat or community, the species that exerts the most influence on the other species that grow there. Where more than one species together exert the most influence they are said to be co-dominant. The term habitat may be used on all scales, from a dead log to a woodland, for example.

*Ecosystem* is a term originally coined by the great ecologist Arthur Tansley (1875–1955) in 1935 to describe the interdependence of all the species in a living community on one another and on their physical environment. Basic principles relating to ecosystems include the flow of energy from one organism to another in a simple linear fashion, as in food chains, or in a more complex manner, as in food webs, and the cycling of nutrients among the living organisms and the soil. Nutrient cycling and recycling occurs when minerals are taken up from the soil by plant roots, passed to herbivores and on into carnivores. All the nutrients, however, are returned to the soil when the plant, herbivore or carnivore dies and is decomposed by bacteria and fungi. Ideally, no nutrients are lost in an ecosystem, so that the pool of nutrients in the soil is continually being renewed and made available for other organisms. In fact, however, nutrients may be lost by leaching into groundwater or streams and rivers. As with a habitat, ecosytem principles may be used on all scales, from a small puddle to a lake, from a dead tree to a forest.

Groups of individuals of a given species within a community, habitat or ecosystem are said to form a population. The individuals within a population will be genetically homogeneous for the major features that characterise the species, but genetically diverse for secondary characteristics such as growth rate or flower colour, unless the individuals comprise a clone, a population of organisms that are near genetically identical in all respects (see Chapters 5 and 10).

Within a community the organisms interact with one another in a dynamic way. They may compete for major resources, most often light but frequently water, major nutrients or pollinators. Such competition may result in one species becoming dominant over its competitors. This may be because the species is able to grow faster or larger than its competitors within the ecosystem, for example a tree species in a forest, or because it is better adapted for dealing with an environmental stress such as extracting nutrients from impoverished soil or excluding salt in a saline habitat such as a salt marsh.

The competition pressure exerted by one species may be so extreme that it excludes all other species. This competitive exclusion is best exemplified by the results of a classic experiment with two species of duckweed, *Lemna polyrhiza* (*Spirodela polyrhiza*) and *Lemna gibba*. When grown alone in the laboratory, *L. gibba* grew more slowly than *L. polyrhiza*. When grown together, however, *L. polyrhiza* was rapidly replaced by *L. gibba*. The reason was that the *L. gibba* plants contained sacs filled with air, enabling them to form a floating mass over other species, thereby cutting off their access to light.

Plants living in the same habitat do not always compete, however. For example, a number of species are able to co-exist because they have evolved to exploit different niches, a *niche* being the position that a species occupies in space and time, within an ecosystem, relative to other species. The ability to exploit particular niches may involve physiological, structural, life-form and life-cycle adaptations (see Chapter 1). For example, in a deciduous woodland the dominant species cast a deep shade for part of the year, the sunny months of the late spring and summer, but cast little shade during the winter when light levels are lower and days shorter. Different woodland species deal with the lack of light during the summer in different ways. Thus butcher's broom (*Ruscus aculeatus*) can survive in shade by virtue of being evergreen, for it is able to grow, albeit slowly, during the autumn, early spring and, to a lesser extent, the winter months. Dog's mercury (*Mercurialis perennis*), in contrast, is a *shade plant* and can grow perfectly well in the deep shade of spring and summer by being physiologically adapted to do so (see Chapter 8). Indeed, in unshaded places it soon dies out as it is unable to compete with the other species, *sun plants*, better adapted to grow in such situations. Other species, the *geophytes* (see Chapter 1), survive the deeply shaded summer months as underground storage organs (see Chapters 1, 2 and 10). Such species, for example wood anemone (*Anemone nemorosa*) grow, flower and set seed in the early weeks of the year, before the leaves appear on the trees. Yet other species, such as foxglove (*Digitalis purpurea*), avoid shade by surviving in seed banks in the soil (see Chapters 8, 9, 15 and 16) until tree death or coppicing (see below) allows light to enter, whereupon the seeds germinate. Others, for example marsh thistle (*Cirsium palustre*), evade shade by changing position as the pattern of shade changes, as when trees become established in clearings and create new shade, and tree death and management elsewhere create new clearings. Some species, such as honeysuckle (*Lonicera periclymenum*), climb to the canopy, thereby reaching the light, while others, such as ferns, grow within the canopy as *epiphytes* (see Chapter 1). Finally, some species, such as the toothworts (*Lathraea* spp.), require no light at all, having evolved to obtain their carbon by parasitising trees.

Sometimes, particular groups of plants tend always to grow together without competing in a habitat or ecosystem, forming what are sometimes called guilds. For example, again in deciduous woodland, bluebell (*Hyacinthoides non-scripta*), wood anemone (*Anemone nemorosa*), primrose (*Primula vulgaris*) and violet (*Viola riviniana*) all form a guild. The five native British woody climbers, honeysuckle (*Lonicera periclymenum*), dog rose (*Rosa canina*), woody nightshade (*Solanum dulcamara*), ivy (*Hedera helix*) and clematis (*Clematis vitalba*), form another guild. This is in part dependent on local soil type and climate, but is also indicative that they all occupy the same niche, but without competition because each exploits the niche in a slightly different way.

In addition to competition, other interactions between species are important in ecosystem dynamics. One of the most significant is *mutualism*, where two species interact to enable both to grow together in an environment or niche that neither could exploit alone. The ubiquitous mycorrhizal associations between certain fungi and the roots of plants are good examples (see Chapters 2 and 6). A mycorrhizal fungus receives carbon from its photosynthetic host, while enabling that host to grow in soils low in soluble phosphates, which means almost any natural soil. Mycorrhizal associations may be even more complex and subtle in some circumstances. For example, seedlings of broad-leaved plants are in part able to become established in the thick sward of a grassland, which shades all below it, because they are supplied with sugars from established plants via an underground web of mycorrhizal fungi. Another example of mutualism is seen in lichens, where the combination of an alga and a fungus (see Chapter 1) enables the lichen to be the first, or primary, coloniser of very exposed habitats such as bare rock surfaces.

Another type of interaction is *parasitism*, where one species acquires its nutrients by killing and enzymatically degrading another organism, the host, or by removing nutrients from a living host, thereby debilitating it (see Chapters 15 and 16). Examples of parasites of plants include bacteria, fungi, viruses and higher plants (see Chapter 15 and Table 15.1). Parasitism may be especially important in natural ecosystems, in limiting the vigour of certain plant populations and therefore any competitive advantage they may have over other species, or by killing long-lived individuals, as when honey fungus (*Armillaria mellea*) kills a woodland tree, thereby allowing other organisms to occupy the site. Parasites may occasionally cause epidemics in natural ecosystems, usually when they are aliens introduced from another country (see Chapter 18). This was the case with the epidemic of Dutch elm disease caused by a new strain of the pathogen, *Ophiostoma novo-ulmae*, from the 1970s onwards.

Related to parasitism, and equally important for plants, is herbivory, where species are eaten or damaged by other, animal, species. Examples include sap-sucking aphids (e.g. *Myzus persicae*), leaf-cutting vine weevils (*Otiorhyncus sulcatus*) and browsing mammals such as rabbits (*Oryctolagus cuniculus*). As with parasitism, herbivory may result in debilitation and loss of vigour, or even death of the host. Introduced alien herbivores may cause epidemics (see Chapter 18). Organisms that feed on plants may also carry disease from one individual to another, as is the case with many viruses that are spread by insects, or may create wounds that allow the entry of pathogens, as when a bird pecks at a ripening apple allowing the entry of *Penicillium* spp.

Allelopathy is another interaction sometimes encountered among plants, although its ecological significance is often disputed. This involves the production of chemicals by one species to inhibit the growth of a competitor species. An often-quoted example is that of black walnut (*Juglans nigra*), which chemically inhibits the growth of seedlings of other plants close to it. Classic examples of allelopathy are the production of antibiotics by some microorganisms to inhibit the growth of others, as when the common soil fungi, *Penicillium* species, produce penicillins for this purpose.

No mention has been made of predator–prey, parasitic and mutualistic relationships among animals, for they are beyond the scope of this book, but they are just as important as interactions among plants in determining ecosystem dynamics.

Finally, it is important to mention the concepts of succession and climax communities. In Chapter 1 it was suggested that ecosystems are relatively stable assemblages of plants, animals and microorganisms. By and large this is true, but where a sudden change in a component of the ecosystem occurs, whether it be climatic or a change in one or more species, or both, for they are often linked, relatively rapid changes in the whole ecosystem are triggered and continue until stability is once more achieved. If a piece of woodland is cleared and left to itself, for example, it is first colonised by a series of pioneer herbaceous species, mainly annuals, then by perennial herbs and shrubs until finally, as the seedlings of tree species become established and grow, it becomes dominated again by trees. This represents a succession, and the woodland dominated by large trees may be regarded as the climax community. Another example might be a pond which gradually becomes silted up by soil washed into it by streams and by the build up of organic matter from dead marginal vegetation. Slowly moisture-loving shrubs such as willows (*Salix* spp.) become established at the margins and further add to and dry out the habitat to the point where trees characteristic of the region become established, again forming a climax community of woodland. Successions may develop at various speeds, depending on circumstances, or may be reversed, as when a woodland slowly becomes flooded because of a change in the water table.

## FULLY NATURAL ECOSYSTEMS IN GARDENS

Few gardens are large enough to enclose a significant area of a natural ecosystem, whether woodland, grassland or wetland. Even those that are large enough rarely enclose an ecosystem that is completely natural. Management of some kind is almost always practised, usually to enhance the diversity and quantities of wild species present, for unless the area is exceptionally large the year-round 'interest' required by most gardeners is rarely seen in unmanaged landscapes. The modifications required to enrich an area with wild flora and fauna is dealt with below.

If the natural ecosystem enclosed is part of a Site of Special Scientific Interest (SSSI), designated as such by one of the statutory conservation agencies – the Countryside Council for Wales, English Nature or Scottish Natural Heritage – or an Area of Special Scientific Interest designated by the Environment and Heritage Service on Northern Ireland, it will require a different kind of management to conserve those characteristics that make it scientifically distinct. In such cases a site-specific management plan will be made available to the garden owner, leaving little scope for individual creativity.

## CREATING HABITATS IN THE GARDEN

Creating representations of natural habitats using native species of plants brings the special beauty of the natural world into the garden. Whether such plantings are more attractive to wildlife than a traditional garden, as is often claimed, however, is now being questioned (see below).

It was pointed out in Chapter 1 and above that natural ecosystems and other natural assemblages of plants, animals and microorganisms are relatively stable, the individual species within them being largely complementary and mutually sustaining, with nutrients constantly being cycled and recycled. In creating garden habitats for native species, a principal aim should be to achieve similar stability. This will not be achieved immediately, for the garden ecosystem will take time to become fully established, but if planned carefully, with successions and competition and food chains and webs thought through, it will settle down in a relatively short time. There are certain basic guidelines and premises, based on ecological principles, that should be followed, and these are set out below.

### Basic guidelines and premises

1. *Study the surroundings*, especially the local climate and soils (see Chapter 6), for these will be the best guide to what kinds of plants grow well in an area, and what kinds of wildlife are most likely to be attracted to the garden. A city-centre wetland garden, unless a canal or river is nearby, will probably be colonised much more slowly by wild species than one in the countryside where ponds and streams occur naturally. A woodland garden richly stocked with ferns, mosses and liverworts will be easier to achieve on the wetter, western side of the country than in the east, where rainfall is generally low. If the local soils are acid, it is likely that the garden soil will be too, so that acid-loving wild plants (*calcifuges*; see Chapter 6) may be grown with ease. If the local soil is alkaline, lime-loving plants (*calcicoles*; see Chapter 6) will not only be most appropriate, but the most easily grown too, while the insects such plants attract are likely to be nearby, feeding on local wild plants.

2. *Then study the garden itself* very carefully, noting the soil pH, the prevailing wind speed and direction, the aspect and degree of exposure, the nature of the soil, whether free-draining or not, clayey, gravelly, loamy or with rocky outcrops (see Chapter 6), and the drainage of the site. All these factors will need to be taken into account when deciding what kind of habitat can most easily be achieved and what sorts of plants are most likely to succeed. The existing plantings in the garden will also be important, especially the trees and shrubs, even if they are exotics. Unless these are unsightly, dangerous or simply in the way, they are best left, even though not strictly appropriate to the planned habitat, for they will supply instant structure, microhabitats for invertebrates and birds, and food for existing wildlife in the garden. If it is decided that a tree should be removed, the local planning authority should be consulted before any work is carried out, for planning laws relating to trees must be adhered to.

3. *Use pesticides, fungicides and herbicides with great care*, for these will militate against diversity of macro- and microorganisms in the habitat, both directly and indirectly. Herbicides, for example, not only destroy particular groups of plants, but also result in loss of the invertebrate species and their larvae that feed on them, and perhaps loss of the species of birds or small mammals that feed on these. The use of insecticides may kill a much wider range of species than the target insect, with knock-on effects on the other species they sustain. Residues from pesticides, fungicides and herbicides may pollute

the soil and groundwater if they are not used according to the instructions on the labels, with unpredictable consequences. Residues in dead insects or plants may enter the food chains and webs, killing the birds and other animals that feed on them. Parent animals may also pass on toxins to their young during feeding. When a new habitat is created in a garden, the organisms commonly regarded as pests, diseases and weeds may initially be a problem, but with time, as the ecosystem comes into balance, the problem should diminish significantly. Chapters 15 and 16 include science-based advice on managing the use of fungicides, pesticides and herbicides in the garden.

4. *Avoid the excessive use of fertilisers and manures*, especially those high in nitrogen. These may favour over-lush or vegetative growth of plants at the expense of flowers, or over-rapid growth of some species rather than others, distorting natural patterns of competition and militating against diversity, unless they are applied with great care. If over-applied, they may find their way into watercourses or ponds, promoting the abundant growth of aquatic weeds such as green algae. The aim should be to mimic natural ecosystems in which all nutrients are cycled through living organisms, eventually being returned to the soil, decomposed by microorganisms and recycled (see Chapter 1 above). Chapters 6 and 7 include science-based advice on managing the use of fertilisers and manures in the garden.

5. *Avoid tidiness and promote untidiness.* Long rather than short grass provides a much greater diversity of habitats for invertebrates and low-growing plants. Grass that is not mown until after seed set provides both a food source for birds and mammals, and seed for the growth and spread of the grass species themselves. Even grass that is mown regularly can be made to provide a wide range of habitats by setting the mower blades higher than normal.

Piles of logs, sticks and other debris provide habitats for insects and other arthropods and animals that feed on them, for mosses, liverworts and ferns, and for mushroom- and toadstool-forming fungi. Piles of stones or rocks may serve similar functions and will also provide niches for some types of plants, especially alpines.

6. *Do not cultivate the soil.* This breaks up soil structure (see Chapter 6) and disturbs the soil microflora, fauna and seed banks, thereby militating against habitat stability. The only circumstance is which cultivation is desirable is when the habitat being created is one in which certain types of annual wild flowers are to be grown, for such species are often adapted to grow in disturbed soil, their seeds often being stimulated to germinate by exposure to light and/or cold (see Chapters 1 and 9). Annual wild flowers have for centuries succeeded as weeds in ground cultivated for agriculture and horticulture.

7. *Never remove plants or seeds from the wild* in order to grow them in the garden. Many species and/or the sites in which they grow are protected by law. It is, however, bad practice to disturb any site or species, for this may have unforeseen consequences. It is tempting to imagine that the small amount of a plant or its seed removed by one person will have little impact, but with large numbers of people now having access to wild sites, even limited collection by each individual can cause massive damage collectively. Indeed, many species, for example snowdrop (*Galanthus nivalis*), have been brought close to extinction in the wild as a result of over collection.

Wild plants or seed for use in creating gardens should be obtained from reputable commercial nurseries or seed merchants. These will ensure that the stocks have been produced in a sustainable way and are correctly named. In cases of doubt, the statutory conservation agencies or local wildlife trusts will provide advice. Whenever possible, seeds of wild plants should be sourced locally, for these are more likely to resemble local plant populations genetically than seeds obtained from elsewhere.

Some wild plants will, of course, find their way into the garden naturally, having been deposited there by wind, or in the faeces or on the coats of birds and other animals. Such natural colonisers should be encouraged, not treated as weeds. Ferns, mosses and liverworts will all enter the garden as blown spores and will establish themselves wherever moisture is present.

The same general rules concerning collection from the wild should be applied to fauna as well as flora. If a garden habitat is appropriate and well-managed, it will be colonised naturally by the wild invertebrates, reptiles, amphibians, birds and mammals present in the area.

Ensuring that these have access to the garden is important.

8. *Seed of wild plants sown direct into a garden habitat rarely produce plants that survive.* Viability is frequently low and seedlings are often eaten by herbivores or killed by pathogens. Seedlings should be raised in pots or trays, using the same procedures and applying the same scientific principles as are set out in Chapter 9. They should then be grown on in pots until they are robust enough to be planted out. Wild plants already growing in the garden may be propagated vegetatively (Chapter 10) or from seed (Chapter 9), but it should be remembered that in the case of seed, hybridisation with cultivated plants may have occurred as a result of crosspollination (see Chapter 5). The viability of home-collected seed is unpredictable (see Chapter 9).

9. *Plan the plantings carefully.* It was pointed out in Chapter 1 that in natural ecosystems the size, shape and form of individual species of plants is complementary, both above and below ground, so that each has access to light and the nutrients required for growth and reproduction. Similarly, the reproductive cycle of each species is coordinated with that of others, such as pollinators, and with the environment, to ensure that germination, flowering and seed set occur at the optimum times for these processes. The mix of species used in planting up a habitat should, therefore, not only be appropriate to that habitat and to the local climate and soil conditions, but should also be chosen to ensure that the vertical, horizontal and temporal matrix of species, life forms and life cycles is appropriate. Moreover, in planting up a habitat, careful account needs to be taken of nutrient flow, food chains and webs, competition, niches, guilds, mutualism, parasitism and successions (see above). Floras, especially local floras, and the publications of conservation bodies and local wildlife trusts, will provide important additional information on the requirements of individual species – sun or shade, wet or dry – and on the types of niche and habitat that they are able to exploit, but these are no substitute for good field observation.

10. *Consult appropriate texts for advice on practical horticultural procedures* (see Further reading, below) for creating garden habitats, for these are beyond the scope of this book.

11. *Take account of the views and wishes of neighbours.* They may object strongly, for example, to the prevalence of wild plants, insects or molluscs in an adjacent garden, which they may regard as weeds and pests, and which may easily spread to their own garden. Neighbours might also object, understandably, to untidiness, or to trees that are too large and/or cast shade or threaten drains and foundations. Many people are also highly sensitive to grass pollen. Failure to take the views of neighbours into account could result in disputes and legal action.

12. *Take account of all planning regulations*, especially with regard to the proposed removal, pruning, pollarding or coppicing of trees. If in any doubt, consult the designated tree officer of the local authority. Failure to take account of such matters could result in legal action. Note that *all major tree work* approved by the local planning authorities should be carried out by a properly qualified, registered and insured tree surgeon.

What follows is a summary of ecological features desirable for creating four types of garden habitat – woodland margins, hedgerows, grasslands and wetlands – together with suggestions of appropriate native species with which to stock them initially (adapted from Baines, 2000). Careful study of plants growing locally in the wild will make it possible to modify the plant lists few individual gardens.

## Woodland margins

The aim might be to create a habitat resembling the margin of an ancient woodland. This would probably have been managed, to greater or lesser extent, by coppicing or pollarding some of the trees to produce firewood and materials for fencing and furniture making, and by felling some larger trees to provide timber for constructing vehicles and buildings. Coppicing involves cutting large shrubs such as hazel (*Corylus avellana*) or trees such as ash (*Fraxinus excelsior*) close to ground level every few years in order to stimulate the production of long, straight branches for harvesting. In pollarding, practised where browsing animals such as deer were a problem, the branches of trees such as oaks (*Quercus* spp.) were removed on a regular basis, leaving the trunk intact. The ecological consequences of both of these management practices was the creation of clearings in the tree canopy, allowing light-loving plant species to regenerate on a regular basis, thereby promoting diversity. In the wild, clearings only occur unpredictably when a tree dies, is struck by lightning or is blown down; such clearings may then, however, persist for a considerable length of time as woodland glades if there are sufficient densities of grazing animals such as deer.

With light entering the woodland margin and with coppicing occurring regularly, the margins of ancient woodlands were rich in herbaceous plant and shrub species, these in turn attracting and supporting an equally diverse fauna. A similar level of diversity may be achieved in a garden woodland margin if the following specific guidelines are applied, in addition to those summarised above.

1. *Choose a south- or south-west-facing site*, to ensure that a significant amount of light will enter the planted area.
2. *Attempt to create a layered structure* when planting, with tall trees, shrubs of different sizes and a herbaceous layer. The mix might include evergreen and deciduous trees and understorey shrubs, shade- and sun-loving species in the herbaceous layer, climbers, and spring-flowering geophytes (see Chapter 1 and above). The size of the garden will determine the number and vigour of trees and shrubs that can be planted, but even the smallest plot can usually accommodate at least one small tree, such as silver birch (*Betula pendula*) or rowan (*Sorbus aucuparia*) and several shrubs, as well as a rich diversity of herbaceous species. When space is limiting, it is usually better to overplant initially, removing some trees and shrubs as they become too large for the space, or carrying out heavy pruning or coppicing of certain shrubs such as *Cornus* spp., and pollarding small trees. Indeed, these two practices add a dynamic component to the garden.
3. *Create log piles*, and promote areas of leaf litter and brushwood. These will all provide habitats for a diversity of organisms. Shredded bark as a mulch may be a useful way of preventing the growth of competitor plants around newly planted trees and shrubs until they become established, and serves as a useful substitute for leaf litter and brushwood until these are produced naturally.

Table 19.1 gives the names of some trees, shrubs and herbaceous species suitable for use in a garden woodland margin. The species listed have been chosen for their ability to thrive in a wide range of conditions.

Finally, a hazard that should be recognised in the garden woodland margin is the threat of infection by honey fungus (*Armillaria* spp; Fig.19.1). There are several species of this pathogen present in Britain, but only two, *Armillaria mellea* and *Armillaria ostoyae*, kill otherwise healthy trees and shrubs. Other species may infect roots, causing limited damage if the specimen is healthy, but hastening death in a tree or shrub already weakened by another infection or by stress such as waterlogging. The spores of *Armillaria* species initially enter plants through wounds, and the infected individual then provides a food base from which the invasive bootlace-like bundles of hyphae (*rhizomorphs*) of the pathogen may grow out to infect other shrubs and trees, and even some herbaceous species. If this disease occurs, the infected specimen should be removed, together with the roots and stump, and all diseased material burnt. It is important to be aware that *honey fungus can spread for considerable distances* and may reach the plots of neighbours.

## Hedgerows

The aim might be to plant a boundary with native hedgerow species, managing the resulting hedge to maintain a dense meshwork of branches attractive to wildlife and capable of serving as a wildlife corridor linking one area with another. If the site allows, the hedge should be planted along an approximately north–south line to give a sunny side and a shaded side, thereby increasing the range of plant species that may be grown. While the shrubs and small trees comprising the hedgerow may be allowed to grow to a large size, it is preferable to use the ancient management practice of laying, also called plashing, to maintain the hedge at a reasonable height and to encourage dense growth at the base. To achieve this the shrubs are first allowed to grow to a height of 2–3 metres, and then the stems and main branches are cut part through close to the ground and laid near horizontally, all in the same direction. Stakes may be used to hold the laid branches in position and, if desired, their effectiveness may be increased by weaving a top layer of branches between them. Side branches will be encouraged to grow upwards by this procedure, albeit more slowly than leaders (see Chapter 11), and these may be cut off at a predetermined height (1.5–2 metres), thereby encouraging the growth of further side shoots. Flailing to maintain the hedge at a desired level is more effective in promoting growth of side shoots than using a hedge trimmer, perhaps because the trauma caused induces greater production of growth-stimulating hormones (see Chapters 2 and 11). It is advisable to flail or cut only one side of the hedge per

**Table 19.1**  Some native UK plants of potential value in creating a woodland margin habitat.

| Type of plant | Examples |
|---|---|
| **Large trees** (too large for most gardens, but most may be grown as coppiced or pollarded specimens | Hornbeam (*Carpinus betulus*), ash (*Fraxinus excelsior*), Scots pine (*Pinus sylvestris*), gean (*Prunus avium*), bird cherry (*Prunus padus*), oaks (*Quercus patrea* and *Quercus robur*), white willow (*Salix alba*), elm (*Ulmus procera*; will only grow reliably in areas where the Dutch elm disease pathogen is absent) |
| **Smaller trees** (some still too large for some gardens, but most may be grown as coppiced or pollarded specimens) | Deciduous species: field maple (*Acer campestris*), alder (*Alnus glutinosa*; prefers wet conditions), silver birch (*Betula pendula*; needs a well-drained, open position), downy birch (*Betula pubescens*; prefers wetter conditions), crab apple (*Malus sylvestris*), aspen (*Populus tremulus*; prefers moist, acid soil), rowan (*Sorbus aucuparia*; perhaps the best tree of all for the very small wildlife garden), wild service (*Sorbus torminalis*) Evergreen species (less useful than deciduous species, for they cast a heavy shade throughout the year): holly (*Ilex aquifolium*), yew (*Taxus baccata*) |
| **Shrubs and shrubby trees** | Strawberry tree (*Arbutus unedo*), dogwood (*Cornus sanguinea*), hazel (*Corylus avellana*), hawthorn (*Crataegus monogyna*), blackthorn (*Prunus spinosa*) field rose (*Rosa arvensis*), dog rose (*Rosa canina*; also a climber), downy rose (*Rosa villosa*; commoner in the north than *R. canina*), wayfaring tree (*Viburnum lantana*), guelder rose (*Viburnum opulus*) |
| **Climbers** | White bryony (*Bryonia dioica*); great bindweed (*Calystegia sepium*); clematis or old man's beard (*Clematis vitalba*); ); ivy (*Hedera helix*); honeysuckle (*Lonicera periclymenum*); dog rose (*Rosa canina*, may also be grown as a shrub); woody nightshade (*Solanum dulcamara*). |
| **Annuals** | Herb Robert (*Geranium robertianum*) |
| **Biennials** | Garlic mustard or Jack by the hedge (*Alliaria petiolata*), foxglove (*Digitalis purpurea*) |
| **Herbaceous perennials** | Columbine (*Aquilegia vulgaris*), nettle-leaved bellflower (*Campanula trachelium*), greater celandine (*Chelidonium majus*), wild strawberry (*Fragaria vesca*), sweet woodruff (*Galium odoratum*), wood cranesbill (*Geranium sylvaticum*), water avens (*Geum rivale*; moist soils), herb bennet (*Geum urbanum*), white deadnettle (*Lamium album*), yellow archangel (*Lamium galeobdolon*), red deadnettle (*Lamium purpureum*), great woodrush (*Luzula sylvatica*), primrose (*Primula vulgaris*), betony (*Stachys officinalis*), red campion (*Silene dioica*), hedge woundwort (*Stachys sylvatica*), green alkanet (*Pentaglottis sempervirens*), common violet (*Viola riviniana*) |
| **Geophytes** (mainly spring-flowering, before the leaves re-form on deciduous trees and shrubs) | Wood anemone (*Anemone nemorosa*), lords and ladies (*Arum maculatum*), lily-of-the-valley (*Convallaria majalis*), snowdrop (*Galanthus nivalis*), stinking hellebore (*Helleborus foetidus*), bluebell (*Hyacinthoides non-scriptus*), stinking iris (*Iris foetidissima*), wild daffodil (*Narcissus pseudonarcissus*), herb paris (*Paris quadrifolia*; needs moist alkaline soil), Solomon's seal (*Polygonatum multiflorum*), lesser celandine (*Ranunculus ficaria*) |
| **Ferns** | Lady fern (*Athyrium felix-femina*), common buckler (*Dryopteris dilatata*), male fern (*Dryopteris felix-mas*), hart's tongue (*Asplenium scolopendrium*), common polypody (*Polypodium vulgare*) |
| **Mosses and liverworts** | These will arrive naturally from wind-blown spores. |

All species have been chosen with the following criteria in mind: number of other species they attract or support; ease of growth over a wide range of garden conditions; diversity. Adapted from Baines (2000).

**Fig. 19.1** Honey fungus (*Armillaria mellea*). *Photograph courtesy of Royal Horticultural Society.*

year to maintain potential nesting sites, and never to cut hedges during the nesting season.

Most farmland hedgerows date from the Enclosures Acts of some 200 years ago, and often contain only a small number of species, sometimes only hawthorn (*Crataegus monogyna*) and blackthorn (*Prunus spinosa*). Older hedges may contain more species, which have accumulated naturally over the centuries. It was once thought that the age of a hedge could be estimated very crudely from the number of species it contained. It is now clear, however, that the situation is much more complex than this, with some relatively recent hedges containing many species and some older ones being made up of a few species only. Ancient hedgerows are rich in species of herbaceous plants at the base and support a very diverse fauna.

A suitable mix of native hedgerow shrubs to form the basis of a garden hedge might be 75% hawthorn (*Crataegus monogyna*), with the following species added in smaller numbers to generate diversity, as space and local conditions permit: field maple (*Acer campestre*), dogwood (*Cornus sanguinea*), spindle (*Euonymus europaeus*), holly (*Ilex aquifolium*), native privet (*Ligustrum vulgare*), wild roses (*Rosa arvensis*, *Rosa canina* and *Rosa villosa*), bramble (*Rubus fruticosus*), raspberry (*Rubus idaeus*), guelder rose (*Viburnum opulus*). The shrubs may be further supplemented with climbers, as listed in Table 19.1. Most of the herbaceous species listed in Table 19.1 may be used to provide a herb layer at the base of the hedge.

If a proposed hedgerow forms a boundary with a neighbour's plot, it is essential that his or her views are taken into account before any work commences. Failure in this regard could lead to disputes and/or legal action.

### Wild-flower meadows

In the UK there is currently less than 2% of the hectorage of wild-flower meadows that existed 50 years ago. The change from hay to silage production, selective herbicides, replanting with monocultures of rye grass (*Lolium perenne*), the heavy use of fertilisers and drainage have all taken their toll. In the garden, the aim might be to recreate the kind of meadow that would have been common before 1950, in which a mixture of native grasses is combined with a range of native perennial flowering plants. By using appropriate mowing regimes, it should be possible to ensure that different parts of the meadow come into bloom at different times of year. Wild-flower meadows (Fig.19.2), if properly planted and managed, provide habitats and food for a very large number of insects, especially bees and butterflies, other invertebrates, reptiles, birds and small mammals. The following specific guidelines should be followed in creating and managing the meadow, in addition to the general guidelines listed above.

1. *Ensure that the ground is nutritionally relatively impoverished*, otherwise growth of some species, especially grasses, will be too lush and will swamp other species. If the area has been heavily fertilised in the past it will be beneficial to grow a vigorous crop such as potatoes for one or two seasons before establishing the meadow. It is important to avoid legume species for this purpose, however, since these fix atmospheric nitrogen (see Chapters 2 and 6), thereby negating the aim of impoverishing the soil.

2. *Try to ensure that different parts of the site have different drainage regimes*, so that wet and dry meadow habitats may be created, for each will support a different range of species (see also Chapter 8).

3. *Avoid ryegrass (Lolium perenne)* in any grass seed mixture used, since this will out-compete less-aggressive species. A suitable mixture might be one containing three or more of the following species, in equal amounts: common bent (*Agrostis tenuis*), meadow foxtail (*Alopecuris pratensis*), red

**Fig. 19.2** Fritillaries (*Fritillaria meleagris* and its white subvariety) growing in the wild-flower meadow at Wisley. *Photograph courtesy of Mike Sleigh/Royal Horticultural Society.*

fescue (*Festuca rubra*), Yorkshire fog (*Holcus lanatus*), timothy (*Phleum pretense*), and smooth meadow grass (*Poa pratense*). A small amount of seed of sweet vernal grass (*Anthox anthum odoratum*) may be added to the mix to ensure a new-mown hay aroma at mowing time (see Chapter 12). The mixture chosen should be sown at about 20% of the density recommended for lawns.

4. *Adopt different mowing regimes for* different parts of the meadow. The exact times for mowing will depend on geographical location and prevailing weather conditions. Baines (2000) suggests the following:

■ SPRING FLOWERS: mow from July onwards, but not before;

■ SUMMER FLOWERS: mow until June and not again until late September;

■ AUTUMN FLOWERS: mow only after mid-to-late September.

It is best to mow with a scythe or sickle, since the blades of most rotary mowers cannot be set high enough. The hay should be allowed to lie for a few days for any seeds to be discharged, but should then be removed, or it will rot and stifle the plants beneath it.

5. *Grow the wild flowers to be incorporated into the meadow in pots or seed trays* and then plant out in early October among the mown grass, by which time they should be strong enough to survive and compete. They may be planted 20–40 cm apart, depending on ultimate size and competitiveness, in patches of single species, or as guilds (see above), or mixed in a more random fashion. If an appropriate mowing regime is used, all will set seed and produce more seedlings or will spread vegetatively. A list of species suitable for early- and late-flowering meadows is given in Table 19.2.

6. *Cornfield flowers are not suitable for growing in meadows*, but require special treatment. Since the advent of modern selective herbicides, such species, most of which thrive in disturbed ground, have all but disappeared from the countryside, but may be seen from time to time on the disturbed ground of construction sites (see Chapters 9, 13, 15 and 16). Some, such as corncockle (*Agrostemma githago*), have become almost extinct in the wild in the UK. As with flowering meadow species, grow in pots before transplanting to the flowering site, which should be freshly cultivated. Many of the seeds will require either exposure to light or cold before they will germinate (see Chapter 9). Ideally, seed should be collected each autumn and the cornfield plot established anew the following spring.

Suitable cornfield species are as follows: corncockle (*Agrostemma githago*), scarlet pimpernel (*Anagallis arvensis*), cornflower (*Centaurea cyanus*), corn marigold [*Chrysanthemum segetum* (*Xanthophthalmum segetum*)], red deadnettle (*Lamium purpureum*), pineapple mayweed (*Matricaria matricarioides*), common poppy (*Papaver rhoeas*) and white campion (*Silene latifolia*). Seed of old cultivars of bread wheat (*Triticum aestivum*) may be added to the mix, or of durum wheat (*Triticum durum*). With their long awns these are most attractive, as well as supplying abundant grain for wildlife.

**Table 19.2** Some native UK species suitable for growing in wild-flower meadows.

| Type of meadow | Examples |
|---|---|
| **Spring meadow** | Bugle (*Ajuga reptans*; prefers moist conditions), daisy (*Bellis perennis*), lady's smock (*Cardamine pratensis*; prefers moist, alkaline conditions), snake's head fritillary (*Fritillaria meleagris*; prefers moist, alkaline or neutral conditions), cat's ears (*Hypochaeris radicata*), rough hawkbit (*Leontodon hispidus*), snowflake (*Leucojum vernum*), wild daffodil (*Narcissus pseudonarcissus*), cowslip (*Primula veris*; prefers some lime in the soil), self heal (*Prunella vulgaris*), yellow rattle (*Rhinanthus minor*; a hemiparasite of grasses), salad burnet (*Sanguisorba minor*), lesser stitchwort (*Stellaria graminea*), dandelion (*Taraxacum officinale*) |
| **Summer meadow** (species marked with an asterisk* are also suitable for autumn-flowering meadows) | Yarrow (*Achillea millefolium*), harebell (*Campanula rotundifolia*), *hardhead (*Centaurea nigra*), *knapweed (*Centaurea scabiosa*), *meadow saffron (*Colchicum autumnale*), lady's bedstraw (*Galium verum*), perforated St John's wort (*Hypericum perforatum*), *field scabious (*Knautia arvensis*), ox-eye daisy (*Leucanthemum vulgare*; requires trampling from time to time or will die out), bird's foot trefoil (*Lotus corniculatus*), *musk mallow (*Malva moschata*), meadow buttercup (*Ranunculus acris*), sheep's sorrel (*Rumex acetosella*; prefers slightly acid soil), *clary (*Salvia verbenaca*), great burnet (*Sanguisorba officinalis*; prefers damp soil), *devil's bit scabious (*Succisa pratensis*, prefers damp conditions), goatbeard (*Tragopogon pratensis*) |

These species have been chosen for their relative ease of growth in a variety of garden conditions, and to ensure flowering at different times of year. Adapted from Baines (2000).

## Ponds and marshes

Wetlands are among the most rewarding of gardens habitats to create, for they attract a very great diversity of invertebrates and amphibians, and provide important sources of food and water for other animals, especially birds (Fig. 19.3). With flexible butyl liners now readily available, ponds in particular are relatively easy to create. They are an immensely valuable resource, for natural and semi-natural ponds and shallow waters in the countryside are disappearing at an unprecedented rate as a result of pollution and/or drainage and climate change, a situation exacerbated by the disappearance of hedgerows and their associated drainage ditches.

In addition to the general guidelines on the construction of natural habitats in the garden, the following apply specifically to ponds and marshes.

1. *Site the pond in an open, relatively sunny site*, ideally facing south-west. Avoid overhanging trees, as shed leaves will pollute the water. While ponds often look best if sited in a hollow, avoid such areas if they are boggy, as they may encourage anaerobic methane production by soil bacteria, causing the liner to bubble up.

2. *The pond or marsh should be kept topped up with rainwater*. This may be facilitated by siting the habitat near a house or other building so that run off from the roof can be used. Tapwater should be avoided because it is a resource that should be conserved (see Chapter 18); also it is often rich in nutrients, especially in hard-water areas, and these promote the overgrowth of algae. Also, tapwater may contain chemicals used in water treatment, and these may affect some species adversely.

3. *Ensure that the pond has a series of levels of water*, grading from at least 60 cm depth at the centre to accommodate rooted species with floating leaves, through to shallow water, and marshy ground.

4. *Line the pond with a thin layer of soil* that is deficient in all major nutrients; subsoil (see Chapter 6) may be used for this purpose.

5. *Raise plants in pots before planting them in the pond*, as with other garden habitats. A range of suitable native plant species is given in Table 19.3. Other wildlife will find its own way to the pond from the nearby ponds, canals, rivers and drainage ditches, often surprisingly rapidly.

6. *Ensure that the pond is safe for children and pets*, by fencing it if appropriate. Also, provide a means

**Fig. 19.3** A wildlife pond with a gently sloping 'beach' area allows birds and other animals easy access to the water. *Photograph courtesy of Tim Sandall/Royal Horticultural Society.*

of escape, such as a ramp, for any small wild mammals that may fall in.

## ADAPTING AN EXISTING GARDEN

Not everyone has the space, time or inclination to establish fully functioning specialised habitats such as those described above. Also, evidence is accumulating to suggest that 'normal' gardens, with their mix of exotic and natural species and cultivars, may contain a wider diversity of wildlife than re-creations of native ecosystems. An alternative is to adapt an existing garden to make it even more attractive to wildlife. This can readily be achieved by following the various guidelines listed

above and adding to the existing plantings native and exotic plant species that provide food sources or habitats for native fauna. Plantings should be dense, designed to achieve a matrix that will be relatively self sustaining and therefore requiring little or no cultivation or maintenance except for the pruning and shaping of trees and shrubs (see Chapter 11), and resowing of annuals (see Chapter 9), as appropriate. Horticultural advice on how to establish a self-sustaining garden is given in the list of Further reading.

If the garden includes a lawn, this should be kept 'long' by mowing at a high setting, thereby creating habitats for insects and other invertebrates, and for rosette forming and creeping plants. As a general rule, herbaceous plants, trees and shrubs that produce abundant berries or seeds, or have flaky or fissured bark, are likely to be attractive to a range of species,

**Table 19.3** Some native UK species appropriate for planting in ponds and associated marshes.

| Type of plant | Examples |
|---|---|
| **Bottom-rooted plants with floating leaves, for deeper water** | White waterlily (*Nymphaea alba*) and yellow waterlily (*Nuphar lutea*; both very vigorous, so need a large pond); fringed waterlily (*Nymphoides peltata*), amphibious bistort (*Persicaria amphibium*), broad-leaved pondweed (*Potamogeton natans*), water crowfoot (*Ranunculus aquatilis*) |
| **Submerged aquatic oxygenators for deeper water** | Water starworts (*Callitriche* spp.), hornwort (*Ceratophyllum demersum*), spiked water milfoil (*Myriophyllum spicatum*; prefers lime-rich water), curly pondweed (*Potamogeton crispus*) |
| **Emergent plants for deeper water** | Flowering rush (*Butomus umbellatus*), bog bean (*Menyanthes trifoliata*), greater spearwort (*Ranunculus lingua*; very invasive so needs regular culling), burr reed (*Sparganium erectum*), lesser reedmace (*Typha latifolia*) |
| **Marginal plants for shallow water** | Water plantain (*Alisma plantago-aquatica*), yellow flag (*Iris pseudacorus*), water mint (*Mentha aquatica*), water forget-me-not (*Myosotis scorpioides*), lesser spearwort (*Ranunculus flammula*), brooklime (*Veronica beccabunga*) |
| **Plants for marshy ground** | Bugle (*Ajuga reptans*), marsh marigold (*Caltha palustris*), lady's smock (*Cardamine pratensis*), hairy willowherb (*Epilobium hirsutum*), hemp agrimony (*Eupatorium cannabinum*), meadow meadowsweet (*Filipendula ulmaria*), water avens (*Geum rivale*), ragged robin (*Lychnis flos-cuculi*), purple loosestrife (*Lythrum salicaria*), meadow buttercup (*Ranunculus acris*), salad burnet (*Sanguisorba minor*), marsh woundwort (*Stachys palustris*), betony (*Stachys officinalis*) |

All species have been chosen for their tolerance of a wide range of garden climates and to provide diverse plantings in all depths of water. Adapted from Baines (2000).

**Fig. 19.4** The bumblebee *Bombus terrestris* feeds on a Michaelmas daisy (*Aster* sp.). *Photograph courtesy of N.A. Callow.*

and these in turn will provide food sources for other species, and so on.

Examples of plants that are particularly attractive to birds and bumble bees (Fig. 19.4) (which are very good indicators of the attractiveness of a garden to wildlife) are given in Tables 19.4 and 19.5 respectively. Many of the plants that are attractive to bees are also attractive to other flying insects. The following, however, are especially attractive to butterflies: *Buddleja* spp. and cultivars, especially *Buddleja davidii*, *Sedum spectabile* and other spp. and cultivars, *Brassica* spp. and cultivars and other members of the Brassicaceae. A diversity of other species will also be required, however, to provide food sources for their larvae.

*Always consult Table 18.1* to ensure that invasive non-native species are not planted.

**Table 19.4** Native UK and exotic plants attractive to birds.

| Common name | Latin name | Value |
| --- | --- | --- |
| **Herbaceous** | | |
| Yarrow | *Achillea millefolium* | Seeds |
| Columbine | *Aquilegia vulgaris* and spp. and cvs. | Seeds |
| Sedges | *Carex* spp., especially *Carex flagellifera* | Seeds |
| Valerian | *Centranthus ruber* | Seeds |
| Sunflower | *Helianthus annuus* | Seeds |
| Love in a mist | *Nigella damascena* | Seeds |
| Iceplant | *Sedum spectabile* cvs. | Seeds |
| Grasses | *Stipa tenuissima* and other Poaceae | Seeds |
| Blessed Mary's thistle | *Silybum marianum* | Seeds |
| **Shrubs and shrubby climbers** | | |
| Barberries | *Berberis* spp. and cvs. | Purple berries, cover, nesting sites |
| Japanese quince | *Chaenomeles japonica* | Fallen fruits, cover, nesting sites |
| Clematis | *Clematis* spp. and cvs. | Seed heads |
| Mezereon | *Daphne mezereum* | Red fruit |
| Ivy | *Hedera helix* and cvs. | Black berries, insects, good cover, nesting sites |
| Himalayan honeysuckle | *Leycesteria formosa* | Red berries |
| Honeysuckle | *Lonicera periclymenum* | Red berries, insects, cover, nesting sites |
| Fly honeysuckle | *L. xylosteum* and cvs. | |
| Oregon grape | *Mahonia aquifolium* | Purple berries |
| Mahonias | *Mahonia × media* 'Buckland' and other spp. and cvs. | Purple berries, cover, nesting sites |
| Photinia | *Photinia davidiana* and cvs. | Red berries, cover |
| Firethorn | *Pyracantha coccinea* and cvs. | Red, orange or yellow berries, cover, nesting sites |
| Buffalo currant | *Ribes odoratum* | Black berries |
| Currants | *Ribes* cvs. | Black, red, or white berries |
| Species roses | *Rosa* spp. | Red hips |
| Blackberry | *Rubus fruticosus* | Black fruits |
| Raspberry | *R. idaeus* | Red fruits |
| Guelder rose | *Viburnum opulus* 'Compactum' | Red berries |
| **Trees and large shrubs** | | |
| Korean fir | *Abies koreana* | Seeds, insects, cover, viewpoint for hunting owls |
| Field maple and other maples | *Acer campestre* and other *Acer* spp. and cvs. | Seeds, perching and nesting sites |
| Common birch | *Betula pendula* | Seeds |
| Hawthorns | *Crataegus* spp. and cvs. | Red haws, perching and nesting sites |
| Hollies | *Ilex* spp. | Red or yellow berries, cover |
| Junipers | *Juniperus* spp. and cvs. | Black berries, cover, nesting sites |
| Crab apples | *Malus* 'John Downie' *Malus × robusta* 'Red Sentinel' *Malus transitoria* *Malus × zumi* 'Professor Sprenger' Other *Malus* spp. and cvs. | Red, yellow or green fruits of various sizes, perching sites |
| Spruces | *Picea* spp. and cvs. | Seeds, insects, cover |
| Pines | *Pinus* spp. and cvs. | Seeds, insects, cover |
| Bird cherry and cherries | *Prunus padus* and *Prunus* spp. | Red fruits, perching sites |
| White willow | *Salix alba* and other *Salix* spp. | Seeds, perching sites |
| Elderberry | *Sambucus nigra* and other *Sambucus* spp. and cvs. | Black or red berries |
| Whitebeams, rowan, etc. | *Sorbus* spp. and cvs. | Red or yellow berries, perching sites |
| Yew | *Taxus baccata* | Red fruits, cover, nesting sites |

Note: Non-native *Cotoneaster* spp. and cultivars, which produce abundant berries, are often recommended as being attractive to birds. They are, however, highly invasive (see Chapter 18) and have, therefore, not been listed in this table.

**Table 19.5** Some UK native and exotic garden plants that provide food for bumble bees.

| Flowering time | Examples |
| --- | --- |
| **Spring and early summer** | Maples (*Acer* spp.), sycamore (*Acer pseudoplatanus*), bugle (*Ajuga reptans*), columbines (*Aquilegia* spp.), perennial cornflower (*Centaurea scabiosa*), winter heaths (*Erica* spp.), yellow archangel (*Lamium galeobdolon*), pieris (*Pieris japonica* and cvs.), bistort (*Persicaria bistorta*), wild and bird cherries (*Prunus avium* and *Prunus padus*), lungwort (*Pulmonaria officinalis*), rhododendrons (*Rhododendron* spp.), flowering currant (*Ribes* spp. and cvs.), willows (*Salix* spp.) |
| **Summer** | Chives (*Allium schoenoprasum*), antirrhinums (*Antirrhinum* spp. and cvs.), buddlejas (*Buddleja* spp. and cvs.), thistles (*Carduus* and *Cirsium* spp.), foxglove (*Digitalis purpurea*), scabious (*Knautia arvensis*), lavenders (*Lavendula* spp. and cvs.), honeysuckles (*Lonicera* spp. and cvs.), purple loosestrife (*Lythrum salicaria*), catmints (*Nepeta* spp. and cvs.), marjorams (*Origanum* spp.), wild roses (*Rosa* spp.), blackberry (*Rubus fruticosus*), raspberry (*Rubus idaeus*), sages (*Salvia* spp. and cvs.), thymes (*Thymus* spp.), clovers (*Trifolium* spp.), stonecrop (*Sedum acre*), vetches (*Vicia* spp.), broad bean (*Vicia faba*) |
| **Late summer and autumn** | Monk's hood (*Aconitum* spp. and cvs.), borage (*Borago officinalis*), tutsan (*Hypericum androsaemum*), iceplants (*Sedum* spp. and cvs.), woundworts (*Stachys* spp.), snowberry (*Symphoricarpus alba*) |

Note: Non-native *Cotoneaster* spp. and cultivars are often recommended as being attractive to bumblebees. They are, however, highly invasive (see Chapter 18) and have therefore not been listed in this table. Adapted from Macdonald (2003).

## ENRICHMENT

All gardens designed to attract and accommodate the natural world may be enriched with artificial structures to provide additional food sources, supplies of water, over-wintering and nesting sites and safe havens. Possibilities include bird tables and feeders, shallow ponds or birdbaths, bird nesting boxes (different designs are required for different species), small mammal feeding, nesting and hibernation structures, nest boxes for bumble bees and solitary wasps, log piles, rock piles and open-jointed dry stone walls. On a large scale, whole buildings, including houses and flats, may be adapted to encourage wild species. Advice on how to make or purchase enrichment structures and to adapt buildings is contained in the books in Further reading, below.

## PERSONAL OBSERVATION AND STUDY

At the outset of this chapter it was suggested that careful study of the local geology, soils, climate, flora and fauna is an essential prerequisite for the creation of a wildlife habitat in a garden. Once established, such habitats and existing gardens adapted to accommodate wildlife provide excellent resources for personal scientific study and research. Possibilities include the study of insect and other invertebrate diversity and life cycles, food sources for particular birds and mammals, hibernation and migration patterns, plant competition and succession, plant dispersal, pollinators and pollination mechanisms, insect galls on plants, the life cycles and dispersal of plant pathogens and the symptoms they cause, and the diversity and ecology of wood-rotting and mycorrhizal fungi. The possession of simple equipment such as forceps, a lens, a simple microscope and binoculars will greatly increase the scope of possible studies. Johnston (1990) and Spedding & Spedding (2003), both on the list of Further reading, illustrate what may be achieved. A current debate that may be informed by the research of private gardeners is whether re-creations of natural habitats with native species attract a richer diversity of wildlife than traditional gardens.

## CONCLUSION

Gardens designed to embrace rather than exclude the natural world are a relatively modern development, fuelled in recent years by the loss of natural landscapes and of habitats for native flora and fauna. Garden habitats that mimic natural habitats may be created from scratch, or existing gardens may be adapted, to include and be attractive to wild species. If the natural history, landscape, climate, soils and ecology of the area in which the garden is situated are studied carefully, and if a small number of basic guidelines are followed, it is relatively straightforward to design and create gardens of and for wild species that are visually attractive, and provide endless possibilities for the observation of the natural world and for private or classroom research and study. Moreover, while not necessarily conservation resources in their own right, such gardens provide important refuges for native flora and fauna, and corridors for their movement, especially in urban and intensively farmed landscapes.

## FURTHER READING

Baines, C. (2000) *How to Make a Wildlife Garden*. Frances Lincoln, London.

Barnes, G. & Williamson, T. (2006) *Hedgerow History*. Windgatherer Press, Macclesfield.

Blamey, M., Fitter, R. & Fitter, A. (2003) *Wild Flowers of Britain and Ireland*. A. & C. Black, London.

Blamey, M. & Grey-Wilson, C. (2003) *Cassell's Wild Flowers of Britain and Northern Europe*. Cassell, London.

Broad, K. (1998) *Caring for Small Woods*. Earthscan, London.

Buczacki, S. (1986) *Ground Rules for Gardeners – a Practical Guide to Garden Ecology*. Collins, London.

Buczacki, S. (2007) *Collins Wildlife Gardener*. Collins, London.

Buczacki, S. (2007) *Garden Natural History*. New Naturalist Series. Collins, London.

Burton, J.A. & Tipling, D. (2006) *Attracting Wildlife to your Garden*. New Holland, London.

Greig, B.J.W., Gregory, S.C. & Strouts, R.G. (1991) *Honey Fungus*. HMSO, London.

Hawthorne, L. & Maughan, S. (2001) *RHS Plants for Places*. Dorling Kindersley, London.

Johnston, J. (1990) *Nature Areas for City People*. London Ecology Unit, London.

Johnston, J. & Newton, J. (1992) *Building Green. Practical Guide to Using Plants on and Around Buildings*. London Ecology Unit, London.

Junker, K. (2007) *Gardening with Woodland Plants*. Timber Press, Cambridge.

Lancaster, R. (2001) *Perfect Plant Perfect Place*. Dorling Kindersley, London.

Mabey, R. (1997) *Flora Britannica*. Chatto & Windus, London.

Macdonald, M. (2003) *Bumblebees*. Scottish Natural Heritage, Inverness.

Milliken, W. & Bridgwater, S. (2004) *Flora Celtica*. Berlin, Edinburgh.

Rackham, O. (1986) *The History of the Countryside*. Dent, London.

Rackham, O. (2006) *Woodlands*. Collins New Naturalist, London.

Spedding, C. & Spedding, G. (2003) *The Natural History of a Garden*. Timber Press, Cambridge.

Thompson, K. (2008) Greater than the sum of their parts. *The Garden*, **133**, 58–61.

Thompson, P. (2007) *The Self Sustaining Garden*. Frances Lincoln, London.

Walters, S.M. (1993) *Wild and Garden Plants*. Collins New Naturalist, London.

Wilson, R. (1981) *The Back Garden Wildlife Sanctuary Book*. Penguin, London.

# 20

# Gardens for Science

## SUMMARY

The history and roles of botanic gardens and the Royal Horticultural Society's garden at Wisley in the curation of plant collections, in the study of botany, horticultural plant science and conservation and in the dissemination of knowledge about plants, are described. A list is provided of publicly funded research institutes also engaged, directly or indirectly, in horticultural science research. The significance of university research is emphasised.

## INTRODUCTION

It is fitting that this book, which deals with the scientific principles and knowledge that relate to the practice of gardening, should end with a short account of the specialist gardens and associated institutions devoted to the cultivation and curation of plant collections, to the study of plant science and conservation, and to botanical and horticultural education and training. They form a very diverse group and have played a major role in advancing knowledge of plants relevant to horticulture, in the introduction and breeding of new garden plants, in training new professionals and in promoting education about plants in schools and among the wider public.

## BOTANIC GARDENS

### Origins

Botanic gardens were the earliest scientific gardens in Europe. The first was set up in Pisa in 1543, followed quickly by another at Padua in 1545. There followed the establishment of gardens at Florence, Bologna, Pavia, Leipzig and then Leiden, in 1590. Botanic gardens did not appear in Britain until 1621, when one was laid out at Oxford, followed by Edinburgh (1670), the Chelsea Physic Garden (1673) and Glasgow (1705). All these gardens, during their early years of existence, were physic gardens, in which medicinal and poisonous plants were grown as reference collections for medical students. This early function, in which the emphasis was on the correct naming and classification of plants, set the pattern of research in botanic gardens thereafter. With the exception of the Chelsea Physic Garden, which was run by the Worshipful Society of Apothecaries, all were attached to the medical schools of ancient universities.

As time went by, the curators of the botanic gardens, perhaps feeling restricted in their gardening by the needs of medicine, began to acquire a much wider range of plants to grow, in part to increase the attractiveness of their gardens, but also to demonstrate plant diversity. 'Rare' and 'exotic' plants were especially prized. With the gradual dissociation of the study of botany from medicine in the late eighteenth and early nineteenth centuries, botanic gardens devoted to the study of plants and plant diversity in their own right were established in other universities:

307

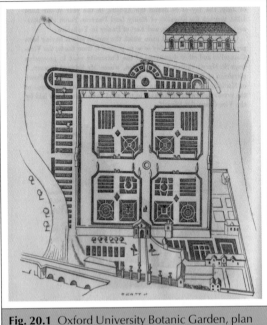

**Fig. 20.1** Oxford University Botanic Garden, plan
c.1670. *Photograph courtesy of the Royal
Horticultural Society, Lindley Library.*

Cambridge in 1760, Glasnevin (Dublin) in 1796 and thereafter in other universities in the UK and around the world throughout the nineteenth and twentieth centuries. The first botanic garden in the UK, at Oxford University (Fig. 20.1) is still involved in teaching and research. The botanic garden in Cambridge is particularly interesting, for it was laid out originally by John Henslow (1796–1861), Charles Darwin's mentor and teacher, to demonstrate plant diversity among and within species. Many of Henslow's plantings are still there to be seen and interpreted, and are currently providing the basis for much research on the thinking of both Henslow and Darwin on plant diversity and its origins. Regrettably, in many places, especially the UK, university botanic gardens are now under financial pressure, partly because they require long-term funding, which universities are often unable to provide. Many, however, continue to thrive and to play important roles in research, teaching and outreach to the wider public. They are all well worth visiting and deserve support.

The Royal Botanic Garden, Edinburgh, together with its three specialist gardens at Benmore in Argyll (Fig. 1.1), Logan in Dumfries and Galloway, and Dawyck in Peebleshire, moved in the 1950s from the care of the University of Edinburgh to that of the Scottish Office (later the Scottish Executive). Now, as the national botanic garden of Scotland, it holds major collections of plants and pursues an international programme of research, conservation, education and training.

The Royal Botanic Gardens, Kew is the largest state-funded botanic garden in the UK. The origins of this garden were quite different from those already mentioned, for it was founded in 1753 by Princess Augusta, and continued for many years as a royal garden, only later being taken into state ownership. The Royal Botanic Gardens, Kew have always had a significant role to play in the formulation and implementation of UK state policies relating to plants. It was especially influential in the development of the natural resources of the British Empire, the introduction of commercial crops to the 'colonies', and in the compilation of colonial floras (see below). Currently, among many other initiatives, it plays a worldwide role in the development of modern approaches to plant classification and evolution (systematics), and in conservation. It has important programmes of botanical and horticultural training and education and plays a leading role in the drafting and implementation of international conservation treaties and conventions.

The nineteenth century saw the establishment in Britain of botanic gardens that were quite different from their predecessors. These were not associated with universities, having been funded by private botanical and horticultural societies whose members were amateurs rather than professionals. Such societies were largely self financing, and were established to promote, through their botanic gardens, botany and horticulture as prestigious recreational activities among wealthy middle-class subscribers such as doctors, lawyers and those engaged in commerce and industry. The gardens were also landscaped, to provide an attractive amenity. The private botanic gardens gave subscribers access to the wealth of new plant introductions flooding into the UK as a result of the activities of plant collectors, as yet unavailable for growing by private individuals other than the very wealthy. Among the private botanic gardens were those at Liverpool (1802), Hull

(1812), Glasgow (1817), Kensington (1818; later Chiswick, the London Horticultural Society Garden), Edinburgh (1820; the Caledonian Horticultural Society Garden), Manchester (1829), Belfast (1829), Birmingham (1832), Bath (1834), Sheffield (1836), Leeds (1840) and Regent's Park (1841; the Royal Botanic Society of London Garden). The Birmingham Botanical Garden is a fine example of this type of botanic garden and, although many of the others founded at the same time still exist, is the only one still thriving in private ownership (see Further reading, below).

State, university and charitably funded botanic gardens have since been set up around the world, and are now flourishing as never before, liaising closely with one another in the fields of plant research, conservation, education and training. Botanic Gardens Conservation International, a charity founded in 1987 and based in Richmond, near London, plays a seminal role in promoting such cooperation (Table 20.1) between botanic gardens worldwide.

The newest botanic garden in the UK is the National Botanic Garden of Wales, at Llanarthne, Caernarvonshire. Established in 2000 as a millennium project, it had a faltering start, but now has a clear focus in the *in situ conservation* of Welsh native plants and the *ex situ conservation* of plants from the southern hemisphere. It is also notable for its sustainable infrastructure, which gives it an excellent environmental footprint.

Another major millennium garden is the Eden Project in Cornwall. Although not a botanic garden in the strict sense, it does have an important education and training role and is developing a programme of conservation work.

## Collections

The importance of botanic gardens lies in the fact that they are the holders of collections of living plants, ideally, but not always, sourced from the wild. These collections have been built up for various purposes – medicine, education, display, conservation and so on – but all are supported by documentation concerning the origins of the plants that comprise them. The importance of such documentation cannot be overemphasised, yet it is not always as complete as it could and should be. Most botanic gardens have glasshouses (Fig. 20.2) and many have outstations

**Table 20.1** Significant recent documents published by Botanic Gardens Conservation International, Richmond, UK.

| Books and manuals |
| --- |
| Akeroyd, J., McGough, N. & Wyse Jackson, P.S. (1994) *A CITES Manual for Botanic Gardens* |
| Akeroyd, J. & Wyse Jackson, P.S. (1995) *A Handbook for Botanic Gardens on the Re-introduction of Plants to the Wild* |
| Anon (2006) *The Gran Canaria Declaration II: Climate Change and Plant Conservation* (with Cabildo de Gran Canaria, Spain) |
| Leadlay, E. & Greene, J. (1998) *The Darwin Technical Manual for Botanic Gardens* |
| Secretariat of the Convention of Biological Diversity (2003). *Global Strategy for Plant Conservation* (with Botanic Gardens Conservation International), Secretariat of the Convention of Biological Diversity, Montreal |
| Waylen, K. (2006) *Botanic Gardens: Using Biodiversity to Improve Human Wellbeing* |
| Willison, J. (1994) *Environmental Education in Botanic Gardens* |
| Willison, J. (2006) *Education for Sustainable Development: Guidelines for Action in Botanic Gardens* |
| Wyse Jackson, P.S. & Sutherland, L.A. (2000) *International Agenda for Botanic Gardens in Conservation* |

| Journals |
| --- |
| Leadlay, E. & Oldfield, S. *BG Journal* (Journal of Botanic Gardens Conservation International) |
| Willison, J. & Kneebone, S. *Roots* (Botanic Gardens Conservation International Education Review) |

with different climatic and soil conditions, thereby increasing the range of plants that may be grown in the collections. Some also have seed banks.

The collections of certain species, notably those represented in the World Conservation Union (IUCN) *1997 Red List of Threatened Plants*, may be important conservation resources. The holding of endangered plants for conservation purposes in a collection at some distance from the natural habitat of the species concerned is called *ex situ* conservation. The value of *ex situ* collections is greatly increased if they include several accessions, obtained from several different sites, thereby representing a high level of genetic diversity. This is rarely the case with older collections, although those of several botanic gardens may, together, include sufficient diversity if they were obtained from different sources. New collections, especially seed banks established specifically for *ex situ* conservation, such as the Millennium Seed Bank at the Royal Botanic Gardens, Kew, have taken great care to build infra-specific genetic variation into their accessions.

Most botanic gardens now have accession and deaccession policies, which ensure a clear focus for their collections (see Table 20.1; Leadlay & Greene, 1998). This prevents overlap and enables individual institutions to hold much more representative collections of their specialisms than would otherwise be the case. Sometimes the specialisation is in plants of particular families or from particular countries, for example *Rhododendron* spp. and plants from China and Asia in the case of the Royal Botanic Garden,

**Fig. 20.3** *Rhododendron praetervisum* Sleumer, from Malaysia, part of the Royal Botanic Garden Edinburgh's extensive living collection of *Rhododendron* species and cultivars. *Photograph courtesy of the Royal Botanic Garden Edinburgh.*

Edinburgh (Fig. 20.3). Alternatively, some collections focus on native plants of the region or country in which they are situated, or on plants of local or international conservation value.

The details of botanic garden collections are now normally held in sophisticated computer databases specifically designed for the purpose. This greatly facilitates the production of informative labels (see Chapter 4), an essential feature of plants displayed in botanic gardens, and the production of plant lists, online and in book form, which can be made available to scientists around the world.

The plants contained in botanic garden collections should not be regarded as being 'owned' by the parent institutions, but are an international resource which can and should be made available to all engaged in legitimate scientific study and conservation. The return (repatriation) of plant or seed collections (*germplasm* collections), and data derived from them, to the countries in which the collections were first made, is now seen as an important activity. Whether, however, they should be exploited for commercial purposes by collection holders or others, and what proportion of the profits arising from such activities should be returned to the countries of origin, is currently a matter of vigorous debate and international legislation. The value of living collections of plants, for breeding purposes, for use in horticulture, or as a source of novel compounds for use in the pharmaceutical industry, is potentially immense.

Living collections may be displayed to provide an amenity, as in the private botanic gardens of the nineteenth century; to show how plants are related and classified, as was traditionally the case with university botanic gardens; to illustrate plants from different countries (phytogeography); or, more recently, to convey ecological or conservation messages by representing plants growing in the wild in particular locations; or some combination of these. However they are displayed, they must be curated by botanists, horticulturists and propagators (both professionals and amateurs under the supervision of professionals), to ensure they are maintained to the very highest standards.

The scientific value of living collections is greatly extended if they are associated with collections of pressed or preserved plants in a herbarium (see Chapter 4), where a much greater range of specimens may be held than is possible in a living collection, and if they are supported by a library of plant science books, journals and illustrations.

The collections of botanic gardens provide the basis for research, education, training and conservation activities, discussed below.

## Research

It was noted above that the origin of botanic gardens for the teaching of *materia medica*, with the plants properly named and classified, set the general direction of research thereafter, and this persists to the present day. The principal areas covered vary greatly from one garden to another, depending on the nature of the collections and the resources and facilities available for their study. Usually they fall into one or more of the subject categories listed in Table 20.2. The best and most effective research is that which draws on a range of approaches, including the use of traditional procedures involving observation and recording by eye, and with the aid of light and electron microscopes, in the laboratory, the herbarium, the garden and the field; genetic experimentation; and the application of modern techniques, especially DNA analyses and statistical analyses of complex data sets.

## Conservation, education and training

It has already been shown that botanic garden collections and the fruits of their research are important *ex situ* and *in situ* conservation resources. Moreover, the horticultural skills and scientific knowledge of those involved in the curation of botanic garden collections may be exploited in the reintroduction of plants to the wild where they have become rare or extinct, and in the restoration or recreation of natural habitats for endangered plant communities (see Table 20.1). The conservation work of botanic gardens is based on an international treaty, the *Convention on Biological Diversity* (*CBD*; see Further reading and Chapter 18).

Botanic gardens also have a major role in training new professional botanists and horticulturists. Many run, or contribute to, training courses in practical horticulture, and those associated with universities are often involved in teaching degree courses in horticultural and plant science and conservation, at all levels.

Most botanic gardens have significant schools outreach programmes which deal with young people direct, and also offer continuing professional development for teachers. And finally, most work to disseminate knowledge about plants and their cultivation, their role in society and their conservation to a wider public (see Table 20.1). Mechanisms to achieve this include 'friends' organisations, horticultural societies, lectures, courses, demonstrations, exhibitions of science and of art that draws its inspiration from the natural world, and retailing.

**Table 20.2** Examples of the categories of research pursued by botanic gardens.

| Category of research | Notes |
|---|---|
| Plant anatomy and reproduction, and chromosome structure (cytogenetics) | |
| Phylogenetics | The study of how plants evolved, and the patterns and lines of descent and relatedness. |
| Systematics | The application of the findings of the above categories in the classification (taxonomy) and naming of individual plants and groups of plants. |
| Floristics | The compilation of lists of various levels of sophistication that document all the plants growing in a particular area, country or region. These provide an invaluable resource for phylogenetic studies and systematics (see above), and for the conservation of plants in their natural habitats (*in situ* conservation). The simplest products of this kind of research are checklists of plants, and the most sophisticated are floras, held in books and electronically. Floras document in great detail definitive accounts of all the plants in a country or region, and may be used in the identification of specimens. Some floras may take decades to complete, and require the combined efforts of botanic gardens around the world, as with the *Flora of China*, which involves many western and Asian botanic gardens and is coordinated from the Chinese National Academy of Science and the Missouri Botanical Garden, USA. |
| Conservation science | Notably studies of the genetics of populations of plants and the application of the findings, together with findings from phylogenetics, systematics and floristics, in informing conservation practice and policy. |
| Ethnobotany and economic botany | The study of the utilisation of plants by individuals, societies and commerce worldwide (including plants as sources of new drugs). |
| Horticultural science | Notably the curation and documentation of living collections. |

## THE GARDENS OF HORTICULTURAL SOCIETIES

The gardens of the Royal Horticultural Society (RHS) at Wisley, Rosemoor, Harlow Carr and Hyde Hall, and the gardens of other horticultural societies in the UK and around the world, notably the USA, have important roles similar to those of botanic gardens in education about plants, the training of new professionals and the conservation of garden plants, especially as the holders of National Collections (see Chapter 18). The RHS garden at Wisley, for example, has the advantage of being associated with a herbarium of cultivated plants (Fig. 4.1), laboratories equipped for modern research and horticultural libraries. The scientific research based on the collections includes the phylogenetics, systematics and taxonomy of cultivated plants, new procedures for

the identification and control of pests and diseases, horticultural physiology, soil science and the conservation of cultivated plants. Much of the research is carried out in collaboration with universities and research institutes (see below).

The RHS garden at Wisley conducts trials (Fig. 20.4) to assess, in addition to 'garden worthiness', the drought, light and low-temperature tolerance of new and existing cultivars. It also maintains an important horticultural database and provides facilities for the registration of cultivated plant names and publishes the *RHS Plant Finder*, the definitive list of the names of plants in cultivation in the UK. All of these inform research on the taxonomy, physiology, pathology and conservation of the plants concerned. Finally, the garden offers horticultural advice, based on its scientific research, to members of the RHS and others, and has an extensive list of publications (Table 20.3).

**Fig. 20.4** The Trials Field at the RHS Garden, Wisley. *Photograph courtesy of Ali Cundy/RHS Herbarium.*

**Table 20.3** Significant publications from the Royal Horticultural Society.

| Journals | |
| --- | --- |
| *The Garden* | Journal of general horticultural interest distributed monthly to RHS members. |
| *The Plantsman* | A specialist journal published quarterly. |
| *The Orchid Review* | Published six times per year. |
| *Hanburyana* | An annual publication by the RHS Botany Department, dedicated to horticultural taxonomy and including a list of standards deposited in the Wisley Herbarium. |
| **Reports, registers, bulletins and leaflets** | |
| *Report of the RHS Science Departments* | Published annually. |
| Plant Registers | The RHS is the registration authority for nine groups of horticultural plants: chrysanthemums, clematis, conifers, dahlias, delphiniums, lilies, narcissus, orchids and rhododendrons. Registration handbooks and supplements are published regularly. |
| Plant Bulletins | Full-colour bulletins give results of major trials, with detailed descriptions of plants receiving the RHS Award of Garden Merit, and background botanical and cultivation information. |
| Gardening Matters | Books, booklets and leaflets that give practical advice and information on various garden-related issues. |
| Garden Guides | There is a guide to each of the RHS gardens, at Wisley, Rosemoor, Hyde Hall and Harlow Carr. |

More information on all RHS publications and research programmes is available on the RHS website: www.rhs.org.uk.

## UNIVERSITIES AND RESEARCH INSTITUTES

In addition to the work of botanic and horticultural society gardens, much significant research on horticultural science is carried out by universities and publicly funded research institutes in many countries. Most have associated gardens and trial plots, and their research is usually based on both field and laboratory experiments. The UK research institutes currently pursuing science of direct or indirect relevance to horticulture are listed in Table 20.4, together with an outline of the scope of their work. All are also involved in education and training.

Much of the fundamental science underlying horticultural practices has also been developed in universities around the world. Departments of horticulture, plant science, biology, environmental biology, genetics and soil sciences, among others, often with associated botanic gardens and/or experimental gardens, have made major contributions to our understanding of how plants grow and develop (Chapters 2, 3, 8–14 and 17), plant genetics, molecular biology and breeding (Chapter 5), plant diversity, taxonomy, ecology and conservation (Chapters 1, 4, 18 and 19), soil structure, composition, diversity and management (Chapters 6 and 7) and the interaction of plants with, and the control of, pests and pathogens (Chapters 15 and 16). The universities, of course, have a major role in the training of professional horticulturists and plant scientists.

*Field trials* are an essential component of most practical horticultural research, whether in research institutes or universities. Basic information on how to lay out a field trial to ensure the results obtained from it are statistically significant is provided in Fig. 20.5. Further information may be obtained from the sources listed in Further reading.

## CONCLUSION

Much important scientific research relevant to horticulture is being carried out in botanic gardens, horticultural society gardens, universities and publicly funded research institutes worldwide. Such

| Row 1 | Row 2 | Row 3 |
|-------|-------|-------|
| A | A | B |
| C | B | C |
| B | C | A |

**Fig. 20.5** The layout of field trials and experiments. When considering the layout of field trials or experiments, or recording their results, it is important to minimise errors arising from faulty design or incorrect analysis. Simply bisecting a plot and allocating different treatments to the two halves does not give valid results because of possible variations in, for example, soil conditions and aspect across the plot. The correct procedure is to use a randomised layout, with as many replications of each treatment as possible. A very simple layout is shown here, incorporating three replicates (rows 1–3) and three treatments (boxes A–C). Each row contains all three treatments, which should be allocated randomly to the rows using tables of random numbers. More complex designs use larger numbers of replicates or treatments. Also, the individual plots may be split up. Recording of results should ideally be carried out without the recorder knowing the treatment, to prevent recorder bias. Finally, the results should be analysed using the proper statistical procedures (see Further reading).

organisations also have significant roles in the breeding of plants of horticultural importance, conservation, education and the training of professionals. Readers are referred to the annual reports and websites of botanic gardens, Botanic Gardens Conservation International, and individual universities and research institutes for further information on current publications.

## FURTHER READING

Ballard, P. (2003) *An Oasis of Delight: The History of the Birmingham Botanical Gardens*, 2nd edn. Brewin Books, Studley.

Bofinger, V.J. & Wheeler, J.L. (1975) *Developments in the Design and Analysis of Field Experiments*. Commonwealth Agricultural Bureau, Slough.

**Table 20.4** Plant research institutes in the UK currently undertaking research directly or indirectly relevant to horticultural science.

| Institute | Description |
|---|---|
| **East Malling Research (EMR)**, New Road, East Malling, Kent ME19 6BJ www.emr.ac.uk | EMR has a broad programme of research focused on top fruit (apples and pears), stone fruit (plums ands cherries), soft fruit (raspberries and strawberries), hops, broad-leaved timber trees (principally wild cherry and ash) and woody ornamentals (especially elderberry, buddleija and lilac). There are also smaller projects on Rosaceae genomics and biodiversity of broad-leaved timber trees. |
| **Forest Research**, Alice Holt Lodge, Farnham, Surrey GU10 4LH, and Northern Research Station, Roslin, Midlothian EH25 9SY www.forestresearch.gov.uk | Forest Research undertakes research on all aspects of trees found in the UK. Its interests include the evaluation of woodland resources, land regeneration, woodland biodiversity, tree diseases and tree improvement and genetics. |
| **Institute of Grassland and Environmental Research (IGER)**, Plas Gogerddan, Aberystwyth SY23 3EB www.iger.bbsrc.ac.uk | The research programme of this institute focuses on grassland-related sustainable agriculture with strands impinging on production, environment, amenity and biodiversity. IGER works in partnership with a commercial company to breed and release new varieties of grass suitable for grasslands. |
| **John Innes Centre (JIC)**, Norwich Research Park, Colney, Norwich NR4 7UH www.jic.ac.uk | JIC undertakes fundamental and strategic research on plant metabolism, nutrition and secondary product synthesis, the cellular basis of growth and development, the genetic and molecular-genetic basis of variation relating to plant breeding, and the causes of, and resistance to, diseases and abiotic stress. Model plants are used in experiments, but all work is related to legumes, brassicas and cereals of commercial importance. The independent **Sainsbury Laboratory** focuses on the cellular and genetic basis of plant pathogenic interaction. |
| **Rothamsted Research (R Res)**, Harpenden, Hertfordshire AL5 2JQ www.rothamsted.ac.uk | R Res undertakes research related to the sustainable management of agricultural land and sustainable production practices which reduce reliance on non-renewable inputs. The research is organised into five centres: bioenergy and climate change, crop genetic improvement, soils and ecosystem functions, sustainable pest and disease management, and mathematical and computational biology. |
| **Scottish Crop Research Institute (SCRI)**, Invergowrie, Dundee DD2 5DA www.scri.ac.uk | SCRI's horticultural research is focused on soft fruits and potatoes. The famous tayberry originated there, but most current research and breeding is on blackberry, blackcurrant and raspberry, with some work also on blueberry, gooseberry and strawberry. Research on all aspects of potato genetics, diseases, quality and crop management feed into a commercial breeding programme. |
| **Warwick Horticultural Research International (HRI) at Wellesbourne**, Wellesbourne, Warwickshire CV35 9EF, and **Warwick HRI at Kirton**, Kirton, Lincolnshire PE20 1NN www2.warwick.ac.uk | Warwick HRI is the principal UK organisation carrying out horticultural research and development and transferring the results to industry. It has a wide range of research on bulbs (especially *Narcissus*) and flower crops, fruits (especially tomato, strawberry and apple), hardy nursery stock, mushrooms, protected edible crops (e.g. tomato, cucumber and lettuce), protected ornamentals and vegetables (especially cabbage, cauliflower and onion), and also strong programmes on plant molecular biology and physiology, in association with the University of Warwick. |

Bown, D. (1992) *4 Gardens in One: The Royal Botanic Garden Edinburgh*. HMSO, Edinburgh.

Desmond, R. (1995) *Kew: the History of the Royal Botanic Gardens*. The Harvill Press, London.

Drayton, R. (2000) *Nature's Government: Science, Imperial Britain, and the Improvement of the World*. Yale University Press, New Haven, CT.

Elliott, B. (2004) *The Royal Horticultural Society. A History 1804–2004*. Phillimore & Co, Chichester.

Faherty, W.B. (1989) *A Gift to Glory: In The First Hundred Years of the Missouri Botanical Garden (1859–1959)*. Harris & Friedrich, Washington DC.

Garbari, F., Tomasi, L.T. & Tosi, A. (1991) *Giardino dei Semplici: L'Orto Botanico di Pisa dal XVI al XX Secolo*. Cassa di Risparmio, Pisa.

Heywood, C.A. & Heywood, V.H. (1990) *International Directory of Botanical Gardens*, 5th edn. Koeltz Scientific Books, Köenigstein.

Kate, K.E. (1999). *The Commercial Use of Biodiversity*. Earthscan, London.

Lord, T. (ed.) (published annually) *RHS Plant Finder*. Dorling Kindersley, London.

Mead, R. & Carnow, R.N. (1983) *Statistical Methods in Agriculture and Experimental Biology*. Chapman & Hall, London.

Minelli, A. (1995) *L'Orto Botanico di Padova, 1545–1995*. Marsilio Editori, Venice.

Oldfield, S. (2007) *Great Botanic Gardens of the World*. New Holland (UK) Ltd., London.

Peterson, R.G. (1994) *Agricultural Field Experiments, Design and Analysis*. Marcel Dekker, New York, NY.

Rae, D. (2003) An introduction to the Global Strategy for Plant Conservation. *Sibbaldia* **1**, 25–8.

Rix, M. (2006) Cambridge University Botanic Garden. *Curtis's Botanical Magazine* **23** (part 1, special part), 1–144.

Walter, K.S. & Gillett, H.J. (1997) *IUCN Red List of Threatened Plants*. IUCN – The World Conservation Union, Gland and Cambridge.

Walters, S.M. (1981) *The Shaping of Cambridge Botany*. Cambridge University Press, Cambridge.

Williams, C., David, K. & Cheyne, P. (2003) *The CBD for Botanists: an Introduction to the Convention on Biological Diversity for People Working with Botanical Collectors*. The Royal Botanic Gardens, Kew, in association with the Darwin Initiative for the Survival of Species, London.

# Glossary

**abscisic acid (ABA)** A plant hormone that is particularly involved in managing the water economy of plants. ABA plays a central role in the control of stomatal closure. *See also* **stoma**.

**abscission** The separation or falling away of leaves, flowers, fruits or other plant organs. *See also* **abscission zone**.

**abscission layer** *See* **abscission zone**.

**abscission zone** A zone at the base of a leaf, flower, fruit or other plant organ in which a layer of cells (the **abscission layer**) disintegrates to aid the falling away of that organ, as in leaf fall from **deciduous** trees.

**accessory pigment** A pigment such as **chlorophyll b** or a **carotenoid**, that absorbs light energy in a different part of the spectrum from **chlorophyll a** and then passes it to chlorophyll a so that it may be utilised in photosynthesis.

**acclimation** The biochemical and/or physiological changes made by plants in response to their environment which improve their survival capacity.

**achene** A small, **indehiscent** fruit, usually with a single seed, surrounded by a thin, dry fruitwall (the **pericarp**), as in members of the Ranunculaceae (e.g. *Clematis* spp.).

**acid rain** The wet or dry deposition of a mixture of secondary atmospheric pollutants with a pH of less than 5.6, formed by the transformation of gases such as sulphur dioxide and nitrogen oxides, emitted principally by coal-fired power stations and vehicle exhausts.

**ADP (adenosine diphosphate)** A chemical molecule formed when **ATP** is broken down to release its energy.

**adventitious** Of buds, shoots or roots that grow from an unusual place on the plant, as when roots arise direct from a stem in a cutting.

**aerenchyma** A tissue consisting of long files of gas-filled spaces that allow oxygen to diffuse through the plant. This tissue is important for plants that grow in waterlogged conditions or in water.

**aerobic respiration** A type of **respiration** requiring atmospheric oxygen in which carbohydrates are oxidised to carbon dioxide and water, with the release of chemical energy. Most plants and animals respire aerobically.

**after-ripening** A period of air-dry storage that some seeds must undergo before they will germinate.

**algae** (sing. **a**) A term used loosely to describe simple, unicellular or multicellular photosynthetic plants that lack a true vascular system and are not differentiated into roots, stems and leaves. They are found in wet places or, more usually, freshwater or marine habitats. *See also* **phytoplankton**.

**allele** Usually two, sometimes more, forms of a **gene** occurring at the same relative position (**locus**) on each of a pair of **homologous chromosomes**. Alleles may be dominant or recessive. The effects of a dominant allele may be seen in the form or growth of a plant (the **phenotype**), whether the allele is paired with an identical dominant allele or with a recessive form of the allele. The effects of a recessive allele are only observed in the phenotype if it is paired with another identical, recessive allele or if the tissue is **haploid**.

**allelic variation** The difference in a particular characteristic (trait) of a plant **phenotype** controlled by two or more **alleles**.

**anaerobic respiration** A type of **respiration** in which carbohydrates are partially oxidised, in the absence of atmospheric oxygen, with the release of chemical energy. Many microorganisms are capable of anaerobic respiration.

**angiosperm** A flowering **seed** plant whose seeds are borne within a mature **ovary** (the **fruit**).

**anion** A negatively charged **ion** such as nitrate ($NO_3^-$) or sulphate ($SO_4^{2-}$).

**anion-exchange capacity** (**AEC**) A measure of the ability of a soil to adsorb (bind) **anions**. To adsorb anions, the surface must be positively charged. In soils of temperate regions, few such surfaces generally occur, but in tropical soils the clay mineral kaolinite and oxides of iron and aluminium are widespread and can have an AEC.

**anther** The terminal portion of a stamen bearing the sacs in which the **pollen grains (microspores)** are borne.

**antheridium** (pl. **ia**) Male sex organ (a type of **gametangium**), within which **gametes** are produced in fungi and lower plants such as **algae**, **liverworts**, **mosses** and **ferns**.

**anthocyanins** Coloured pigments belonging to a group of chemicals called **flavonoids**. They accumulate in **vacuoles** and are particularly important in determining the colours of flowers, although they also occur in stems, leaves and fruits.

**antiflorigen** *See* **florigen**.

**apical dominance** The term used for the process in which the apical bud prevents the growth of the lateral buds below it.

**apical meristem** *See* **meristem**.

**apomixis** The production of a seed by a non-sexual process.

**apoplast** The space, permeable to water and dissolved substances, comprising the non-living, unthickened **cell walls** and **intercellular spaces** of a plant tissue.

**arbuscular mycorrhiza** *See* **vesicular-arbuscular mycorrhiza**.

**ATP (adenosine triphosphate)** A chemical molecule of fundamental importance in plant cells as a carrier of energy. *See also* **ADP**.

**autotrophic** Able to synthesise the complex organic substances required for nutrition from simple molecules, usually carbon dioxide and water. All green plants are autotrophic.

**auxins** The first category of plant hormones to be recognised. The major naturally occurring auxin is

**indole-3-acetic acid** (**IAA**), but several synthetic auxins are used in horticulture. They have several effects in plants including the promotion of root initiation.

**available water capacity** (**AWC**) It is the water held between field capacity (the water content at which a soil ceases draining) and permanent wilting point (the water content at which a plant will wilt and not recover). This water is held in storage pores and is available to plant roots.

**axil** The upper angle between a leaf **petiole** or small branch and a stem.

**axillary bud** A bud occurring in an **axil**.

**bacterium** (pl. **ia**) A very simple, usually unicellular **microorganism** that reproduces by dividing or by forming spores. Bacteria do not have a clearly defined nucleus and the genetic material usually takes the form of a single, circular chromosome. They may occasionally be **photosynthetic** (*see* **Cyanobacteria**) but most gain their nutrients by breaking down dead organic matter or by developing a **symbiotic** association with a plant or animal host.

**betanin** The pigment which gives the colour to red beets.

**bilateral symmetry** Exhibited by a flower that is symmetrical in one plane only and therefore may appear flattened (e.g. the flowers of sweet peas, *Lathyrus* spp.).

**binomial** A system of naming plants, formalised by Carl von Linné (Linnaeus), in which two names are given in Latin: the **genus** (with a capital initial letter), followed by the **species**, with a lower-case initial letter (e.g. *Bellis perennis*, the common daisy). Both names are written in italics.

**biocide** A substance that kills living organisms.

**biome** A large ecological community of plants, animals and microorganisms, or a complex of communities, that extends over a large geographical area and is characterised by a dominant type of vegetation. The characteristics of a biome are determined by many factors, but especially climate.

**bioregulant** A substance that, when applied to a plant, changes or controls its growth. *See also* **plant growth regulator**.

**biotroph** A plant **pathogen** that obtains its nutrients from the living cells of its host and usually has little ability to live apart from it.

**bitter pit** A condition of apple fruit, caused by calcium deficiency, in which the skin develops sunken brown spots less than 1 mm in diameter and the flesh exhibits numerous pale brown spots. Affected apples may have a bitter taste.

**blue-greens** *See* **Cyanobacteria**.

**botanical key** A device for identifying plants in which a series of paired, mutually exclusive, statements each leads to a further pair of statements and so on, and eventually to the identity of the plant.

**boundary layer** A layer of air between the leaf surface and the surrounding air. This normally contains more water vapour than the surrounding air.

**bract** A reduced or modified leaf with a flower in its **axil**. Bracts may function as petals, as in *Euphorbia* spp. (spurges).

**bud** An immature shoot or flower protected by a casing of **scale leaves**.

**buffering** The ability of a soil to maintain the concentration of an **ion** in solution. Because soils contain charged surfaces, exchange of adsorbed (bound) **cations** from these surfaces with the soil solution can maintain the concentration despite, for example, uptake by plant roots. Generally, clayey soils are better buffered than sandy soils.

**bulb** A storage organ consisting of fleshy modified leaves (often called **scales**) that are attached to a base plate, which is a compressed stem from which the roots grow.

**bundle sheath** A single layer of cells ensheathing a vascular bundle.

**C-3 plants** Plants in which the first product of photosynthesis contains three carbon atoms.

**C-4 plants** Plants in which the first product of photosynthesis contains four carbon atoms.

**carbon/nitrogen ratio (C/N ratio)** The ratio of the weight of carbon to that of nitrogen in plant material or soil. Soils typically have a ratio in the range 10–15:1, but in plant materials the ratio varies considerably (from 10:1 to 100:1) depending on the age and nature of the material.

**calcicole** A plant that prefers chalky or calcarous soils to acid ones.

**calcifuge** A plants that is adapted to growing on acid soils and does not grow well on chalky soils, often showing yellowing of the leaves.

**callus tissue** The undifferentiated mass of cells produced at the site of a wound.

**CAM** *See* **crassulacean acid metabolism**.

**cambium** A layer of cells, between the **xylem** and **phloem** of stems and roots, capable of dividing to produce secondary xylem and phloem, thereby increasing the girth of the organ. A specialised cambium, the **cork cambium (phellogen)**, arises in the outer layers of older stems or roots to produce a protective, corky layer of cells.

**carotenes** **Carotenoid** pigments consisting of only carbon and hydrogen atoms. They are mainly orange in colour.

**carotenoids** A group of pigments that are orange or yellow in colour. Some are **accessory pigments** in photosynthesis, whereas others are responsible for some of the yellow and orange colours in flowers and fruits.

**carpel** A female reproductive organ comprising an **ovary**, enclosing the **ovules**, usually with a terminal **style** tipped by a **stigma**. The ovules mature to become the seeds and the carpel wall thickens and matures to form the fruit.

**cation** A positively charged **ion** such as potassium ($K^+$) or calcium ($Ca^{2+}$).

**cation-exchange capacity (CEC)** A measure of the ability of a soil to adsorb (bind) **cations**. To adsorb cations, the surface must be negatively charged. There are many such negatively charged surfaces in soils, including various clay minerals, living roots and dead organic matter.

**catkin** A pendulous flower **spike** in which the individual flowers are usually of one sex only.

**cell** The basic unit of the body of a living organism. Organisms may be unicellular (e.g. some **algae**) or multicellular, as in most plants. Plant cells have a **cell wall**, **plasmalemma**, **cytoplasm**, **nucleus**, **organelles** and **vacuoles**.

**cell wall** The rigid wall, comprised principally of **cellulose microfibrils**, proteins and sometimes **lignin** or **suberin** embedded in a **matrix** of **hemicelluloses** and **pectins**, that surrounds the plant cell.

**cellulose** A **polysaccharide** molecule composed of chains (**polymers**) of glucose molecules that is the principal structural material of the plant **cell wall**.

**cellulose microfibril** An aggregation, usually crystalline, of **cellulose** molecules found in plant **cell walls**.

**centromere** A structure which joins the two halves of a **chromosome** (the **chromatids**). The centromere becomes attached to the nuclear spindle during nuclear division.

**chelated** Referring to an organic compound (a compound containing carbon) containing a metal ion.

**chemical dwarf** A plant that has been dwarfed by the application of a chemical known as a growth retardant.

**chimera** A plant made up of genetically different cells.

**chipping** A method used to propagate **bulbs** by dividing them vertically into small pieces (chips). Chipping can also refer to the practice of removing a small part of a seed coat to facilitate water uptake and germination of hard seeds.

**chlorophyll a** The green pigment in plants capable of absorbing the radiant energy of sunlight (mainly from the red and blue regions of the spectrum) during the first stages of **photosynthesis**. Chlorophyll normally occurs in the **chloroplasts**.

**chlorophyll b** An **accessory pigment** that absorbs light energy and passes it to **chlorophyll a** for use in photosynthesis.

**chloroplasts** Microscopic, membrane-bound **organelles** in the cells of plants that contain a complex system of membranes (**thylakoids**) and **chlorophyll** pigments. Chloroplasts are the sites of **photosynthesis**.

**chromatids** The two daughter strands of a **chromosome** that has undergone division. As a **nucleus** divides, the chromatids are pulled apart, so that one ends up in each of the two daughter nuclei.

**chromatophore** A specialised type of cell containing **carotenoid** pigments.

**chromophoric group** The light-absorbing region of a pigment that is responsible for its colour.

**chromosome** A thread of **DNA** and protein that carries genetic information in the form of **genes** arranged in a linear manner. Chromosomes are located in the **nuclei** of all plant cells. They occur in **homologous** pairs and each species has a characteristic number of chromosomes.

**circadian clock** An internal clock that keeps time by going through a cycle that returns to the same point at approximately 24-hour intervals. The term circadian means 'about a day' and is used because the cycles are not exactly 24 hours long.

**CITES** *See* **Convention on International Trade in Endangered Species of Wild Flora and Fauna**.

**cladistics** An approach to taxonomic analysis in which organisms are classified according to their evolutionary relationships. *See also* **taxonomy**.

**cladode** (**phylloclade**) A specialised, photosynthetic stem that resembles a leaf.

**clamp** A traditional structure for storing vegetables (such as potatoes, beet, turnips, swedes and carrots). A clamp consists of a pile of the vegetables to be stored, encased with layers of straw and soil to provide protection from frost and rain, and with ventilation holes plugged with loose straw to prevent the build-up of carbon dioxide and the depletion of oxygen.

**climate change** Any change in climate, whether natural or induced, local or worldwide. *See also* **global warming**.

**clone** A group of genetically identical genes, cells or individual plants derived from a single, common ancestor by non-sexual means.

**codon** A group of three **nucleotides** within a molecule of messenger **RNA** that acts as a unit specifying the code for making a specific amino acid during protein synthesis.

**co-evolution** The synchronous **evolution** of adaptations in two or more populations of living organisms that interact so intimately that each exerts a selective force on the other, as with flowering plants and their pollinators.

**coleoptile** A sheath surrounding the apical meristem of a grass seedling that provides protection as the shoot grows towards the surface of the soil following germination.

**colour wheel** A wheel of colours that begins with yellow and circles through orange, red, purple, blue, green and back to yellow. Contrasting colours lie opposite to each other, whereas harmonious colours lie next to each other.

**companion cells** Specialised, metabolically active cells, each linked to a **sieve cell** in **phloem** tissue.

**composite flower** A structure resembling a true flower that is in fact made up of a large number of smaller flowers, tightly pressed together on the surface of a **receptacle**.

**compost** Organic residues (often with soil materials) that have been piled in a moist condition and allowed to decompose. The term is also used to describe a medium used to grow plants in pots and containers.

**conidium** (pl. **ia**) An asexual, non-motile, usually wind-dispersed **spore** produced by a **fungus**.

**controlled-atmosphere storage** A form of storage used mainly for fruit in which the atmosphere is controlled by increasing the concentration of carbon dioxide and, in some cases, also reducing the concentration of oxygen.

**Convention on Biological Diversity (CBD)** An abbreviated form of the United Nations Framework Convention on Biological Diversity which was agreed at the **Earth Summit** in 1992 and which provides a framework for international action to protect biodiversity.

**Convention on International Trade in Endangered Species of Wild Flora and Fauna (CITES)** An international agreement, dating from 1975, with the purpose of regulating the trade in endangered plants and animals.

**convergent evolution** The evolution of similar characteristics by unrelated organisms adapted to grow in a similar environment.

**core flush** A physiological disorder of stored apples, resulting in a pink discolouration of the core, especially in fruit exposed to dull, sunless summer weather before harvest.

**cork** A tissue made up of **cells**, impregnated with the fatty substance **suberin**, that have a protective function, especially in woody plants.

**cork cambium (phellogen)** *See* **cambium**.

**corm/cormel** A corm is a storage organ consisting of a compressed stem covered with dry scales. A cormel is a small corm.

**corpus** The cells comprising the central region of an **apical meristem**.

**cortex** A layer of tissue in a plant stem or root, bounded on the outside by the **epidermis** and enclosing the **stele**.

**corymb** An **inflorescence** in which the stalked flowers arise from a single axis, with the stalks of the flowers being progressively shorter from the base to the tip of the axis.

**cotyledon** A seed leaf, attached to the **embryo**. Cotyledons usually have a storage function in the seed and in some species act as the first photosynthetic organs following germination. In the Monocotyledoneae only one cotyledon is normally present, while in the Dicotyledoneae there are normally two and sometimes more.

**cover crop** A crop, such as kale or mustard, grown to cover the soil in winter to prevent erosion and to take up nitrogen mineralised in the autumn which might otherwise leach. Or, a crop, such as kale or mustard, grown to cover the soil in winter to prevent erosion and to take up nitrogen mineralised in the autumn which might other wise leach. Or, a crop, usually mown grass, grown primarily to suppress weeds in permanent plantations.

**crassulacean acid metabolism (CAM)** A type of **photosynthesis** in which plants open their **stomata** at night and temporarily store carbon dioxide in the form of malic acid. During daylight, the stomata are closed and the malic acid is broken down to release carbon dioxide, which is then re-fixed into the usual products of photosynthesis, namely sugars.

**critical day length** The daily duration of light above which short-day plants do not flower or are delayed in flowering, or below which long-day plants do not flower or are delayed. The critical **day length** differs between plants but is a constant for any one **species** or **cultivar**.

**cultivar (cv)** The name used for a plant variety that has been raised in cultivation, to distinguish it from a wild species.

**cuticle** A thin, non-cellular, waterproof, waxy layer containing **cutin** that covers the outer surfaces of the leaves and stems of higher plants.

**cutin** A polymer of fatty acids that is the major component of the **cuticle** and gives it its waterproofing properties.

**Cyanobacteria** (sing. **um**). A group of **bacteria** which possess **chlorophyll a** and carry out **photosynthesis**. Cyanobacteria are sometimes called blue-green algae, or **blue-greens**.

**cyclic lighting** A schedule in which flowering is controlled by giving light intermittently (rather than continuously) during a night-break.

**cytokinins** A group of plant hormones mainly originating in the roots. They have many effects, including the promotion of cell division and the inhibition of **senescence**.

**cytoplasm** The living part of a **cell** enclosed by the **plasmalemma**, but excluding the **nucleus**.

**cytoplasmic male sterility** (**CMS**) A condition in which functional pollen is not produced, determined by genetic information contained within **organelles** (usually **mitochondria**) in the **cytoplasm**.

**dark-dominant plants** Plants in which flowering depends on whether or not a critical duration of darkness is exceeded. Most are short-day plants.

**day length** *See* **photoperiod**.

**day-neutral plants** Plants in which flowering is not controlled by **day length**.

**deciduous** Used to describe parts of a plant, usually leaves, that are shed seasonally, for example at the onset of winter or of a dry season. Also used to describe a plant that sheds its leaves seasonally.

**de-etiolation** Changes that occur in dark-grown plants when they are exposed to light, such as a reduction in stem elongation and the expansion of leaves.

**dehiscent** Of a structure, usually a fruit, that bursts or splits open as it matures to release its contents, usually seeds.

**dehydrins** Proteins that accumulate in plants in response to the experimental application of **abscisic acid** and any environmental influence that has a dehydration component, such as freezing, salinity and drought.

**de-vernalisation** The process by which the vernalising effect of a previous exposure to cold is reversed by subsequent exposure to high temperatures. *See also* **vernalisation**.

**differentially permeable membrane** A membrane that allows the passage of some molecules, such as those of water, but not of others, such as those of dissolved minerals. *See also* **semi-permeable membrane**.

**diffusion** A process in which different substances become mixed as a consequence of the random movement of their component atoms, ions or molecules.

**diffusion pathway** The path followed in the diffusion of a gas or a solute from a region of relatively high concentration to a region of lower concentration.

**diploid** (**2$n$**) Of a tissue or organism in which the majority of cells contain two complementary sets of chromosomes (the normal state). *See also* **haploid**.

**discontinuous variation** Clear differences in a character that can be observed in a population of organisms. Such differences are brought about by **isolating mechanisms** that for plants may be reproductive (plants cannot hybridise because they flower at different times), ecological (plants cannot hybridise because they have become adapted to separate environments) or distributional (plants cannot hybridise because they have become stranded on different mountain tops or in different valleys).

**DNA** (**deoxyribonucleic acid**) The genetic material of most living organisms, comprising a double helix of **nucleotides** in which the principle sugar is deoxyribose, linked by the hydrogen bonds between complementary pairs of bases (cytosine and guanine, and thymine and adenine). The sequence of the bases constitutes the genetic code.

**dominant allele** *See* **allele**.

**dormancy** Seeds or buds that fail to grow when the environmental conditions are favourable for growth are termed dormant.

**dormancy: double** A type of **dormancy** in which seeds have first to be exposed to a warm temperature followed by a chilling treatment, a second exposure to a warm temperature and finally a further exposure to cold.

**dormancy: enforced** Seeds or buds that fail to grow because the environmental conditions are unfavourable. This is not considered to be true dormancy.

**dormancy: epicotyl** A type of seed dormancy in which the root emerges in warm temperatures but the shoot apex is dormant and fails to grow until after the seed plus root system is exposed to cold.

**dormancy: induced** This occurs when seeds that are capable of germination are exposed to conditions that lead to the dormant state.

**dormancy: innate** A type of dormancy that is under genetic control but can be influenced by the environmental conditions during maturation.

**Earth Summit** The popular term for the United Nations Conference on Environment and Development held in Rio de Janeiro, Brazil, in June 1992.

**ecosystem** A biological community, together with its non-living (physical) environment.

**ectomycorrhiza** A **mutualistic symbiosis** involving a **fungus** and the roots of a plant, in which the fungus is restricted to a sheath around the outside of the root and the spaces between the surface cells.

**embryo** The rudimentary **diploid** plant which develops after fertilisation and is contained within the seed; or a rudimentary plant which develops from a plant cell grown in **tissue culture**.

**endodermis** The innermost layer of cells of the root **cortex**, surrounding the **stele**. The cell walls are partially or completely thickened with **suberin**, restricting the free diffusion of water and solutes into and out of the stele, thereby allowing the cells of the endodermis to regulate the passage of such substances.

**endomycorrhiza** A **mutualistic symbiosis** involving a fungus and the roots of a plant in which the fungus grows within the tissues of the root and may penetrate the cells. *See* **vesicular-arbuscular mycorrhiza**.

**endosperm** A nutritive tissue, derived from nuclei in the **embryo** sac, within the developing **seed**.

**entropy** A measure of the disorder in a system: the higher the entropy, the greater the disorder.

**enzyme** A protein that acts in minute amounts in biological systems to promote chemical changes without itself being changed.

**epidermal hair** (**trichome**) A hair-like structure that is an outgrowth of an epidermal **cell** on a leaf or other plant organ. *See also* **epidermis**.

**epidermis** The outermost layer of **cells** in a plant.

**epinasty** The bending downwards or rolling under of a leaf caused by the differential growth of the upper part of the petiole. **Auxins** and **ethylene** are usually involved in this phenomenon.

**epiphyte** A plant which derives its physical support, and thereby access to light, by growing on another species, usually a tree, but does not derive nutrients from it.

**ethylene** A gaseous hormone that is important in **senescence** and fruit ripening. It also causes **epinasty** in plants growing in waterlogged soils.

**etiolation** Plants growing in the dark do not produce **chlorophyll** and are said to be etiolated. Dark-grown seedlings are also elongated, have few fibres and reduced leaf expansion.

**eukaryote** A cellular organism in which the genetic material is contained within a membrane-bound nucleus.

**evocation** A term used for changes at the shoot tip that result in flowering.

**evolution** The process by which the diversity of living organisms arose, over a period of more than 3 billion years, from ancestral forms. **Natural selection** appears to be the principal mechanism of evolution. Natural selection is the process in which, for a given population of organisms, the proportion of individuals that are better adapted to the prevailing environment increases over many generations, relative to the proportion of individuals that are less well adapted. This occurs because not all **gene** combinations confer equal advantage in an environment, so individuals differ in their reproductive success (biological fitness). For evolution to occur, there must be variation in heritable characteristics within a population, and the variation must result in dif-ferential reproductive success.

*ex situ* **conservation** The conservation of biodiversity (genetic diversity and/or species) outside its natural habitat, usually in a collection such as a zoo, botanic garden or seed, gene or spore bank.

**explant** A small piece of tissue cut from a plant for propagation or establishment of a **tissue culture**.

**F1** The first (filial) generation of plants following cross-pollination of two parental lines. *See also* **F1 seeds**.

**F1 seeds** The first offspring of a cross between two distinct cultivars. Plants produced from F1 seeds do not breed true. *See also* **F1**.

**F2** The second (filial) generation of plants, produced by self-pollinating or inter-pollinating individuals of the **F1** generation.

**false fruit** A fruit in which the **receptacle**, rather than the ripened **ovary** and its contents, constitutes the most conspicuous part, as in a strawberry.

**fan and pad cooling** An evaporative cooling system used by commercial growers in which the outside air is drawn over wet pads into the greenhouse.

**fasicular cambium** A **cambium** involved in the production of secondary **xylem** and **phloem** that occurs within the vascular bundles, between the primary xylem and phloem.

**feeder roots** Delicate, short-lived, lateral roots that are actively involved in the uptake of minerals and water.

**ferns** (Filicinophyta) The most advanced and numerous **pteridophytes**, in which there is in the life cycle an alternation between a large, leafy **sporophyte** generation (the 'fern'), which reproduces by means of asexual, **haploid spores** and a very small, simple, free-living **gametophyte** generation that bears the sexual structures and, after fertilisation, gives rise to a new **sporophyte**.

**fertiliser** Any organic or inorganic material of natural or synthetic origin that is added to the soil/plant system to supply nutrients essential for plant growth. Most fertilisers are applied to soil, but some can be applied direct to foliage.

**fertiliser: compound** A **fertiliser** containing more than one nutrient; for example Growmore.

**fertiliser: controlled-release** A **fertiliser** that is treated so that it does not dissolve rapidly and the nutrients are released in a controlled manner over a prolonged period.

**fertiliser: straight** A **fertiliser** containing a single compound; for example, muriate of potash.

**field experiment** An experiment conducted out of doors, either in a plot set aside for the purpose or under natural conditions.

**field heat** (or **sensible heat**) The heat held by a crop at the time of harvest.

**field trial** The cultivation of a plant or plants in an outdoor plot for the purpose of assessment and/or comparison with other plants.

**fitness (biological)** *See* **evolution**.

**flavonoids** A group of water-soluble pigments responsible for the colours of many flowers and some fruits. They include anthocyanins, flavones and flavonols.

**flora** All the plants in a given area or ecosystem, or a book or database designed to list and assist in the identification of individual plants from a specific area, region or country.

**floral stimulus** An unknown chemical stimulus that is exported from the leaves and causes flowering at the stem apex and lateral buds.

**florigen/antiflorigen** Florigen is the name that has been historically given to the unknown chemical substance that is exported from the leaves and causes flowering. In contrast, antiflorigen is the name that has been given to an unknown chemical substance that is exported from the leaves and prevents flowering. *See also* **floral stimulus**.

**fogging** The injection of water vapour into the air to maintain a high humidity around plants in a greenhouse.

**fruit** In **angiosperms**, a mature, ripened **ovary** or fused group of ovaries.

**fruits: climacteric** Fruits that produce a burst of **ethylene** gas (the climacteric) during ripening.

**fruits: non-climacteric** Fruits that do not produce a burst of **ethylene** gas during ripening.

**fungus** (pl. **i**) Organisms (often **microorganisms**) that lack chlorophyll, usually grow as filaments (**hyphae**) and reproduce by means of **spores**. Fungi obtain their nutrients by breaking down dead organic matter or by forming a **symbiotic** relationship with a plant or animal host.

**gametangium** (pl. **ia**) A structure in which **gametes** are formed, as in **mosses**, **ferns** and **fungi**.

**gamete** A specialised, **haploid** sex cell, which fuses, during fertilisation, with another gamete of the opposite sex or mating type to form a **diploid zygote**.

**gametophyte** The **haploid** stage in the life cycle of a plant, during which **gametes** are produced.

**gemmae** (sing. **a**) Asexual (i.e. non-sexual) reproductive structures, usually comprising small balls of cells, produced by some **thallose liverworts** in gemmae cups.

**gene** A unit of heredity comprising a sequence of **DNA nucleotides** coding for a single function or several related functions.

**genetic code** The series of triplets of bases of **DNA** that controls the processes leading to the synthesis of a specific protein or proteins.

**genetic engineering** *See* **genetic modification**.

**genetic locus** The position of a specific gene on a chromosome (*see* **allele**).

**genetic modification (GM)** A process whereby **genes** specifying a particular characteristic are identified and extracted from the cells of a plant species and transferred in the laboratory to the cells of another individual (which may be of a different species) where they are incorporated into the genetic material and expressed. *See also* **transgenic**.

**genome** All the genes carried by a single set of **chromosomes** of an individual.

**genome sequencing** The determination by research of the entire **genetic code** of an organism.

**genotype** The genetic information contained within an organism, as opposed to its physical appearance (**phenotype**).

**genus** (pl. **era**) The rank in the taxonomic hierarchy in plants, occurring between family and species. In a **binomial** plant name the genus occurs first, has a capital initial letter and is italicised (e.g. *Bellis* in *Bellis perennis*, the common daisy).

**geophyt**e Any species of land plant adapted to survive unfavourable conditions such as a cold and/or dry period by means of an underground storage organ such as a **rhizome**, **bulb**, **corm** or **tuber**.

**germ cell** A cell that produces a **gamete**.

**germplasm** The hereditary material of an individual that is transmitted to the **germ cells** of offspring during sexual reproduction. Often used loosely to describe seeds or plants held in a collection such as a gene bank.

**gibberellins** A group of plant hormones that is widespread throughout the plant kingdom. A major function is the control of stem elongation but they also have many other effects including the modification of flowering and the promotion of germination.

**global warming** An increase in the temperature of the troposphere (the part of the atmosphere closest to the surface of the Earth) which in the past occurred as a result of natural processes but is now accelerating as a result of the increased emission of **greenhouse gases**, largely as a result of the burning of fossil fuels. The current rate of global warming will probably lead to changes in worldwide climate patterns.

**glycoprotein** A complex chemical substance made up of a carbohydrate linked to a protein.

**graft** A union between two different plants.

**granum** (pl. **a**) Part of the internal contents of a **chloroplast**, consisting of a stack of membranous discs (**thylakoids**). There may be up to 40 to 80 grana in a single chloroplast.

**greenhouse gases** Gases that trap heat in the atmosphere, especially water vapour, carbon dioxide, methane, nitrous oxide and chlorofluorocarbons (CFCs). *See also* **global warming**.

**growth retardant** A generic term for a group of synthetic chemicals that reduce stem length.

**guard cell** A specialised leaf or stem **epidermal** cell, pairs of which surround each **stomatal pore**. Guard cells have differentially thickened walls and changes in the **turgor** of the cells cause the opening or closing of the **stomata**. *See also* **epidermis**.

**gymnosperm** A **seed** plant whose seeds are not enclosed in an **ovary**. The conifers and their relatives are gymnosperms.

**habitat** The place in which an organism or group of organisms lives.

**haploid** (*n*) Used for a **nucleus** or **cell** containing only a single set of chromosomes, formed as a result of **meiosis** of a **diploid** nucleus or cell, as in the production of **gametes**.

**hard seeds** Seeds with hard coats that inhibit germination by preventing the uptake of water. Germination cannot occur until the seed coat is rendered permeable by abrading, **chipping** or natural breakdown in the soil.

**Hartig net** The network of ectomycorrhiza fungal hyphae that permeates the outer layers of root cells *See also*, **ectomycorrhiza**.

**heartwood** The wood occupying the central region of a tree trunk or branch, consisting mainly of **xylem** vessels and fibres, impregnated with oils, gums and resins that may give it a dark appearance. The xylem of the heartwood provides support but is not normally involved in the transport of water.

**hemicelluloses** A miscellaneous group of chains (**polymers**) of various sugars, mainly glucose and xylose, that occur in the **matrix** of the plant **cell wall**.

**herbarium** (pl. **ia**). A collection of dried or otherwise preserved plants, together with detailed information concerning their collection from the field, used in the study of plant classification and evolution.

**heterotrophic** Derives energy from the digestion of complex organic substances, normally the tissues of plants and/or animals. Animals and most microorganisms are heterotrophic.

**heterozygous** Having different genetic information at a particular point (**locus**) on the complementary **chromosomes** of a **diploid cell**.

**hexaploidy** (**6n**) A form of polyploidy in which an individual possesses three times (**6n**) the **chromosome** complement of a normal **diploid** (**2n**) individual.

**homologous chromosomes** In **diploid** organisms, a pair of **chromosomes** having the same pattern of **genes** along their length, one chromosome being derived from the female parent and the other from the male.

**homozygous** Having identical genetic information at a particular point (**locus**) on the complementary chromosomes of a **diploid** cell.

**hormone** *See* **plant hormone**.

**horsetails** Primitive, spore-forming land plants belonging to the Phylum Sphenophyta, a group that together with the ferns (Filicinophyta) is often placed in the informal group **Pteridophyta**.

**hybrid vigour** The increase in vigour resulting from a cross (hybridisation) between two, often inbred, lines of a species or cultivar. It is always accompanied by increased heterozygosity.

**hybridisation** To produce seed by crossing two genetically different individuals. A hybrid is a plant resulting from such a cross.

**hydrolysis** A chemical reaction between a compound and water, resulting in the splitting of the compound which is then said to have been hydrolysed.

**hypertrophy** Excess swelling (usually of shoots at the base of the stem) as a result of treatment with hormone weedkillers or in response to water-logged soils. In both cases, the gaseous hormone, **ethylene**, is probably involved.

**hypha** (pl. **ae**) The tube-like filament that is the basic body form of most **fungi**.

**hypocotyl** In an embryo or developing plant, the region of the axis between the cotyledons and the root.

**IAA** *See* **indole-3-acetic acid**.

**illuminance** A measurement of light in terms of the sensitivity of the human eye.

**imbibition** A physical process through which water can enter cells. The first stage of water entry into dry seeds is by imbibition.

**immobilisation** The conversion of an element (commonly nitrogen, phosphorus or sulphur) from an inorganic to an organic form by microorganisms, so that the element is not readily available to plants or other organisms.

*in situ* **conservation** The conservation of biodiversity (genetic diversity, species and ecosystems) in their natural surroundings.

**inbreeding depression** The loss of vigour, fertility and yield often associated with inbreeding (fertilisation of the flowers of a plant with its own pollen, resulting in an increase in the number of homozygous recessive genes).

**incompatibility** The inability of one plant to fertilise another successfully; or of a particular **scion** to form a successful **graft** with a **rootstock**; or of a **pathogen** to infect a particular host. *See also* **incompatibility** (**graft**).

**incompatibility** (**graft**) When two different plants are unable to produce a stable graft union.

**indehiscent** Not **dehiscent**.

**indeterminate** Describing the growth of a flower spike in which extension continues unchecked as the lower flowers open; or of any other plant or fungal structure in which growth continues unchecked.

**indole-3-acetic acid** (**IAA**) The major naturally occurring auxin in plants. It is synthesised mainly in leaf primordia, young leaves and developing seeds. Moves by polar transport in living cells or bidirectional transport in phloem. Promotes apical dominance, vascular tissue differentiation and fruit growth; induces formation of adventitious roots; stimulates ethylene synthesis; delays abscission.

**inflorescence** An arrangement of flowers on a stem of a plant.

**inoculum** Spores, cells or particles of a pathogen (e.g. **fungus**, **bacterium** or **virus**) with the potential to infect a plant or a batch of soil.

**integrated pest and disease management** The integrated use of all methods available for the control of pests and diseases that minimises the use of chemicals.

**integuments** The outer coats (usually two) of the **ovule** which, after fertilisation, develop into the **testa** (seed coat).

**intercellular spaces** The spaces between cells, occupied by (for example) water vapour, air, carbon dioxide and oxygen, depending on circumstances.

**interfasicular cambium** A **cambium** involved in the production of secondary **xylem** and **phloem** that arises between the vascular bundles in the unthickened stem.

**intermediate-day plant** A plant that will only flower, or will flower most rapidly, when the **day length** is neither too long nor too short.

**intermittent mist system** A system used in propagation that employs an artificial leaf to switch on spray jets of water each time its surface dries out.

**interstock** A piece of another plant inserted between two incompatible plants during grafting. In some cases, the interstock overcomes the incompatibility and enables a successful **graft** to be established.

**ion** An atom or group of atoms (a molecule) with an electric charge. *See also* **anion** *and* **cation**.

**irradiance** The amount of light energy received at any one time.

**isolating mechanisms (distributional, ecological** and **reproductive)** *See* **discontinuous variation**.

**juvenile phase** The period during which plants exhibit **juvenility**.

**juvenility** The name given to an early phase of growth (often lasting for many years in trees) during which flowering cannot be induced by any treatment. Plants sometimes have a characteristic morphology during their juvenile phase of growth, as in ivy (*Hedera helix*) and many *Eucalyptus* spp.

**Koch's postulates** A series of requirements that must be fulfilled if a specific **pathogen** is to be identified unequivocally as the cause of a particular disease.

**lamina** The flat blade of a leaf.

**leaf primordium** (pl. **ia**). The group of **cells** at the stem apex or in a **bud** that is the precursor of a leaf.

**leafy liverworts** Members of the primitive phylum of spore-producing land plants called the Hepatophyta in which the **gametophyte** is a small plantlet with rows of thin flattened leaves on either side of and below a delicate stem.

**lenticel** A pore, filled with spongy cells, in the bark of a woody plant that allows air to reach the underlying tissues.

**lichen** A composite organism formed by the **symbiotic** association of a **fungus** with a photosynthetic **alga** or **Cyanobacterium** (blue-green). *See also* **symbiosis**.

**light-dominant** Plants in which flowering depends not only on the duration of the dark period but also on the kind of light during the day. Most are **long-day plants**.

**light integral** The total amount of light received during a period of time.

**light saturation point** The rate of photosynthesis increases with increasing light levels until, above a certain value, there is no further increase. This value is known as the light saturation point.

**lignin** A tough, aromatic polymer deployed as a strengthening material in plant cell walls, particularly in the **xylem**. It is the major component, therefore, of wood.

**liverwort** The common name for a primitive, flowerless land plant belonging to the group Hepatophyta. *See* **leafy liverwort** *and* **thallose liverwort**.

**locus** *See* **genetic locus**.

**long-day plant** A plant that will only flower, or will flower more rapidly, when the **day length** is longer than a particular value, known as the **critical day length**.

**low-temperature breakdown** Deterioration of apple fruits during low temperature storage following a cool, wet summer.

**lux** A unit of light (**illuminance**) that is related to the sensitivity of the human eye.

**macronutrient** A chemical element that is essential for plant growth and generally needed in relatively large quantities if the plant is to grow well. The macronutrients are nitrogen, potassium, phosphorus, sulphur, calcium and magnesium.

**manure** The excreta of animals, often mixed with either bedding materials or litter or both, in varying stages of decomposition.

**marker-assisted breeding** Plant breeding in which a **phenotype** that may be difficult to identify by eye (e.g. disease resistance) is identified by the presence of a particular DNA sequence always associated with the presence of that **phenotype**.

**matrix** The amorphous mass of **hemicellulose** and **pectic polymers** in which the **cellulose microfibrils** of the **cell wall** are embedded.

**maturity (commercial)** The stage of maturity at which plant structures are at their optimal stage of development for commercial use.

**maturity (physiological)** This refers to progress through a series of developmental stages that typically include the growth, maturation, ripening and **senescence** of a plant or plant organ.

**medullary rays** *See* **rays**.

**meiosis (reduction division)** Two nuclear divisions, one after the other, resulting in the reduction of the chromosome complement of the **nucleus** from **diploid** to **haploid**, as in the production of **gametes**. In meiosis, one diploid nucleus gives rise to four haploid nuclei, each in a new daughter cell.

**membranes** Sheet-like, living structures that cover, line or occur in cells. Plant membranes consist of two layers (a bilayer) of lipid (fatty molecules) and associated proteins. *See also* **plasmalemma, tonoplast** *and* **thylakoid**.

**meristem** A plant tissue composed of **cells** capable of dividing and thereby giving rise to new growth. Such tissues occur principally at the stem and root apices, in the tissue between the **xylem** and **phloem** (**cambium**), and in developing leaves.

**meristem culture** A type of **micropropagation** involving the growth and proliferation of excised stem **apical meristems** on a culture medium in a laboratory to produce large numbers of new, clonal plants. Meristem culture is used frequently to rid vegetatively-propagated plants of viruses.

**methyl jasmonate** A gaseous chemical messenger that may be involved in the triggering of resistance to pathogens in plants, including those adjacent to an infected plant.

**micro-cuttings** Shoot tissues that are removed from **tissue cultures** and rooted in a normal rooting medium.

**micronutrient** A chemical element that is essential for plant growth but generally needed in only small quantities. The micronutrients are iron, zinc, chlorine, copper, manganese, boron, molybdenum and nickel.

**microorganisms** Organisms such as **bacteria, fungi** and **viruses** that are so small that they can only be observed with the aid of a microscope.

**micropropagation** **Vegetative propagation** of plants by induction of the proliferation in culture of new plantlets from small **explants** of parental tissues. *See* **meristem culture**.

**micropyle** A pore in the developing **ovule** through which the pollen tube passes during fertilisation; and later, when the seed matures and begins to germinate, through which water enters.

**microspore (pollen grain)** In seed plants, a cell, usually haploid, that develops into a male gametophyte. Microspores are produced in the **anthers** of Angiosperms or the male cones of Gymnosperms.

**middle lamella** The gluey material, mainly **pectins**, between adjacent plant **cell walls** that serves to cement them together.

**midrib** The main, central vein of a leaf or leaflet.

**mineralisation** The conversion of an element from an organic (i.e. containing carbon) form to an inorganic (i.e. not containing carbon) form as a result of decomposition by microorganisms. The elements nitrogen, phosphorus and sulphur are commonly made available to plants by this process.

**mitochondrion** (pl. **ia**) A membrane-bound, subcellular **organelle** in which **respiration** occurs in a living cell.

**mitosis** Nuclear division, normally followed by cell division, in non-sexual cells of living organisms, resulting in two identical daughter nuclei, each containing the same number of chromosomes, identical to those of the mother nucleus.

**monopodial** The pattern of growth in which the apical growing point does not die back at the end of the season. *See also* **sympodial**.

**morphogenesis** Changes in the shape and form of the plant during development or in response to changes in the environment.

moss The common name for a flowerless, primitive land plant belonging to the Phylum Bryophyta. Mosses have an alternation of a leafy **gametophyte** generation and smaller, non-leafy **sporophyte** generation attached to the gametophyte.

mutation Any change in the **genetic code**, often leading to a changed **phenotype**, that may occur by chance or may be induced by certain chemicals, radiation, **transposons** or the activity of **viruses**. Natural mutations are known to gardeners as **sports**.

mutations: germ cell Mutations occurring in the **cells** that give rise to **gametes**. They may thus be transmitted to the next sexual generation.

mutations: somatic Mutations occurring in cells that comprise the body (leaves, stems, roots, etc.) of the plant. They are not transmitted from one sexual generation to another, but may be perpetuated by **cloning**.

mutualism The living together (**symbiosis**) of two or more organisms in close association (usually both physical and physiological) that is mutually beneficial. *See also* **mutualistic symbiosis**.

mutualistic symbiosis A **symbiosis** in which both partners benefit.

mycorrhiza A **mutualistic symbiosis** in which a **fungus** associates with the roots of a plant.

mycotoxin A toxin produced by a **fungus**.

National Plant Collections Scheme Under the auspices of the National Council for the Conservation of Plants and Gardens (NCCPG), a collaboration involving UK national institutions concerned with horticulture, professional horticulturists and taxonomists, and amateur gardeners to establish National Collections, each with a clear focus and comprising as complete a representation of a genus or section of a genus as possible of a plant of horticultural importance. Currently, there are approximately 650 National Collections, located throughout the British Isles.

natural selection *See* **evolution**.

niche The position that a **species** or other **taxon** occupies in space and time, relative to other species, within an **ecosystem**.

night-break lighting A term used to describe schedules in which the onset of flowering is controlled by giving a relatively short exposure to light (a night-break) at a particular time in the night.

nitrogen fixation The process whereby some **microorganisms** incorporate nitrogen gas ($N_2$) from the atmosphere into water-soluble nitrogenous compounds. Nitrogen-fixing microorganisms include free-living soil **bacteria** and **blue-greens** (**Cyanobacteria**), **symbiotic bacteria** (e.g. *Rhizobium*) that form nodules on the roots of plants in the Papilionaceae (e.g. peas and beans) and symbiotic bacterium-like Actinomycetes that form nodules on the roots of alders (*Alnus* spp.) and *Elaeagnus* spp.

nitrogen-fixing bacteria *See* **nitrogen fixation**.

node The point on a stem or branch at which one or more leaves, shoots, flowers or branches are attached.

nucellus The tissue within the **ovule** of the plant that contains the **embryo** sac.

nucleotide A unit structure of a nucleic acid (**DNA** or **RNA**), made up from a nucleoside (a molecule composed of ribose or deoxyribose sugar bound to a base) bound to a phosphate group.

nucleus The membrane-bound **organelle** that in **cells** contains the genetic material.

offset A young plant, **bulb** or **corm** produced vegetatively and attached to its parent, but easily detached and then capable of giving rise to a new independent plant.

oogonium (pl. ia). The female sexual organ (**gametangium**) of some **algae**, and **fungi** in the group Oomycota (e.g. *Phytophthora* spp.). Each oogonium contains one or more egg cells which, after fertilisation, form resting **spores** called oospores.

oospore *See* **oogonium**.

organelles Structures within a cell that have specialised functions and are usually enclosed by a membrane (e.g. **chloroplasts** and **mitochondria**).

organic Materials containing carbon. *See also* **soil: organic matter**.

organogenesis The organisation of an undifferentiated mass of cells into new **meristems** that can produce shoots and roots.

osmosis The flow of water (a solvent) across a semipermeable or differentially permeable membrane from a region of a relatively low concentration of dissolved substances (solutes), and therefore of high concentration of water (high **water potential**), to a region of higher concentration of solutes, and therefore of a lower concentration of water (low water potential).

**ovary** Of a flower, the female reproductive structures.

**ovule** A structure within an **ovary** that, after fertilisation, develops into a **seed**.

**oxaloacetic acid** The first product of photosynthesis in **C-4 plants**.

**ozone (O₃)** A colourless gas produced in the stratosphere by the action of high-energy **ultraviolet (UV) radiation** on oxygen. It thus acts as a screen for UV radiation. It is also a **greenhouse gas**.

**ozone layer** The natural layer of **ozone** in the stratosphere that absorbs some of the **UV radiation** from the sun, thereby reducing the amount of harmful radiation reaching the surface of the Earth.

**palisade mesophyll** A tissue, usually on the upper side of the leaf, made up of columnar cells containing many chloroplasts. In most plants it is the principal tissue in which **photosynthesis** occurs.

**panicle** A complex, branched **inflorescence**.

**parasitism** A **symbiosis** in which one organism (the parasite) invades another (the host) and derives nutrients from it, often causing disease and debilitation. *See also* **pathogen**.

**parenchyma** A tissue made up of large, thin-walled, relatively unspecialised **cells**.

**pathogen** A parasite such as a **bacterium**, **fungus** or **virus** that causes disease in its host. *See also* **parasitism**.

**pattern genes** The genes that control the variegation patterns in some plants. In this case, the plants will breed true from seed.

**pectic polymer (pectin)** A molecule made up of long chains of simple sugars. Pectic polymers usually have a gel-like consistency and are major components of the **matrix** of the plant cell wall and of the **middle lamella**.

**pericarp** The wall, sometimes three layered, of a ripe **ovary** or **fruit**.

**periclinal chimeras** Plants in which an outer layer of mutated cells overlays an inner core of the original **genotype**. *See also* **chimera**.

**periderm** An outer, protective layer of woody stems and roots, consisting principally of corky cells. A component of bark.

**perisperm** A tissue derived from the remains of the **nucellus** that provides the main store of nutrients in some seeds.

**petals** Leaf-like organs, usually brightly coloured, borne in a tight spiral or whorl in a flower.

**petiole** The leaf stalk.

**pH** Negative logarithm of the hydrogen **ion** activity of a solution or cell content. *See also* **soil: pH**.

**phellogen** *See* **cambium**.

**phenetics** An approach to taxonomic analysis that brings together a wide range of characteristics (e.g. leaf length) of each plant in a broad comparison of similarity. *See also* **taxonomy**.

**phenolics** A group of aromatic, often bitter compounds in plants that in the presence of air may be oxidised to dark compounds, as when a cut apple browns.

**phenotype** The visible manifestation of a **genotype** (i.e. the actual appearance of a plant) resulting from the interaction between its genes and the environment in which it grows.

**pheromones** Chemicals that are the sexual attractants for many insects.

**phloem** The living tissue, consisting mainly of **sieve cells**, **companion cells** and **fibres**, which transports organic substances (principally sugars) in vascular plants.

**photometric units** Units that measure light in terms of the sensitivity of the human eye.

**photon** A **quantum** of energy in the visible spectrum between wavelengths 400 and 700 nm.

**photon flux density** The number of **photons** falling on a unit area at any one time.

**photoperiod** The duration of light (**day length**) in a 24-hour period.

**photoperiodic induction** The detection of **day length** and the subsequent changes that take place in the leaf.

**photoperiodism** A response to the length of day that enables an organism to adapt to seasonal changes in the environment.

**photorespiration** A light-induced form of **respiration** in the chloroplasts of **C-3 plants** that may prevent the build up of toxic superoxide during periods of high photosynthetic activity.

**photosynthesis** The process whereby green plants capture the physical energy of sunlight (light reaction) and use this in the chemical reactions (dark reaction) in which carbon dioxide and water are used to make organic compounds (i.e. those containing carbon), notably sugars and other carbohydrates. Photosynthesis is dependent upon the green primary pigment **chlorophyll a**, which absorbs light energy in the blue and red regions of the spectrum, reflecting green light. In plants it occurs in the **chloroplasts**.

**photosynthetically active radiation** The number of quanta per unit area in the visible spectrum between 400 and 700 nm. This is the most useful measurement of the amount of light available for **photosynthesis**.

**photosystems (I and II)** The systems of photosynthetic reactions involved in the capture of light energy by **chlorophyll**.

**phototrophy** The synthesis of foods using light energy (*see* **photosynthesis**).

**phototropism** A process in which plants grow towards the direction of the incoming light.

**phylloclade** *See* **cladode**.

**phyllomorphs** These are found in some members of the family Gesneriaceae and consist of the leaf blade plus its **petiole**.

**phyllotaxy** The arrangement of leaves in pairs, spirals or whorls on a stem.

**phytochrome** A light-sensitive pigment that controls many developmental responses of plants to light. These include germination, stem elongation and flowering. Several different types of phytochrome are known.

**phytoplankton** The photosynthetic component of the **plankton**, comprising minute, mainly microscopic **algae**; they are confined to the surface layers of water, where light is available, and are responsible for 40–50% of the Earth's photosynthetic productivity.

**phytotoxicity** Of a substance that is toxic to plants.

**pith** The tissue, usually composed of large, rounded **cells**, occupying the centre of a stem.

**pits** Regions in a plant cell wall through which strands of cytoplasm pass (the **plasmodesmata**), linking the cell to adjacent cells.

**placenta** The fused margins of the **carpel** or carpels on which the **ovules** are borne.

**plankton** Minute, mainly microscopic organisms that float and drift with the currents in the surface layers of lakes or the sea. There are many millions of such organisms per cubic metre of water, and a significant proportion, the **phytoplankton**, is photosynthetic.

**plant growth regulator** A synthetic chemical that regulates growth in a similar way to a naturally occurring **plant hormone**.

**plant hormone** An organic compound that is synthesised by the plant and causes a physiological response at very low concentrations. In many but not all cases, the hormone is translocated and causes a response in a target region remote from the site of synthesis.

**plasmalemma** The outer membrane of a plant **cell**. It is a living structure consisting of two lipid (fatty) layers (a bilayer) and associated proteins.

**plasmodesmata** *See* **pits**.

**ploidy** The number of sets of **chromosomes** contained within a cell. *See* **haploid, diploid, triploid, tetraploid** *and* **hexaploid**.

**plumule** An embryonic shoot.

**polar nuclei** Two **haploid nuclei** contained within the **embryo** sac of a flowering plant which fuse with another haploid nucleus (a sperm nucleus) to form the triploid (i.e. having three sets of chromosomes) **endosperm**.

**polarity** Plant tissues have a polarity, which means that some substances, such as **auxins**, move unidirectionally within the plant, away from the shoot or root apices.

**pollen grain** *See* **microspore**.

**poly-embryony** An unusual situation in which there are many **embryos** in the same seed, which may have arisen spontaneously from body cells of a fruit (as in *Citrus* spp.) or by repeated division of a sexually derived embryo (as in many orchids).

**polymer** A chemical substance made up of chains of repeated molecules.

**polyploidisation** To render polyploid (i.e. having more than the normal two sets of chromosomes), either as a result of hybridisation or using a chemical (e.g. colchicine) that disrupts **mitosis**.

**polysaccharide** A substance made up of chains (**polymers**) of sugar molecules.

**polytunnel** A hooped supporting frame covered with a flexible plastic film inside which plants may be grown.

**procambial strand** A strand of cells in a bud that will ultimately give rise to the **xylem** and **phloem** tissues of the stem.

**prokaryote** An organism such as a **bacterium** or **Cyanobacterium** in which the genetic material is not enclosed in a membrane-bound **nucleus**.

**propagule** The name given to any part of a plant that is used to start a new plant.

**prothallus** (pl. **i**) A small, flattened structure that is the **gametophyte** (sexual) generation in the life cycle of a **fern** or **horsetail**.

**protoplast** The living contents of a plant **cell**, bounded by the **plasmalemma**.

**protoplast fusion** A laboratory process whereby naked **protoplasts** of plants are released from cells (either mechanically or using enzymes), induced to fuse in pairs and then induced to regenerate a new plant. The process may be used to produce hybrids of plants that cannot be crossed by normal pollination (e.g. in making interspecific hybrids).

**pteridophytes** (**Pteridophyta**) A collective term used informally to refer to **ferns**, **horsetails** and their relatives.

**pulvinus** A swollen region at the base of the **petiole** of some species capable of causing leaf movement.

**quantum** (pl. **a**) A discreet unit of electromagnetic energy. **Quanta** within the wavelengths of the visible spectrum are known as **photons**.

**raceme** An **inflorescence** in which flowers with stalks of equal length arise from a single axis.

**radial symmetry** Exhibited by a flower that is circular in outline (e.g. the flowers of *Primula vulgaris*, primrose) and in which the flower parts radiate from a central point.

**radicle** An embryonic root.

**rate of metabolism** The rate at which a cell or tissue is respiring and carrying out metabolic functions (the chemical reactions of a living cell).

**rays** Sheets of living cells extending radially across the secondary **xylem** and **phloem**. Those rays that link the **pith** with the **cortex** are called medullary rays, whereas those confined to the xylem and phloem are called vascular rays.

**receptacle** The enlarged, elongated, flat, concave or convex end of a stem that bears the flower or flowers and later the **fruit**.

**recessive allele** *See* **allele**.

**recombination** The process during sexual reproduction whereby **genes** in an offspring are rearranged in combinations that are different from those in either parent.

**relative humidity** The amount of water vapour present in the air expressed as a percentage of that which would be present if the air were saturated.

**replication: semi-conservative** A form of replication of **DNA** that uses an existing string of bases to create another double strand.

**respiration** The process, which normally takes place in the **mitochondria**, whereby organic molecules (those containing carbon), usually sugars, are oxidised (combined with oxygen) to release energy for metabolism (the chemical processes of a living cell). Carbon dioxide is a by-product of respiration. *See also* **aerobic respiration** *and* **anaerobic respiration**.

**respiration rate** The rate at which **respiration** is taking place.

**rhizoid** In, for example, **mosses** and **liverworts**, a thread-like multicellular or unicellular structure that serves to anchor the plant and to absorb water and minerals.

**rhizome** A specialised underground stem, which also serves as a storage organ.

**rhizosheath** The soil attached to plant roots.

**rhizosphere** The soil surrounding and influenced by plant roots.

**rhodopsin** *See* **visual purple**.

**ribosomes** Minute structures that are the sites of protein synthesis in a **cell**.

**RNA** (**ribonucleic acid**) A nucleic acid characterised by the presence of the simple sugar ribose and the base uracil. It may occur in three forms, messenger RNA, ribosomal RNA and transfer RNA, all of which are involved in protein synthesis.

**root cap** The tissue, consisting mainly of large **cells**, many of which secrete mucilage, that covers the tip of the root, protecting it and facilitating its progress through the soil as the root grows.

**rootstock** The plant upon which is grafted a shoot or bud (the **scion**) from another specimen, thereby providing the root system of the new combined plant.

**ruderal** A plant with a short life cycle that is able to grow, flower and set seed rapidly if the ground is disturbed thus reducing competition. Many weeds are ruderals.

**sapwood** The outer wood of a tree trunk or branch, consisting principally of xylem tissues which both transport water and provide support. *See also* **heartwood**.

**scale** A modified leaf or leaf base that performs a storage function in a **bulb**.

**scale leaf** A reduced leaf.

**scaling/twin scaling** Methods of propagating bulbs from their scales.

**scion** A term used in grafting for a piece of shoot placed in contact with a root system from another plant (the **stock**).

**scooping** A method of propagating **bulbs** in which the basal plate (compressed stem) and growing point of the **bulb** are scooped out from below, leaving the outer rim of the basal plate intact.

**scoring** A method of propagating **bulbs** in which deep cuts are made in the basal plate (compressed stem) and wedges of tissue are removed.

**secondary thickening** The formation of new layers of tissues (mainly **xylem** and **phloem**) in woody plant stems and roots by the repeated lateral division of the cells of the **cambium**.

**seed** A multicellular structure that develops from the ovule of an Angiosperm or Gymnosperm following fertilisation. It has an outer coat, the **testa**, a food store and an **embryo** capable of giving rise to a new plant.

**seed: orthodox** Seeds that can be dried to a low moisture content and stored at $-18°C$ for long periods.

**seed: recalcitrant** Seeds that are killed if their moisture content is reduced below some relatively high level (12–31%).

**seed bank** The population of **dormant** seeds in a given sample of soil; or a place where seeds are stored for porposes of *ex situ* **conservation**, usually at low temperature.

**seed vigour** A measure of seed quality which relates to the ability of the seed to germinate and establish under a wide range of environmental conditions.

**self-fertilisation** The fusion of male and female gametes from the same individual. *See also* **selfed**.

**self-incompatibility** (**SI**) A genetically determined system to prevent or deter **self-fertilisation**.

**selfed** Pollinated with pollen from the same individual, leading to **self-fertilisation**.

**semi-conservative replication** *See* **replication: semi-conservative**.

**semi-permeable membrane** A membrane whose structure allows the passage only of solvent molecules (water in the case of plants) and prevents the passage of dissolved substances. *See also* **differentially permeable membrane**.

**senescence** The deteriorative processes leading to death of a plant or a plant organ.

**sensible heat** *See* **field heat**.

**sepals** Leaf-like organs, usually green but sometimes coloured, that form the outer tight spiral or whorl of structures in a flower.

**shade avoider** A term used for plants that elongate in response to shady conditions. This response enables them to outgrow their neighbours and reach better light conditions. Many arable weeds are shade avoiders.

**shade plant** A plant species or other **taxon** physiologically adapted to grow and succeed in shady conditions.

**shade tolerator** A plant, often from woodland habitats, that does not elongate in response to shade. Such plants have mechanisms that improve their ability to collect light under shady conditions.

**sheath** *See* **bundle sheath** *and* **ectomycorrhiza**.

**short-day plant** A plant which only flowers, or flowers earlier when the **day length** is shorter than a particular duration known as the **critical day length**.

**sieve cells** Long cells that lack a nucleus and have perforated end walls (the **sieve plates**) and are joined together end to end to form **sieve tubes**. They are a major component of the **phloem** and transport sugars and other organic molecules such as **hormones** about the plant.

**sieve plates** *See* **sieve cells**.

**sieve tubes** *See* **sieve cells**.

**sink** A tissue or organ in which a transported substance (e.g. a sugar) or a diffused substance (e.g. $CO_2$) is consumed or sequestered, thus reducing its concentration at that place.

**soil: aggregate** A group of soil particles cohering in such a way that they appear and behave for many purposes as a single unit.

**soil: horizons** Horizontal layers in a soil that are usually distinguishable from each other by changes in soil colour or texture.

**soil: organic matter** The organic materials (i.e. those containing carbon) in soils which are a mix of live and dead animal, microbial and plant materials.

**soil: pH** The negative logarithm of the hydrogen **ion** activity of a soil. A neutral soil has a pH of 7 while acidic soils have values less than this and alkaline soils have greater values. The typical range of soil pH is from 4 to 9.

**soil: pores** The space in a soil not occupied by solids.

**soil: structure** A description of the results of the ways in which the individual clay, silt and sand particles interact with each other and are bound together by other materials such as organic matter and various oxides into larger units. The arrangement of these secondary, structural, units and of the gaps between them allows a description of structure.

**soil: texture** A description of the proportions of clay-, silt- and sand-sized particles in a soil.

**solarisation** Light-dependent inhibition of **photosynthesis** followed by bleaching of leaves. This can occur when plants are transferred from shady conditions to bright sunlight.

**solutes Ions** present in the soil solution or cell contents.

**somatic hybridisation** *See* **protoplast fusion**.

**spadix** A fleshy spike of flowers, often enclosed by an ensheathing leaf-/petal-like structure (strictly a **bract**) called the spathe, as in *Arum* spp.

**spathe** *See* **spadix**.

**species** (**sp.**, pl. **spp.**) A taxonomic category or rank occurring below the level of the **genus** and referring to a group of related, morphologically similar, usually interfertile individuals. May be subdivided into subspecies, varieties and forma. It is a basic unit of nomenclature, being written after the genus, in italics, with a lower-case initial letter, in a **binomial** (e.g. *perennis* in *Bellis perennis*, the common daisy).

**spike** An **inflorescence** in which flowers without stalks arise from a single axis.

**spiral phyllotaxy** The arrangement of the leaves in a spiral around the stem.

**spongy mesophyll** A tissue located below the **palisade** in a leaf and consisting of large, irregular cells with wide air spaces between them.

**spore** A minute propagule of a **fungus** or **bacterium**, functioning as a seed but without a pre-formed **embryo**.

**sporophyte** The **diploid**, vegetative, often **spore**-bearing stage in the life cycle of plants (as opposed to the **gametophyte** or sexual stage). In the **fern** life cycle, the large, familiar leafy 'fern' is the **sporophyte** stage. In flowering plants the **sporophyte** and **gametophyte** generations are combined.

**sport** *See* **mutation**.

**stamens** The male organs of a flower, each comprising a stalk (filament) and **anther**.

**standard** A plant specimen chosen by horticultural **taxonomists** to serve as the basis for naming and describing a new **cultivar**. The **standard** exhibits the precise characteristics of that **cultivar**.

**stele** The vascular tissues of a root or stem, comprising the **xylem**, **phloem**, **pericycle**, **pith** and, in a secondarily thickened stem, the **rays**.

**stigma** The surface on which pollen grains germinate in the female part of a flower.

**stipule** A leafy structure, usually paired with another, at the base of the **petiole** in some species.

**stock** *See* **rootstock**.

**stolon** A specialised stem that may arise above or below ground and grows outwards to colonise new ground.

**stoma** (pl. **ata**) A pore whose size may be varied (*see* **guard cell**) in the **epidermis** of a leaf or stem, allowing the controlled exchange of gases (usually carbon dioxide, oxygen and water vapour) between the plant and the atmosphere.

**stomatal pore** *See* **stoma**.

**stooling** The repeated cutting back of parent plants required for hardwood cuttings, to maintain their juvenility.

**stratification** Layering seeds in moist soil or sand and maintaining them at low temperatures in order to break **dormancy**.

**stroma** The matrix of a **chloroplast**, in which the membranes comprising the **grana** are embedded.

**style** An extension of a **carpel** that supports the **stigma**.

**suberin** A fatty, hydrophobic, waterproofing substance deposited in the cell walls of the **endodermis** and of **cork** tissue.

**substrate** (a) A substance acted on by an enzyme; (b) the soil or other medium to which an organism is attached and on which it grows.

**sun plant** A plant species or other **taxon** physiologically adapted to grow and compete most successfully in full sunlight.

**superficial scald** A physiological disorder of apples in which the death of surface cells leads to skin browning during storage following growth in hot dry conditions. In the UK it is a serious problem only on 'Bramley's Seedling'.

**symbiosis** Two organisms living together, either in a **mutualistic** or **parasitic** relationship.

**symplastic** Relating to the interconnected, living part of a plant. The symplast is a unit because protoplasts of adjoining cells are connected by the **plasmodesmata** passing through the **pits**.

**sympodial** A type of branching in which the main axis is formed by a series of lateral branches, each having arisen from the one before. It occurs because the apical bud withers at the end of each growing season.

**tap root** A large central root that is the main axis of a root system.

**taxon** (pl. **a**) A named group of organisms of any taxonomic rank, such as a **genus** or **species**. *See also* **taxonomy**.

**taxonomist** One who studies the scientific classification of organisms.

**taxonomy** The scientific classification of organisms.

**teliospore** A thick-walled, **diploid** resting **spore** in the life cycle of a rust fungus (Uredinales).

**telomere** The end of a **chromosome**, consisting of repeated sequences of **DNA** that ensures that during nuclear division each cycle of DNA replication has been completed.

**temperature inversion** The term used to describe the situation when frost occurs inside a **polytunnel** although the outside temperature remains above freezing.

**terminator gene technology** A type of **genetic modification** that prevents germination of the seeds produced by the modified plant.

**testa** The seed coat, which develops from the **integuments**.

**tetraploid** (**4n**) A form of polyploidy in which a plant possesses twice the **diploid** number (2n) of chromosomes.

**thallose liverwort** A **liverwort** (Hepatophyta) in which the **sporophyte** consists of a flattened, lobed structure (the thallus) growing flat on the ground.

**thigmo-seismic effect** The physiological responses of plants to shaking or stroking.

**thylakoid** One of the membranous discs that makes up a **granum** in a **chloroplast**.

**tipburn** Browning and death of leaf tips of lettuce and stored cabbage, resulting from calcium deficiency.

**tissue culture** Plant tissue grown in the laboratory on a sterile, synthetic medium containing sugars and **hormones**.

**tonoplast** The membrane enclosing a **vacuole** in a plant **cell**.

**topophysis** A term used to describe cuttings taken from horizontally growing branches that retain a prostrate habit of growth.

**totipotent** A term used to describe a plant **cell** that can be induced to undergo division and produce a complete new individual.

**tracheid** An elongated, spindle-shaped dead cell with its walls thickened with bands of **lignin**.

**transcription factors** Enzymes involved in **transcription**; that is, the copying of **DNA** to **RNA** prior to protein synthesis.

**transgenic** Of a species whose genome has incorporated and then expressed genes from another species. *See also* **genetic modification**.

**transpiration** The movement of water vapour from a plant to the atmosphere, mainly through the **stomata** of the leaves and stems.

**transport proteins** Proteins involved in the transport of substances across **membranes** in the **cell**.

**transposon** A naturally occurring fragment of **DNA** that moves around the **genome** by cutting itself out of one region of the **DNA** and splicing itself in, at random, elsewhere. In so doing it may disrupt the function of a **gene** or genes, as in *Rosa mundi* where the red spots and stripes on the petals are caused by the disruption of pigment formation by transposon activity.

**trichome** *See* **epidermal hair**.

**triploid (3n)** A form of polyploidy in which a plant possesses three sets of chromosomes.

**true breeding** Term used to describe a plant which, when self-pollinated, produces progeny similar to the parent.

**true fruit** A fruit in which the **pericarp** (fruit wall) is derived from the wall of the **ovary**.

**tuber** A root tuber consists of a swollen portion of a root, while a stem tuber consists of a swollen portion of a stem. Both have a storage function.

**tunica** The layer of cells overlaying the **corpus** in the **apical meristem** of flowering plants.

**tunicate** Enclosed by layered coats, the outermost being the tunic, as in an onion.

**turgid** Term used to describe a **cell** so full of water that further uptake is prevented by the hydrostatic pressure (wall pressure) exerted by the cell wall. *See also* **turgor pressure**.

**turgor pressure** The pressure exerted by the **cell** contents on the **cell wall**. When water is lost from the leaves at a greater rate than it can be supplied from the roots, turgor is lost and the plant wilts.

**twin scaling** *See* **scaling**.

**ultraviolet (UV) radiation** Electromagnetic radiation having wavelengths between 400 and 4 nanometres (i.e. between the wavelengths of violet light and X-rays).

**umbel** An **inflorescence** in which the stalked flowers all arise from the tip of the axis.

**vacuole** A **membrane**-bound sac, filled with liquid, contained within the **cytoplasm** of a plant cell.

**vapour pressure** The absolute amount of water vapour present in the air. Water vapour moves along a concentration gradient from a region of higher vapour pressure to a region of lower vapour pressure.

**vascular bundle** A long strand largely composed of **xylem** and **phloem** and therefore responsible for the transport of water, sugars and other substances in the plant.

**vascular rays** *See* **rays**.

**vector** Carrier (often an insect or other arthropod, but sometimes a fungus or human) of a virus from one host to another.

**vegetative propagation** The propagation of a plant by non-sexual means such as division or by taking cuttings.

**vein** In a plant, a visible strand of conducting tissue (**xylem** and **phloem**) in a leaf, petal or other organ.

**velamen** A tissue comprising layers of dead cells that surrounds the aerial roots of certain tropical orchids and involved in the absorption of water vapour.

**vernalisation** The exposure of seeds or young plants to a period of low temperature, thereby inducing or increasing flower formation.

**vesicular-arbuscular mycorrhiza** A form of **endomycorrhiza** in which the fungus may form specialised feeding branches called arbuscules and swollen storage hyphae called vesicles. Now often simply called **arbuscular mycorrhiza**.

**vessel** *See* **xylem** *and* **phloem**.

**virus** A **microorganism**, capable of replication (reproduction), but only in association with another, more complex organism, and having no cellular structure, consisting only of nucleic acid (**RNA** or **DNA**) with a protein coat.

**virus infection** An infection by a **virus** causing a disease.

**visual purple** The light-sensitive pigment, also called **rhodopsin**, in the rod cells of the eye that enables people to see.

**wall pressure** The inwardly directed pressure exerted by the **cell wall** that balances the **turgor pressure** in a **turgid** cell.

**water potential** A description of the potential of water to do work. By definition, the potential of pure water is zero, so the water potential in unsaturated soils or plant cells is usually negative.

**water stress** Stress in a plant resulting from an insufficient supply of water to the roots or a too rapid loss from the leaves. Many **cell** functions are depressed when plants are subjected to water stress.

**xanthophylls** A group of **carotenoid** pigments that have an oxygen atom in addition to carbon and hydrogen. They are responsible for the yellow colours of flowers and fruits.

**xylem** A plant tissue involved in the transport of water and in support, in which most of the cells have walls thickened with **lignin**. These cells may be elongate, dead, without end walls and joined end to end, forming **vessels** for the transport of water; or they may be spindle-shaped, dead **tracheids**, involved in the transport of water; or they may be dead fibres, involved in support only.

**zeatin/zeatin riboside** Naturally occurring **cytokinins**.

**zygote** The cell containing a **diploid nucleus**, resulting from the fusion of two **haploid gametes** in sexual reproduction, from which the embryo develops.

# Index